AF782753

GENETICS - RESEARCH AND ISSUES

GENETIC DIVERSITY

NEW RESEARCH

GENETICS - RESEARCH AND ISSUES

Additional books in this series can be found on Nova's website
under the Series tab.

Additional E-books in this series can be found on Nova's website
under the E-book tab.

ENVIRONMENTAL SCIENCE, ENGINEERING AND TECHNOLOGY

Additional books in this series can be found on Nova's website
under the Series tab.

Additional E-books in this series can be found on Nova's website
under the E-book tab.

GENETICS - RESEARCH AND ISSUES

GENETIC DIVERSITY

NEW RESEARCH

JULIAN A. CERVANTES AMAYA
AND
MIGUEL M. FRANCO JIMENEZ
EDITORS

NOVA BIOMEDICAL

New York

Copyright © 2012 by Nova Science Publishers, Inc.

All rights reserved. No part of this book may be reproduced, stored in a retrieval system or transmitted in any form or by any means: electronic, electrostatic, magnetic, tape, mechanical photocopying, recording or otherwise without the written permission of the Publisher.

For permission to use material from this book please contact us:
Telephone 631-231-7269; Fax 631-231-8175
Web Site: http://www.novapublishers.com

NOTICE TO THE READER

The Publisher has taken reasonable care in the preparation of this book, but makes no expressed or implied warranty of any kind and assumes no responsibility for any errors or omissions. No liability is assumed for incidental or consequential damages in connection with or arising out of information contained in this book. The Publisher shall not be liable for any special, consequential, or exemplary damages resulting, in whole or in part, from the readers' use of, or reliance upon, this material. Any parts of this book based on government reports are so indicated and copyright is claimed for those parts to the extent applicable to compilations of such works.

Independent verification should be sought for any data, advice or recommendations contained in this book. In addition, no responsibility is assumed by the publisher for any injury and/or damage to persons or property arising from any methods, products, instructions, ideas or otherwise contained in this publication.

This publication is designed to provide accurate and authoritative information with regard to the subject matter covered herein. It is sold with the clear understanding that the Publisher is not engaged in rendering legal or any other professional services. If legal or any other expert assistance is required, the services of a competent person should be sought. FROM A DECLARATION OF PARTICIPANTS JOINTLY ADOPTED BY A COMMITTEE OF THE AMERICAN BAR ASSOCIATION AND A COMMITTEE OF PUBLISHERS.

Additional color graphics may be available in the e-book version of this book.

Library of Congress Cataloging-in-Publication Data

Genetic diversity : new research / editors Julian A. Cervantes Amaya and Miguel M. Franco Jimenez.
p. cm.
Includes index.
ISBN 978-1-62081-598-4 (hardcover)
1. Variation (Biology) 2. Variation (Biology)--Research. I. Cervantes Amaya, Julian A. II. Franco Jimenez, Miguel M.
QH401.G466 2011
576.5'4072--dc23

2012013580

Published by Nova Science Publishers, Inc. †New York

Contents

PREFACE

Genetic diversity plays a very important role in survival and adaptability of a species because when a species's environment changes, slight gene variations are necessary to produce changes in the organisms' anatomy that enables it to adapt and survive. A species that has a large degree of genetic diversity among its population will have more variations from which to choose the most fit alleles. In this book, the authors discuss topics that include genetic diversity of human and animal papilloma viruses; conservation genetics of the lion; genetic diversity for quality traits in wild and ancient wheats; the genetic history of Morocco; different approaches to assess structure and genetic diversity in forest species; new genetic provisions for the reinstatement of ginseng natural populations; genetic and epigenetic variations in disease and drug response; genetic diversity and genetic rescue within wildlife populations; and genetic variability and advances in taxonomy.

Chapter 1 - The papillomaviruses form a highly diverse group that infects mammals, birds and reptiles. There is a huge veterinary and public health importance associated with these viruses. Because of this, human and bovine papillomaviruses are the most studied viruses of this group. Infection with human papillomavirus (HPV) represents the most important risk factor in the genesis of cervical cancer. Bovine papillomavirus (BPV) infects basal epithelial cells leading to the formation of benign and malignant tumors, when in synergism with action of environmental co-factors. Brazil is a country with continental dimensions; therefore the understanding of the genetic variability and the distribution of these viruses is essential for an active surveillance program and the design of more efficient vaccines and diagnostic methods. In what concerns the association of HPV and cervical cancer, HPV-16 is the most frequent type worldwide, with the exception of Indonesia where the HPV-18 happens to be the most common type. Although the HPV-16 is the most frequent, the second most important type differs between different regions in the world. So, this difference in the incidence of HPV genotypes needs to be elucidated, along with the genetic variability of these isolates that could lead to distinct oncogenic potential among different variants and subtypes. Oncogenic variants of HPV-31 were found and some of the mutations were observed in regions involved in host immune recognition. Changes in amino acid sequences between variants from different geographical regions evidence the possibility of functional divergence between them, suggesting that the variants detected may have different adaptive fitness. On the other hand, twelve BPV types are known so far, but only three of them are associated with cancer. There is a great diversity of BPV types among cattle and high co-infection was found. The presence of infection with multiple viral types could lead to suppression of the immunological system making the clinical state of the animals worse. So, further molecular epidemiological investigations on the incidence and diversity of papillomavirus infection will assist in establishing a more accurate view of the distribution of these viruses. This could help the development of accurate and specific diagnostics, prophylactics and therapeutic requirements.

Chapter 2 - Maize (*Zea mays* subsp. *mays*) is a member of the grass family *Poaceae* (*Gramineae*), together with many other important agricultural crops. The wild species commonly known as teosintes are the closest relatives of maize. Teosintes are wild grasses with a native distribution area from Mexico to Nicaragua (Mesoamerican region) and are an important genetic resource for maize improvement. Teosinte from Nicaragua (*Zea nicaraguensis*) is to a large extent an unutilized genetic resource and its properties in terms of desirable traits, such as water-lodging adaptation and disease resistance have to be phenotypically and genetically characterized. This chapter includes results from cytological, genetic, morphological and phylogenetic studies. Cytogenetic studies were based on C-banding techniques. Microsatellite (SSR) markers and DNA sequences of cpDNA regions were used for genetic diversity and phylogenetic studies respectively, whereas morphological studies involved various quantitative morphological traits. The C-banding pattern revealed that *Z. nicaraguensis* is more similar to *Z. luxurians* than other teosintes and cultivated maize. Among the Meso-American teosintes, *Z. diploperennis, Z. perennis* and *Z. nicaraguensis* showed the highest values in the number of rare and unique alleles and the results also indicated that the gene flow between *Z. nicaraguensis* and *Z. luxurians* has been more frequent than between other teosintes in the Mesoamerican region. The morphological characterization revealed that the traits of the number of lateral branch nodes bearing ears, the glume width, the number of tiller nodes bearing ears and the number of tillers per plant were the most important in discriminating between taxa and the principal component analysis grouped many traits around the *Z. nicaraguensis, Z. luxurians* and *Z. mays* subsp. *huehuetenangensis*. Finally, the variation among cpDNA sequences need more analysis to give a definitive phylogenetic resolution among the five *Zea* species, but even so the results support the idea that *Z. nicaraguensis* could be treated as a separate species distinct from *Z. luxurians*.

Chapter 3 - As an ecological and cultural umbrella species, the lion has drawn a lot of attention from researchers from a variety of disciplines. Due to the decrease in suitable habitat and increasing pressure by a growing human population, many lion populations show a declining trend. When it comes to species conservation, initiatives often follow existing taxonomy. This implies that taxonomic distinction reflects the diversity of a taxon sufficiently to preserve it on the long term. In the view of this, it is remarkable that all lion populations in Africa are considered to be the same subspecies, while for other large mammals with a similar geographical distribution, often several subspecies are being distinguished. The following chapter will review the published literature dealing with the genetic diversity of the lion and illustrate how upcoming molecular techniques can contribute to the field of species conservation.

Because the lion populations from West and Central Africa were only rarely included in previous studies on genetic diversity, the position of these populations within the species is still unclear. Recent phylogenetic research showed that the West and Central African lion is genetically distinct from their conspecifics in East and southern Africa. Because of geographic barriers and the climatic history of this region, these populations have formed a unique genetic make-up. This provides a strong argument to increase conservation efforts in West and Central Africa. In this region in particular, anthropogenic pressures have led to small, fragmented, and therefore vulnerable populations.

This chapter will also describe how different types of genetic markers can be combined to create a complete picture and come to a well-designed management plan. With the current development in next generation sequencing and genotyping techniques, these approaches come into reach for conservation studies. The resulting data can be used for a wide variety of research questions, for example on taxonomic uncertainties, defining management units, the design of (captive) breeding programs, forensics, and the genetic basis of morphological differences or disease vulnerability.

Despite the fact that the focus will be on the case of the lion, the resulting conclusions are also applicable to other species. In species conservation one of the main targets is to halt or to minimize the loss of genetic variation, in order to preserve the species as a dynamic entity with enough evolutionary potential to survive on the long term. Therefore it is of utmost importance to have a detailed picture of the present genetic diversity, both on the phylogenetic level, as on the population genetics level, to properly manage the species under consideration.

Chapter 4 - Global Change, a result of human action on ecosystems, is one of the major problems to face humanity at the present time. For crops, this means that there will be a need to find or create new cultivars able to adapt to a changing environment, without a reduction in nutritional standards or industrial quality. The quality characteristics of wheat depend on many factors, including gluten levels, starch levels and hardness, which are affected by seed storage proteins, starch synthases and puroindolines, respectively. There is little diversity among modern cultivars for the abovementioned characters, mainly due to the strong pressure exerted by wheat breeding programmes to adapt cultivars to different climatic conditions. Consequently, the search for allelic variants of these genes in local cultivars or closely related species could serve as a basis to increase the genetic pool of these components in modern wheat. In this context, a large programme has been developed to firstly, evaluate and characterize ancient wheats and some close relatives that could be used in wheat breeding and secondly, to improve wheats in general within the context of sustainable agriculture.

Chapter 5 - Morocco is a country located in northwest Africa. It has a long coastline on the Atlantic Ocean that reaches past the Strait of Gibraltar into the Mediterranean Sea. This layered country has a large history that left a mark on its current makeup. It's spread initially from Carthaginians and Romans to the coming of Islam that introduced the dominant religion and language, which are still both religious and political issues. This chapter reflects that layering, through population genetic studies. A wide range of molecular markers have been used to study the Moroccan population as well as its three ethnical groups (Arabic, Berbers and Sahrawis) and to compare it to other populations, in order to well understand and reconstitute the history of this challenging area.

The results lead to an important conclusion about the origins of the Moroccan population. Thus the male gene pool, studied by means of the Y chromosomal STR and SNP, seems to have a Middle East provenance while the feminine lineage studied through the mitochondrial DNA present a European and North African kinship. Those results were supported by some studies on the X chromosome wish spends two thirds of its life time in females. The differentiated contribution of males and females to the genetic landscape of the Moroccan population was observed with a higher migration rate in males than in females.

The autosomal markers shows a Middle East and North African predominance in the genetic pool of the Moroccan population, wish reflect the importance of the Gibraltar Strait as a genetic Barrier. In the other hand, the results show that there is a small difference between Arabs, Berbers and Sahrawis. The Arabic population is closer to Middle East and North-African populations than Berber, the autochthonous population, which is closest to Europeans.

Chapter 6 - Forest systems provide timber, food, non-food goods and externalities, being one of them the contribution to the genetic diversity safeguard. Conservation of forest genetic resources is essential for sustaining the environmental and productive values of forests. The knowledge of such diversity is particularly relevant in the case of woody species, which have been named "ecosystem engineers". The objective of the study is to show the way to establish the safeguard of genetic diversity through landscape genetics discipline. This chapter present how the results obtained in some forest species using different approaches of this discipline has provided a better understanding on the species genetic resources.

Chapter 7 - Ginseng (*Panax ginseng* C.A. Meyer) belongs to the family Araliaceae and is one of the most important medicinal plants in East Asia. There are almost no wild ginseng plants left in China and Korea, and Russia is the only country where *P. ginseng* is still growing in its natural habitats. All natural populations of *P. ginseng* are now mostly exhausted; therefore, reintroduction is essential to restore this plant to its former range. One hundred sixty-seven (167) *P. ginseng* individuals were collected from the ten administrative areas of Primorsky Krai and studied using AFLP method. A high level of intraspecific genetic diversity ($P = 55.68\%$, $H_S = 0.1891$ at population level; $P = 99.65\%$, $H_t = 0.2857$ at species level) and significant population-genetic differentiation ($\theta_B = 0.363$ from the Bayesian full model; $F_{ST} = 0.360$ from the non-hierarchical AMOVA) were detected for fluorescent AFLPs. The populations with the highest and lowest levels of genetic diversity were determined, and these populations should be conserved for the first time. Two divergent population groups were identified with hierarchical AMOVA, by the UPGMA method, by principal coordinates analysis (PCoA), in minimum spanning tree (MST) and using the STRUCTURE program. These two groups should be recognized as management units (MUs) because the low level of gene flow among them indicates contemporary demographic independence. Using an MST analysis and assignment test, the content of *P. ginseng* populations was verified to differentiate the true population plants from the plants that are not inherent to the population under study. This differentiation can identify the "pure" plants that could be chosen to serve as a stock material for reintroduction.

Chapter 8 - DNA sequence variations (0.1%) between two randomly selected individuals are considered the major factors that determine genetic diversity among individuals, disease susceptibility and response to medication. The most common form of genetic variation is called single nucleotide polymorphism (SNP). Because of their frequency of distribution, stability and simplicity they represent unique genetic markers. They can be used to identify and map complex, common diseases such as high blood pressure and diabetes. SNP markers are also employed to identify drug response, ecology, evolutionary biology, diagnostic test and heterogeneity testing.

Recent progress in genetic technology has also identified a new form of genetic variation known as copy number variation (CNV). They are known to be associated with population diversity and several complex and common diseases including nervous system disorders. Because of their location and nature, many CNVs have functional consequences such as gene dosage alteration, disruption of genes and modulation of activities of other genes. Therefore, these genetic variations will have an influence on phenotypes, susceptibility of an individual to disease, drug response and human genome evolution.

Recently, whole-genome association studies also suggest that epigenetic variation along with nucleotide variation may account for all the heritable phenotypic variation in health and disease. Epigenetics refers to the study of inheritable changes in gene expression without any changes in DNA sequence. These heritable changes include DNA methylation, histone modifications and microRNA expression. Epigenetic control mechanisms may differ among tissues and individuals and they may also change in time during aging. Many genes including enzymes, drug transporters receptors and drug targets are under epigenetic control. It is because of this reason, the importance of epigenetics in clinical medicine has been recently recognized but little is known regarding the role of epigenetics in the pathogenesis of diseases. In the near future, many additional rare polymorphisms, structural variants and tissues specific epigenetic features of the human genome may become significant contributors to disease pathogenesis and medicinal drug response.

Chapter 9 - Rhizobia form nitrogen fixing nodules in legumes but they are not the unique microbial nodular inhabitants. In the last years the endophytic bacteria of different legume nodules have been studied and the results revealed a high genetic diversity. Members of Proteobacteria, Firmicutes and Actinobacteria have been identified in the inner of nodules from different legumes. Moreover several of these endophytes are classified into novel genera and species of bacteria, revealing nodules as an important source of diversity that should be further studied, with special emphasis in their role in the formation and functionality of legume nodules.

Chapter 10 - The importance of genetic diversity within populations has been debated since the study of genetics began. This debate was brought to a head when O'Brien et al., 1985 suggested that population viability problems within the cheetah (*Acinonyx jubatus*) could be linked to low genetic diversity thus supporting the balanced school. Since then there has been a significant increase in conservation genetics studies within threatened and endangered species focusing on genetic variability and it's importance in wildlife conservation. Populations that experience bottlenecks through extirpation, isolation or fragmentation are thought to lose genetic diversity through genetic drift and inbreeding. Small population sizes may lead to inbreeding where related individuals produce offspring. Inbreeding may lead to the buildup of deleterious recessive genes, termed inbreeding depression, that may cause decreased fecundity, increased mortality, slowed growth, developmental defects, increased susceptibility to disease, decreased ability to withstand stress, and decreased ability to compete. Genetic load is the longer term loss of population fitness over time due to an increase in detrimental mutations becoming fixed in small populations through inbreeding. Genetic load may be documented as low estimates of fitness related variables when compared to other populations that have not experience bottlenecks or have no evidence of loss of fitness associated with loss of genetic diversity. Genetic rescue is a management technique that has been used to manage critically endangered populations with high genetic load. This strategy uses unrelated individuals from one population that are introduced into the population with apparent low fitness. This introduction of new genetic material acts to reduce genetic load by lowering or eliminating the frequency of detrimental gene variants in the population. This management technique has been used successfully within both captive and wild critically endangered populations. Due to the rapid increase in the study of genetic variation within wildlife populations there is a growing body of evidence linking fitness variables with genetic methods, suggesting meaningful trends and management practices like genetic rescue that may affect long-term population survival and viability. In this chapter we highlight case studies that describe the consequences of low genetic diversity, inbreeding, genetic load and genetic rescue within several endangered wildlife populations.

Chapter 11 - Several native species are used in South American folk medicine justifying the scientific surveys and earning increasing attention from pharmacological and cosmetic industries among others. *Cunila* is an important aromatic and medicinal genus with increasing researches regarding chemical and genetic variability due its interesting essential oil composition. Along with the increasing scope of chemical studies, concerns about the maintenance of natural populations and its genetic variability have raised toward to the sustainable and rational use of these plants. The current chapter encompasses an actual review about genetic variability among South American *Cunila* species and an introduction to the genetic relationship among South and North American species.

Chapter 12 - Japanese stock enhancement has been carried out since the early sixties by releasing artificially bred seed of several economically important fish and shellfish species. Even though the aim of stocking is to increase the amount of resources in the wild, breeding part of the native population in a captive environment, to achieve this goal also implies the reduction of the genetic variability even in first generation offspring, due to the small genetic contribution of just a few breeders. Consequently, stocking may comprise a potential risk to the wild population if no action is taken with respect to the loss of the genetic variability of the hatchery stock. Hence, in the last two decades, when referring to conservation programs of rare and endangered species, strong attention has been paid on conservation genetics of the captive population, in addition to the primary protection of the species and the maintenance of an adequate natural habitat.

Specifically, in the case of Japanese marine finfish stocked species, such as barfin flounder, spotted halibut or red sea bream -which are migratory species, and thus not easily traceable- it is especially complicated to obtain feasible genetic population data to establish the genetic variability of the native population. Consequently, Japanese marine finfish stock enhancement becomes a novel model for marine finfish genetic diversity conservation programs.

For this purpose, computational modelling simulation of selective breeding scenarios is being applied in Japan in order to minimize genetic diversity loss of especially rare and endangered finfish species and thus to conserve the genetic variability of the native population.

In: Genetic Diversity: New Research
Editors: J. A. C. Amaya and M. M. F. Jimenez
ISBN: 978-1-62081-598-4
© 2012 Nova Science Publishers, Inc.

Chapter 1

GENETIC DIVERSITY OF HUMAN AND ANIMAL PAPILLOMA VIRUSES

Antonio Carlos de Freitas[1,*] Ana Pavla Almeida Diniz Gurgel,[1] Bárbara Simas Chagas,[1] Valdir de Queiroz Balbino[2] and Marcus Vinicius Aragão Batista[2]

[1]Laboratory of Molecular Studies and Experimental Therapy,
Department of Genetics, Center for Biological Sciences,
Federal University of Pernambuco, Pernambuco, Brazil
[2]Laboratory of Bioinformatics and Evolutionary Biology,
Department of Genetics, Center for Biological Sciences,
Federal University of Pernambuco, Pernambuco, Brazil

ABSTRACT

The papillomaviruses form a highly diverse group that infects mammals, birds and reptiles. There is a huge veterinary and public health importance associated with these viruses. Because of this, human and bovine papillomaviruses are the most studied viruses of this group. Infection with human papillomavirus (HPV) represents the most important risk factor in the genesis of cervical cancer. Bovine papillomavirus (BPV) infects basal epithelial cells leading to the formation of benign and malignant tumors, when in synergism with action of environmental co-factors. Brazil is a country with continental dimensions; therefore the understanding of the genetic variability and the distribution of these viruses is essential for an active surveillance program and the design of more efficient vaccines and diagnostic methods. In what concerns the association of HPV and cervical cancer, HPV-16 is the most frequent type worldwide, with the exception of Indonesia where the HPV-18 happens to be the most common type. Although the HPV-16 is the most frequent, the second most important type differs between different regions in the world. So, this difference in the incidence of HPV genotypes needs to be elucidated, along with the genetic variability of these isolates that could lead to distinct

* Corresponding author: Antonio Carlos de Freitas, Department of Genetics, Federal University of Pernambuco, Cidade Universitária, CEP 50.670-901, Recife – Pernambuco – Brazil. E-mail: acf_ufpe@yahoo.com.br.

oncogenic potential among different variants and subtypes. Oncogenic variants of HPV-31 were found and some of the mutations were observed in regions involved in host immune recognition. Changes in amino acid sequences between variants from different geographical regions evidence the possibility of functional divergence between them, suggesting that the variants detected may have different adaptive fitness. On the other hand, twelve BPV types are known so far, but only three of them are associated with cancer. There is a great diversity of BPV types among cattle and high co-infection was found. The presence of infection with multiple viral types could lead to suppression of the immunological system making the clinical state of the animals worse. So, further molecular epidemiological investigations on the incidence and diversity of papillomavirus infection will assist in establishing a more accurate view of the distribution of these viruses. This could help the development of accurate and specific diagnostics, prophylactics and therapeutic requirements.

Introduction

Papillomaviridae Family

Papillomaviruses (PVs) are a large group of viruses that infects a wide variety of vertebrate species. Nowadays, it is widely known that PVs are involved with epithelial and mucosal lesions of several mammalian and non-mammalian groups, such as birds and reptiles [1]. However, the link between the infection caused by PVs and skin diseases was discovered only in the mid-of 20th century. The first article that reported the PVs and their association with papillomatosis was published in 1937, when the researcher Richard Shope observed that the Cottontail rabbit papillomavirus (CRPV, now SfPV1) suspensions might actively immunize rabbits against papillomatosis disease [2]. Furthermore, the researchers Peyton Rous and J. Beard published an article describing the transformation of papilloma lesion in squamous cell carcinoma in rabbits, thus concluding that papillomavirus is involved in skin cancer development [3]. Peyton Rous was awarded the Nobel Prize in Physiology or Medicine (1967) in recognition of his efforts to demonstrate the relationship between virus and cancer. In the early 1970s and 1980s, research groups published several articles demonstrating the involvement of Human papillomavirus (HPV) in cervical smear, genital warts and laryngeal papillomas [4]. However, only in 1983, zur Hausen and co-workers isolated HPV-16 in biopsies from cervical cancer [5]. To date, 151 HPVs have been described (http://pave.niaid.nih.gov/#home) and classified according to the oncogenic potential in High Risk (HR) HPV and Low Risk (LR) HPV [4]. Harald zur Hausen was recently awarded the Nobel Prize in Physiology or Medicine (2008) due to the discovery that HPV is "necessary but not sufficient to develop cervical cancer".

According to the World Health Organization (WHO), cervical cancer represents 13% of female cancer and it is the second leading cancer in women worldwide [6]. HPV is a major sexually transmitted infection (STI), since this viral group is involved in several cancers such as cervical, penis, vagina, vulva, anus, head-neck and non-melanoma skin cancers [7,8]. Furthermore, HPV infection is also involved in benign diseases, such as anogenital warts, laryngeal papillomas, and psoriasis [7,8]. In 2006, HPV vaccines against HPV-6, -11, -16 and -18 were approved. However, persistency and efficacy analysis, as well as selection of other HPV types, should be considered in future studies concerning PVs. STI is not exclusive to human and PV DNA was also detected in genital papillomas of cetacean and macaque species [9,10]. Thus, the medical importance of PVs infections and their relationship with skin diseases lead us to question how the molecular events of host-virus could contribute to diseases caused by this infection.

Papillomaviruses Characteristics

PVs are non-encapsuled and icosahedral virus group, measuring approximately 50-55 nm in diameter, and they contain a circular double-stranded DNA with approximately 8 kilobases (kb) in size. As mentioned previously, PVs infect the epithelial cells of several mammalian and non-mammalian species and can cause benign and malignant diseases [11–14]. Currently, PVs were detected in cats, dogs, cows, cattle, deer, macaques, turtles, snake, goat, rabbits and other animals (Figure 1).

Because their strong cell tropism, PVs are considered epithelialtropic and mucosotripic agents, although several studies have detected PVs DNA in non-epithelial sites, such as blood, semen (spermatozoa) and placent [15–22]. Also, PVs are normally considered host-specific, although some PVs can be found in several hosts [12–14]. For example, *Bovine papillomavirus* 1 (BPV-1) and BPV-2 can infect cow, cattle, buffalo and giraffe.

Figure 1. Some examples of PVs and their hosts, including mammalian, bird and reptile groups. *Human papillomavirus* (HPV) infects Humans; *Feline papillomavirus* (FdPV) infects domestic cats; *Bovine papillomavirus* (BPV) infects domestic cows and cattle; *Alces alces papillomavirus* 1 (AaPV1) infects European elk; *Capreoulus capreolus Papillomavirus* 1 (CcaPV1) infects Western roe deers; *Macaca fascicularis papillomavirus* (MfPV) infects Cynomolgus macaques; *Chelonia mydas papillomavirus* 1 (CmPV1) infects turtles; *Morelia spilota papillomavirus 1* (MsPV1) infects snakes; *Capra hircus papillomavirus* 1 (ChPV1) infects domestic goats; *Cottontail rabbit papillomavirus* (CRPV) infects rabbits.

Papillomaviruses Genome Diversity

PVs genome typically consists of 8 Open Reading Frames (ORFs), in which generally six ORFs are classified as early genes (E1, E2, E4, E5, E6 and E7) and two as late genes (L1 and L2), according to the expression of these genes in the viral development into keratinocytes cells. Early genes are expressed in the surface of epithelium and late genes in suprabasal layer epithelium. PVs genome are structurally conserved, although the E3, E4, E5/E5a and E5b, E8 and E9 genes may be present or absent in these genomes. Furthermore, PVs possess a non-coding region named Long Control Region (LCR), which is located in the upstream of the early and late genes (Figures 2, 3, 4 and 5).

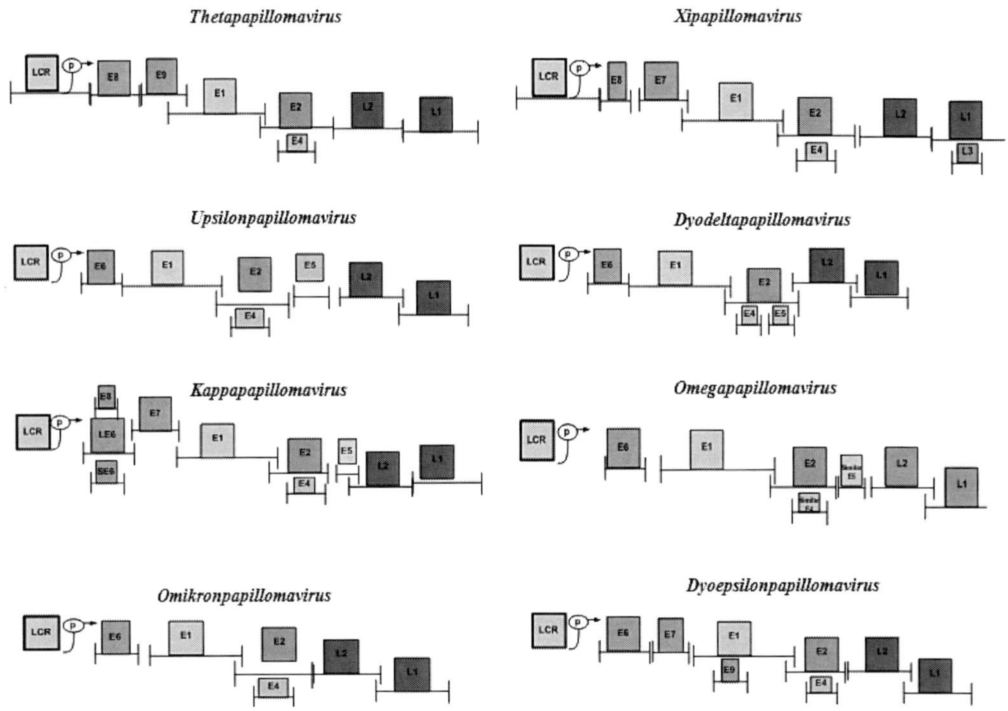

Figure 2. PVs genomes in a liner format of *Thetapapillomavirus, Xipapillomavirus, Upsilonpapillomavirus, Dyodeltapapillomavirus, Kappapapillomavirus, Omegapapillomavirus, Omikreonpapillomavirus and Dyoepsilonpapillomavirus genera.* Concerning the *Thetapapillomavirus* genus, do not was observed the both E6 and E7 genes. *Xipapillomavirus* genus do not have E6 gene. *Upsilonpapillomavirus, Dyodeltapapillomavirus, Omegapapillomavirus, Omikronpapillomavirus* genera do not have E7 gene in their genomes. The E9 gene are present in *Thetapapillomavirus* and *Dyoepsilonpapillomavirus* genera.

Classification of Papillomaviruses

Initially, Papillomaviruses and Polyomavirus were grouped into *Papovaviridae* family due to their small size, non-enveloped capsids and circular double-stranded genome features. However, during the 8th International Committee on Taxonomy of Viruses (ICTV), the division of the *Papovaviridae* in *Papillomaviridae* (constituting of PVs) and *Polyomaviridae* (constituting of Polyomavirus – PVD) families was determined due to their nucleotide sequence distance.

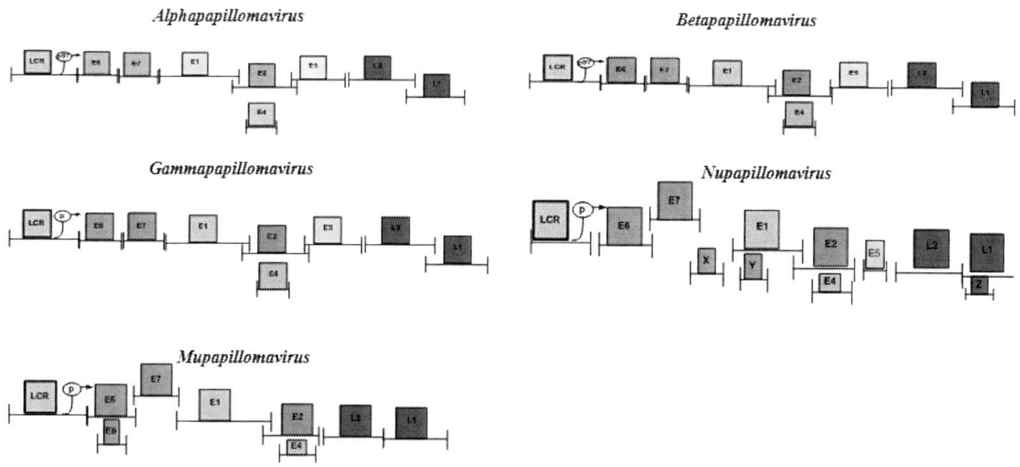

Figure 3. PVs genome in a liner format of *Alphapapillomavirus, Betapapillomavirus, Gammapapillomavirus, Mupapillomavirus and Nupapillomavirus* genera. These genera of virus infect the humans and monkeys and commonly presents E1, E2, E4, E5, E6, E7, L1 and L2 genes. However, the Nupapillomavirus present coding region called X, Y and Z, which possess na unknown function. The Mupapillomavirus possess E8 gene and do not possess the E5 gene.

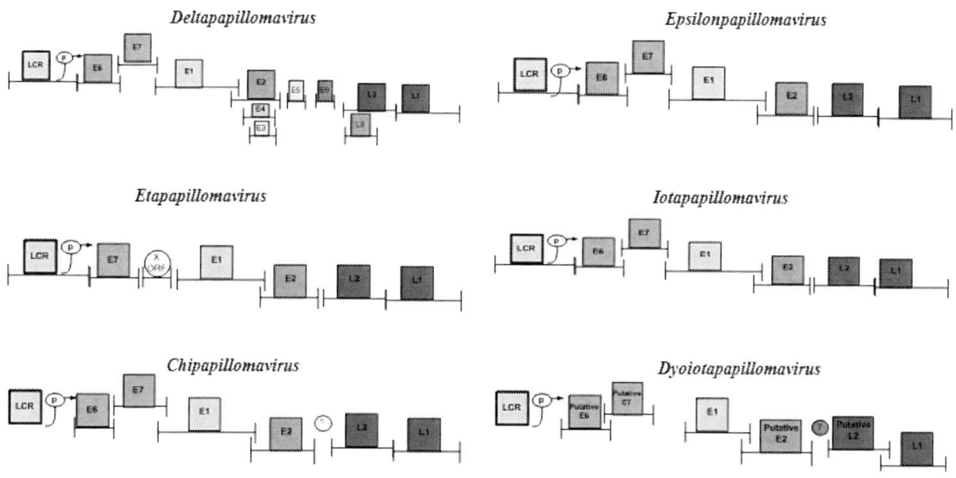

Figure 4. PVs genomes in a liner format of *Deltapapillomavirus, Epsilonpapillomavirus, Etapapillomavirus, Iotapapillomavirus, Chipapillomavirus, Dyoiotapapillomavirus* genera. PVs E3, E9 and L3 genes are present in *Deltapapillomavirus* genus and they do not still possess a known function. *Epsilonpapillomavirus, Etapapillomavirus, Iotapapillomavirus, Chipapillomavirus and Dyoitopapillomavirus* genera do not have E4 gene.

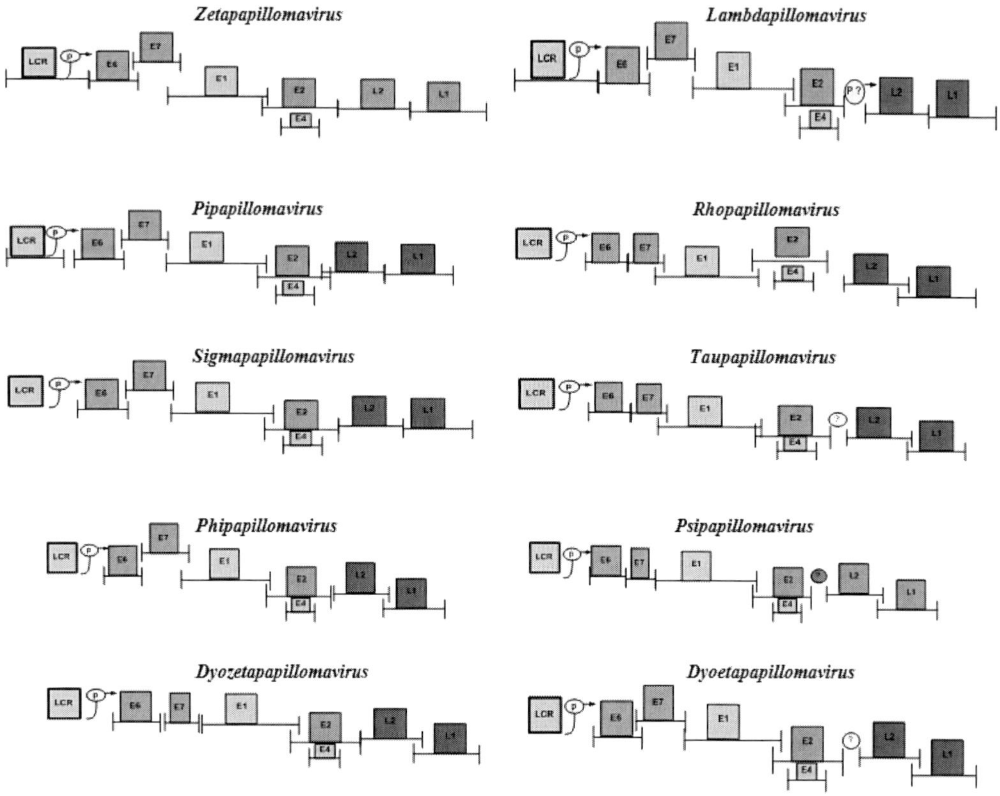

Figure 5. PVs genomes of *Zetapapillomavirus, Lambdapapillomavirus, Pipapillomavirus, Rhopapillomavirus, Sigmapapillomavirus, Taupapillomavirus, Phipapillomavirus, Psipapillomavirus, Dyozetapapillomavirus and Dyoetapapillomavirus* genera. These genera groups do not have E5 gene (Other genera do not present E5 gene, such as *Thetapapillomavirus, Omikronpapillomavirus, Dyoepsilonpapillomavirus, Mupapillomavirus, Epsilonpapillomavirus, Iotapapillomavirus, Etapapillomavirus, Chipapillomavirus and Dyoiotapapillomavirus).* The *Lambdapapillomavirus*, *Taupapillomavirus*, *Psipapillomavirus* and *Dyoetapapillomavirus* possess a non-coding region between E2 and L2 genes.

Although these viral groups have been divided, two papilloma/polyoma virus hybrids were detected, in which there are homologous sequences in early genes to polyomavirus and late genes to papillomavirus. However, these viral groups were not included in Papillomavirus [13].

Actually, the ICTV categorized Papillomavirus into family, genus, species, strain and type-species [13]. Furthermore, PVs were classified through the L1 nucleotide sequence identity, since there was not a recognition culture or serological methods to classify PVs yet. In 2010, *Papillomaviridae* family comprised 29 genera, 189 PVs complete genome deposited in GenBank, being 69 non-human PVs, three infecting birds and two infecting reptiles PVs [13]. To date, there are 151 HPV complete genomes deposited in GenBank http://www.ncbi.nlm.nih.gov/genbank/. PVs genera are written in Greek Alphabet, comprising the following: *Alphapapillomavirus*, *Betapapillomavirus*, *Gammapapillomavirus*,

Deltapapillomavirus, Epislonpapillomavirus, Zetapapillomavirus, Etapapillomavirus, Thetapapillomavirus, Lotapapillomavirus, Kappapapillomavirus, Lambdapapillomavirus, Mupapillomavirus, Nupapillomavirus, Xipapillomavirus, Pipapillomavirus, Rhopapillomavirus, Sigmapapillomavirus, Taupapillomavirus, Upsilonpapillomavirus, Phipapillomavirus, Chipapillomavirus, Psipapillomavirus, Omegapapillomavirus, Dyodeltapapillomavirus, Dyoepsilonpapillomavirus, Dyozetapapillomavirus, Dyoetapapillomavirus, Dyothetapapillomavirus, Dyoiotapapillomavirus and *Omikronpapillomavirus* [13]. However, it is well known that the PV diversity is underestimated and a large number of PVs will be detected in the next years. To avoid the confusion caused by the nomenclature, the ICTV still recommends the use of Greek Alphabet to new employed genera, by adding the prefix Dyo, such as *Dyodeltapapillomavirus, Dyoepsilonpapillomavirus, Dyozetapapillomavirus, Dyoetapapillomavirus, Dyothetapapillomavirus* and *Dyoiotapapillomavirus*.

Evolution of Papillomaviruses

The PV diversity can be explained by multiple evolutionary mechanisms [30]. Virus host-divergence is an important evolutionary force. However, this force solely cannot explain the evolution of PVs and their diversity. Thus, alternative mechanisms such as within-host virus duplication, recombination, viral sorting, or viral adaptation after a host switch, may contribute to explain the PV diversification [31–33].

Although it is not clear how these evolutionary mechanisms act on PVs diversity, some hypotheses were raised through the use of robust methods. Multiple evolutionary mechanisms such as co-evolution, recombination, adaptive radiation and interspecies transmission should drive the PVs diversification [30]. At least six PVs lineages should be associated with an ancestral mammalian host, thus giving rise to the PVs types that infect today's mammalian hosts [31]. The importance of alternative mechanisms other than co-divergence was suggested after the use of different statistical approaches to assess topological and branch-length congruence [32]. Batista et al. [34] used another entropy-based statistical approach to assess the evolution of PVs, showing that the genome has regions that could be used as markers to infer PVs phylogeny.

The understanding of PVs diversity is still limited, probably underestimated. Therefore, more new PVs types should be sequenced to increase our knowledge about PV evolution. The diversity of subtypes and variants could show a more detailed and refined scenario of PVs diversification, thus increasing insights into the representativeness of each PVs type [33].

Papillomaviruses Infection Cycle

The infectious cycle of PVs depends on the differentiation of keratinocytes [11]. The infection starts in the basal layers through micro-lesions or abrasions in the stratified epithelium, which occur in the skin or mucosa by direct contact [11]. From there, the infection could follow three ways: latent infection, clinical or subclinical manifestations [35]. After the entry of the virus into the cell, the viral genome is stabilized in the form of extra-chromosomal elements (episomes) in the nucleus and the number of copies is increased to approximately 50 to 100 per cell. When these infected cells divide, the viral DNA is equally distributed between the daughter cells. One of the infected cells migrates from the basal layer towards the upper layers of the epithelium in order to monitor the process of cellular differentiation of keratinocytes, while the others continue dividing in the basal layer, serving as a reservoir for viral DNA for subsequent cell divisions [36,37]. Although PVs infections occur in the basal layers, the production of the virus is restricted to the suprabasal layer cells, which are not lysed by the production of virions, thus continuing the proliferation. This dependent differentiation promotes the infection and the maintenance of PVs in the basal layers for periods up to several years [38].

From the time of the infection to the release of viral particles goes approximately three weeks, corresponding to the period required for differentiation and desquamation of the basal keratinocytes [39]. The period between initial infection and the appearance of papillomas may be related to the concentration and type of the PVs that infect the tissue [40]. These viruses can cause lesions on the surface of mucosal and cutaneous tissue. Typically, these lesions regress spontaneously although persistent infection of a subset of these viruses can lead to the formation of malignant tumors [41–44].

Due to the fact that HPV and BPV have a great medical and veterinary importance, this chapter will focus on the recent insights into HPV and BPV diversity.

HUMAN PAPILLOMAVIRUS

Human papillomavirus (HPV) are a diverse group of virus that infect humans and monkeys and they are associated with several benign and malign diseases [13,41], notably cervical cancer. Cervical cancer is the second cause of gynecological cancer worldwide [45]. According to WHO, more than 500.000 new cases arise and 250.000 women die per year. Among the cases of cervical cancer, 85% are squamous cell cancer, followed by adenocarcinoma and small cell neuroendocrine tumor [7]. Actually, the prevention strategies to cervical cancer include papanicolaou analysis and excision of precancerous lesions. In the last years, the screening programs allowed the decrease in the number of cases worldwide, however cervical cancer remains as one of the major causes of death in female worldwide. In the developing countries, there is a high frequency of cervical cancer (86%) when compared to developed countries, probably due to inefficient screening programs [6].

Several studies have demonstrated that the HPV infections are associated with cervical cancer [6,41,46]. Also, HPV infection seems to be an important cofactor involved in other anogenital cancers. In 2010, statistical analysis of WHO showed that there are HPV infections in 75% of anal cancer patients [6]. Moreover, HPV-16 and HPV-18 were the most frequent type, detected in 73% of the anal cancer cases. Similarly, the HPV infection is involved in 40% of vulvar cancer, in which 36% of the patients are infected with HPV-16 and HPV-18 [6]. 66% of vaginal cancer patients are infected with HPV, notably by the HPV-16 and HPV-18. Penile cancer seems to have a relationship with HPV infection, since 46% of patients are infected with HPV and 35% could be infected with HPV-16 or HPV-18.

As mentioned previously, 151 HPV types were divided into five genera: *Alphapapillomavirus, Betapapillomavirus, Gammapapillomavirus, Mupapillomavirus* and *Nupapillomavirus* [http://pave.niaid.nih.gov/#home] [12]. In 2010, it was reported that the

Alphapapillomavirus genus was constituted by 74 HPV types, *Betapapillomavirus* genus has 41 HPV types, *Gammapapillomavirus* has 16 HPV types, *Mupapillomavirus* has two HPV types and *Nupapillomavirus* has only one specie (see Table 2) [13]. *Alphapapillomavirus* and *Betapapillomavirus* can be found in humans and monkeys. *Aphapapillomavirus*, *Betapapillomavirus* and *Mupapillomavirus* are involved in benign and malign tumors, while the *Gammapapillomavirus* and *Nupapillomavirus* seem to be involved in benign diseases only [13]. All HPV types are distributed worldwide and can be transmitted by direct contact.

Among these genera, 40 HPV types infect the genital tract, being classified according to the oncogenic potential in High Risk (HR) HPV: 16, 18, 31, 33, 35, 39, 45, 51, 52, 56, 58, 59, 68, 73, 82; and Low Risk (LR) HPV: 6, 11, 42, 44, 51, 53, 83 [4]. The most common HR-HPV types worldwide are the 16 (57%), 18 (16%), 58, 33, 45, 31, 52, 35, 59, 39, 51 and 56 [47]. About 90% of the LR-HPV infections are caused by the HPV-6 and HPV-11 (see Table 1).

HR-HPV infection is commonly transmitted by sexual intercourse and can cause lesions in epithelial tissues. These lesions can regress in 6-12 months [4]. However, if not treated, these lesions can progress from pre-malign to cervical cancer, such as: cervical intraepithelial neoplasia grade 1 (CIN1) (or mild dysplasia); intraepithelial neoplasia grade 2 (CIN2) (or moderate dysplasia); intraepithelial neoplasia grade 3 (CIN3) or *in situ* carcinoma; and squamous cell carcinoma or adenocarcinoma [4]. The Bethesda System is based on papanicolaou analysis and it classifies these pre-malign lesions into: atypical squamous cells of undetermined significance (ASCUS); and squamous intraepithelial lesion (SIL), which comprises the low-grade SIL (mild dysplasia) and high-grade SIL (moderate severe dysplasia and *in situ* carcinoma) [48]. The cervical intraepithelial neoplasia grade 2/3 regress in 28% of the patients in a short period of time [49].

HPV Genome

Similarly to the PVs, HPV DNA has eight open reading frame (ORFs), namely E1, E2, E4, E5, E6, E7 (expressed in early phase of infection), L1, L2 (expressed in late phase of infection) and a control region denominated long control region (LCR) [4,11] (Figure 3). E1 and E2 ORFs are involved in viral DNA replication. The E4 gene is implicated in maturation of HPV virion. The E5, E6 and E7 genes are involved with transformation activities, in which E6 and E7 are involved with integration of HPV DNA within the cell DNA. Thus, HPV DNA can immortalize human keratinocytes due to the E6 and E7 oncogene interactions with p53 and pRb suppressor tumor proteins, respectively. These interactions inhibit the apoptosis process (see above).

L1 and L2 are involved in capsid formation. Moreover, HPV DNA possesses a LCR, which regulates the transcription of the E6 and E7 viral oncogenes through transcription factors of the virus and the host cells [11] (Figure 6).

Table 1. Characteristics of HPV from *Alphapapillomavirus*, *Betapapillomavirus*, *Gammapapillomavirus*, *Mupapillomavirus* and *Nupapillomavirus* genera. All HPV are found worldwide and can infect mucosa and epithelium through contact. *Gamapapillomavirus* and *Mupapillomavirus* HPV can cause begin diseases, such as warts and papillomas. On the other hand, *Alphapapillomavirus*, *Betapapillomavirus* and *Nupapillomavirus* can cause benign and malign diseases, notably cervical cancer, anal cancer, penile cancer and other

Genus	Geografical localization	Tissue tropism/Host	Transmission	Diseases associated	Types/Reference
Alphapapillomavirus	Worldwide	Epihtelium and mucosa Humans, Monkeys	Sexual contact, auto-inoculation	Ginecological malignt and benign tumors	PpPV-1, HPV-2, HPV-3, HPV-6, HPV-7, HPV-10, HPV-11, HPV-13, HPV-14, HPV-16, HPV-18, HPV-28, HPV-31, HPV-26, HPV-27, HPV-29, HPV-30, HPV-31, HPV-32, HPV-33, HPV-34, HPV-35, HPV-39, HPV-40, HPV-42, HPV-43, HPV-44, HPV-45, HPV-51, HPV-52, HPV-53, HPV-54, HPV-56, HPV-57, HPV-58, HPV-59, HPV-60, HPV-62, HPV-66, HPV-67, HPV-68, HPV-69, HPV-70, HPV-71, HPV-72, HPV-73, HPV-74, HPV-77, HPV-78, HPV-81, HPV-83, HPV-84, HPV-86, HPV-87, HPV-89, HPV-90, HPV-82, HPV-85, HPV-91, HPV-94, HPV-97, HPV-102, HPV-106, HPV-117, MmPV-1, MfPV-3, MfPV-4, MfPV-5, MfPV-6, MfPV-7, MfPV-8, MfPV-9, MfPV-10, MfPV-11. [13]
Betapapillomavirus	Worldwide	Epihtelium and mucosa	Contact	Skin benign and malign	HPV-5, HPV-8, HPV-9, HPV-12, HPV-14, HPV-15, HPV-17, HPV-19,

Genus	Geografical localization	Tissue tropism/Host	Transmission	Diseases associated	Types/Reference
		Humans Monkeys		tumors	HPV-20, HPV-21, HPV-22, HPV-23, HPV-24, HPV-25, HPV-36, HPV-37, HPV-38, HPV-47, HPV-49, HPV-75, HPV-76, HPV-80, HPV-92, HPV-93, HPV-96, HPV-98, HPV-99, HPV-104, HPV-105, HPV-107, HPV-100, HPV-110, HPV-111, HPV-113, HPV-115, HPV-118, HPV-120, HPV-122, HPV-124, CgPV-2, MfPV-1. [13]
Gammapapillomavirus	Worldwide	Epihtelium and mucosa Humans	Contact	Skin benign tumors	HPV-4, HPV-48, HPV-50, HPV-60, HPV-65, HPV-88, HPV-95, HPV-101, HPV-103, HPV-108, HPV-109, HPV-112, HPV-116, HPV-119, HPV- 121, HPV-123. [13]
Mupapillomavirus	Worldwide	Epihtelium and mucosa Humans	Contact	Skin benign tumors	HPV-1, HPV-63 [13]
Nupapillomavirus	Worldwide	Epihtelium and mucosa Humans	Contact	Skin benign and malign disease	HPV-41[13]

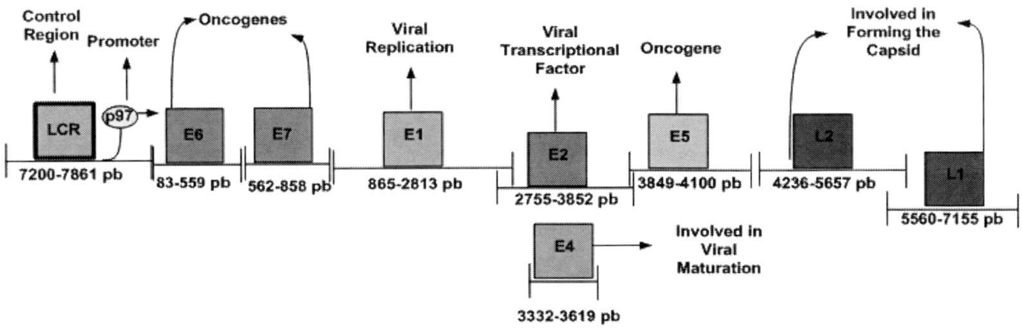

Figure 6. HPV-16 complete genome is shown in a linear format (based in RefSeq NC_001526.2). After the break in stratified epithelium tissue, the HPV can infect basal cell layer, thus starting an infection process. The *entry of HPV* into the *host cell* is initiated by the binding of the virus particles into the cell surface receptors, such as the heparan sulfate (not shown). Subsequently, there are expression of E1, E2 (to replication of viral DNA), E4 (probably involved in maturation of virion), E5, E6 and E7 genes (involved with transformation activity and immortalization), which results in replication of the HPV DNA as an *episome*. L1 and L2 genes are expressed in a late stage, thus resulting in viral capsid formation. Finally, the HPV virion is released at the cell surface, which can result in infection of other sites.

Integration of Human Papillomavirus in Host Genome

Sexually active women are at risk for genital infection by HPV. Most genital HPV infections regress and only a minority of women will develop persistent HPV infection that could cause cervical intraepithelial neoplasia (CIN) which is a precursor of invasive cervical cancer. The integration of high oncogenic risk HPV genome into the host chromosome is considered to be a key event in the progression of CIN to invasive cervical cancer [11], occurring downstream of the early genes E6 and E7, often in the E1 or E2 region. This disruption results in the loss of negative feedback control of oncogene expression by the viral regulatory E2 protein [10].

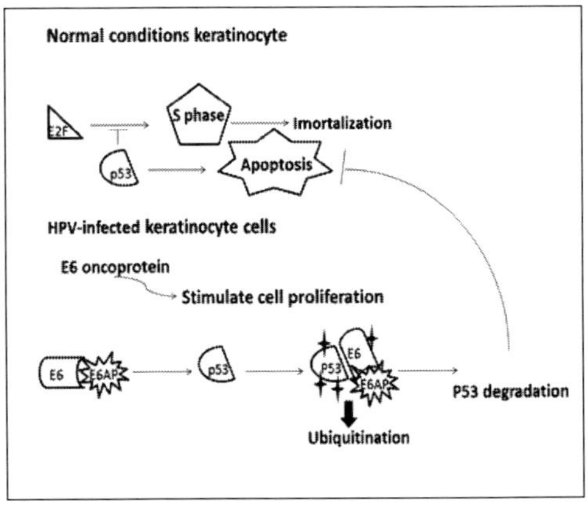

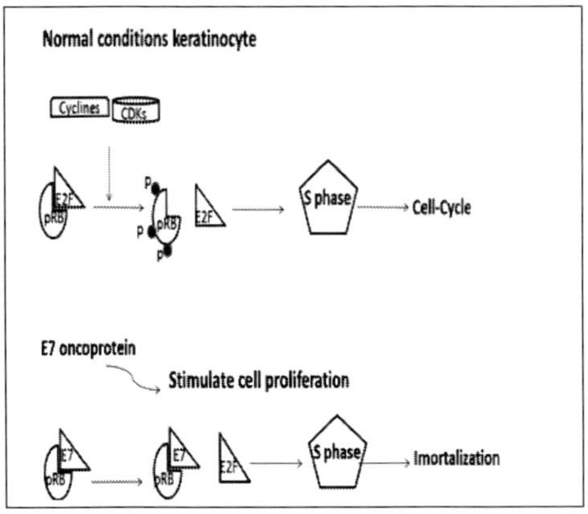

Figure 7. Transformation and immortalization effects of HPV E6 and E7 oncoproteins in keratinocyte cells. Briefly, HPV-E6-AP complex is able to *target p53* for *degradation* via the *ubiquitin-proteasome* system, leading to indeterminate division and immortalization of cells. Furthermore, E7 oncoprotein also leads to immortalization of the cell due to the regulation of pRB, releasing the E2F activity.

This event is often followed by over-stimulation of the transcription of genes E6 and E7 that disrupt the cell cycle through interaction with different proteins, such as the product of the tumor suppressor gene (p53) and the retinoblastoma protein (pRb) inducing viral immortalization capacity, and a high level of chromosomal instability (Figure 7) [50,51].

Several studies of HPV DNA integration have focused mainly on HPV-16 since it is considered the type that confers the highest risk of cervical cancer and also the type most commonly detected in women with normal cervical cytology [52]. In low-grade cervical lesions, the majority of HPV genomes persist in episomal state, whereas in high-grade lesions and invasive carcinomas have integrated HPV DNA [53].

HPV Distribution

HPV-16, HPV-18, HPV-31 and HPV-52 are the four most frequent types found in health women (without lesions) worldwide [6]. However, the frequency of HPV types change according to grade lesion cases, cervical cancer cases, and the where the patients live. For example, in low grade lesion, the fourth most frequent HPV types are 16, 51, 31 and 52, while in high grade lesion the fourth most frequent HPV types are 16, 31, 58 and 33 [6]. On the other hand, the HPV-16, 18, 58 and 33 are the most frequent types in patients with cervical cancer worldwide [6]. Furthermore, the HPV types associated to cervical cancer in developing countries and developed countries are different. In the developing countries, the most frequent types are the HPV-16, 18, 58 and 33, while in developed region the most frequent types are the HPV-16, 18, 33, and 31 [6]. The contrast in the prevalence between the HPV types seems also to occur concerning the cancer type. Indeed, the most common HPV types in squamous cell cancer are the species 16, 18, 58 and 33. Moreover, in adenocarcinoma cancer, the most frequent HPV types are 18, 16, 45 and 33 [6].

According to WHO, the prevalence of HPV in health women in Latin American and Caribbean is 17%. In addition, the prevalence of HPV in women with low grade, high grade lesion and cervical cancer are 22%, 46% and 66%, respectively [6]. In Brazil, the prevalence of HPV in women with normal cytology is 14%. Concerning the cancer type, the four most prevalent HPV types in squamous cell carcinoma in Brazil are HPV-16, 18, 35 and 45 [54]. On the other hand, HPV-18, 16, 33 and 35 types are the most frequent in adenocarcinoma [54]. Also in Brazil, HPV-16 is the most prevalent virus involved in carcinogenic activity; however, the second most important type changes according to the Brazilian regions. For example, in Northeastern Brazil the four most commonly HPV types are 16, 31, 33 and 18 [55]. In contrast, in Midwest region, the four most commonly HPV types are 16, 18, 31 and 33 [56]. Thus, differences on the prevalence of HPV types might have negative impact in adoption of public policy, concerning mass vaccination.

HPV Genetic Variability and Evolution

An HPV type differs from another in at least 10% nucleotide sequence of the L1 gene. The additional term "subtype" is used to identify HPV genomes with L1 nucleotide sequences that differ between 2-10% of that of the closest type, and the varying types differ in less than 2% in nucleotide sequence of L1, and 5% in LCR [12].

Substantial data demonstrate the existence of a natural diversity of variants of a determined type of HPV [56,58]. Different HPV types have distinct oncogenic potential and, the same way, intratypic variants may also have differential oncogenicity causing the development of cervical cancer [60–62]. Moreover, these differences may contribute to disparities in the incidence of cervical cancer worldwide [63].

Studies related to genetic variability revealed that different variants of HPV-16 and HPV-18 co-evolved with the three major human phylogenetic branches: African, Caucasian and Asian. However, HPV-16 variants were grouped into five distinct groups distributed in different geographical regions, such as European (E), Asian (As), Asian-American (AA), an African 1 (Af-1) and African 2 (Af-2) [64–66]. The variants of HPV-18 were properly grouped into three distinct groups: European (E), Asian-American (AA) and African (Af) [67–69]. In relation to HPV-31, it is suggested that HPV-31 variants do not exhibit the same patterns above mentioned with respect to the co-evolution with human ethnic groups [58,70]. However, Northeastern Brazil is characterized by strong ethnical admixture, possibly explaining isolates distribution in different clades of the phylogenetic trees. Another explanation would be that these HPV-31 variants might have existed before the spread of humans and the clusters specific for the ethnic groups arose subsequently, as suggested by Calleja-Macias et al. [63,70].

The possible variations in the nucleotide sequences of HPV have been used as an important tool to assist in epidemiological studies of transmission and viral persistence [71]. The differences in nucleotide sequences between the variants of one type of HPV can result in changes in amino acids encoded and can lead to distinct oncogenic potential between the different variants. Variations in the nucleotide sequence of the LCR region, which contains sequences important for the regulation of viral replication and transcription of early genes could result in loss of the binding capacity of a specific transcriptional factor, as well as changing its recognition site and the affinity of a specific transcription factor [72]. With respect to the L1 region, one or more amino acid changes could lead to a conformational change in the capsid protein, interfering with the conformation of epitopes relevant for viral neutralization and interference in vaccine strategies [73]. It can also be said that the importance attached to nucleotide variations in E6 and E7 genes is directly related to the occurrence of variations in critical regions for the degradation of p53 and pRb, respectively, and the regions involved in immune recognition of the host [74].

Due to the high prevalence of HPV-16 and HPV-18 in the world, not only among asymptomatic women, but also in samples of cervical cancer, several epidemiological studies have been conducted to describe the molecular variants of HPV types of high oncogenic risk [75–83]. Studies suggest that European variants are associated with a lower risk of developing high-grade lesions, whereas non-European variants are associated with an increased risk of incidence and prevalence of these lesions [79,82,84–86]. Moreover, it has been demonstrated an association between non-European variants, both HPV-16 and HPV-18, and an increased risk of persistent infections [73,85].

Genetic variability studies performed on HPV-16 suggested that each viral variant is associated with risk of cervical neoplasia. Londesborough et al. [78] studying the genetic variability in HPV-16 E6 gene in women from UK reported a nucleotide substitution of thymine for guanine at the position 350 of the gene, leading to a substitution of valine for leucine at position 83 regarding the amino acid sequence of this protein. It was suggested that there was an association between this change and an increased risk of persistent infection and cytological progression to cervical intraepithelial neoplasia of grade 2/3 [78]. In studies performed by Kammer et al. [87] and Zehbe et al. [83], this type of change in E6 was associated with an increased risk of cervical disease progression compared with the prototype strain. de Boer et al. [88] found a high frequency of nucleotide changes in the virus type 16 in cervical cancer patients from Indonesia. Among those changes stands the nucleotide variation adenine for guanine at the position 276 of E6 ORF, which promotes the change of the amino acid asparagine to serine at the position 58 of E6 protein [89,90], describing the same change in cervical intraepithelial neoplasia and cervical cancer. Another study reported a change in the nucleotide position 647 in the E7 gene of HPV-16, with a change in the amino acid asparagine to serine at the position 29, and this variation in E7 was more frequent in samples of cervical cancer compared with precursor lesions, suggesting alterations in the biological properties of this variant [91].

With respect to genetic variability of E6 of HPV-18, studies have shown the nucleotide change C491A, which revealed the amino acid change N129K. This codon 129 is highly conserved within genotypes oncogenic HPV [56,92]. However, in vitro analysis has shown that this change to E6 does not interfere with their ability in promoting p53 degradation [93]. Variations in LCR of HPV-18 were also observed, among these variations A41G and T104C [56]. Both variations in this region are associated with high activity of the E6/E7 promoter by modulating Sp1 and YY1 activities [94]. The substitution observed in the codon 149 of L1 gene in HPV-18, is located near a linear epitope, which is present in the surface of the capsid and in common among many HPV genotypes [56,95]. Changes in the L1 gene may compromise the fitness of the virus and may be important in discriminating the infectious potential of different variations and in defining epitopes relevant to vaccine design [73,96,97]. Studies report that specific variants of HPV-18 may increase the oncogenic potential, which may represent an additional increase in the number of patients with cervical cancer in different regions [69,84,86,98,99].

Considering the genetic variations of HPV-31 E6, Chagas et al. [70] detected nucleotide changes at positions 213 and 413, which were previously described [58,101], however changes at positions 141, 190 and 368 were described for the first time in his study, where changes at positions 190, 368 and 413 caused amino acid exchanges in codons 64 (T to A), 123 (K to R) and 138 (A to V), respectively. Still in this work, all mutations found in E7 gene of HPV-31 (positions 67, 111, 136 and 184) were described for the first time, among which 67, 136 and 184 were classified as non-synonymous, causing amino acid changes in codons 23 (H to Y) 46 (E to Y) and 62 (K to E), respectively. Many of observed genetic variations in E6 and E7 HPV-31 were located in T-cell and B-cell epitope sites [70]. These variations may influence the presentation of viral peptides to the T-cell, being this an important mechanism for the control of infection and development of cervical lesions. The recognition of T-cell determinants by T cells (particularly T helper cells) contributes significantly to the cell-mediated immune response against infectious organisms, as T cells cooperate with B cells in the induction and maintenance of an effective antibody response and contribute to the maturation of cytotoxic T cells by interacting with macrophages [102]. Nucleotide and amino acid analysis were also performed to L1 gene of HPV-31, where it was revealed the presence of variants nucleotide. Of these, nine led to amino acid changes in L1 protein. The mutations identified with higher prevalence were the two synonymous A6943C and T6949A. This last mutation was associated with high grade cervical lesions [103]. Analysis of HPV-31 LCR sequence variations performed by Cento et al. [103] revealed some substitutions (G7449A,

G7457A, C7474T, G7525A and T7575C) which potentially affected biding sites for transcription factors. In particular, the nucleotide substitution (G to A) at position 7749 may cause a gain of YY1 and a loss of SRY binding sites. An SRY binding site may be also affected by the T7659G mutation. Similarly, the mutation A7450C may cause the loss of the TATA binding site. Other variations at nucleotide positions 7457 (G to A) and 7474 (C to T) may result in a gain of YY1 biding site and in a loss of E2 biding site, respectively [103].

One early PV lineage, probably with mucosal tropism, originated the supertaxon comprising by *Alphapapillomavirus* and *Omicronpapillomavirus*, which includes viruses infecting primates, carnivores, swine and cetaceans. *Alphapapillomavirus* have viruses associated with cervical cancer in humans [105]. A second early PV lineage gave rise to a clade of cutaneotropic PVs including *Betapapillomavirus* and *Xi-papillomavirus*, that comprise primates, rodents, carnivores, ruminants, hedgehogs and marsupials [14]. Many viruses of this supertaxon have found in healthy skin and are apparently not associated with lesions [106]. Others cause malignant transformation depending on the genetic susceptibility of the host, for example, infections by HPV-5 or HPV-8 are associated with the development of squamous cell carcinoma in individuals suffering from the hereditary disease epidermodysplasia verruciformis [107]. A fourth PV lineage gave rise to the supertaxon including *Kappapapillomavirus* and *Lambdapapillomavirus*, with primates, rodents, lagomorphs and carnivores but not humans as hosts [14].

BOVINE PAPILLOMAVIRUS

Bovine papillomaviruses (BPVs) are a heterogeneous group of viruses that infect bovine cattle and other close related animals. They cause bovine papillomatosis, a worldwide distributed veterinary disease, which leads to important economic losses. The world cattle population has exceeded 1 billion head of cattle. India, Brazil and China have the largest cattle population in the world, whereas USA, Brazil and China are the largest beef producer, respectively [108]. India has around 320 million head of cattle and Brazil has about 200 million head of cattle, however the majority of Indian cattle population is non-commercial. In 2011, it was produced nearly 57 million tons of bovine meat. USA produced about 12 million tons of bovine meat and Brazil produced around 9 million tons of bovine meat [108]. The world's cow milk production is around 583 million tons in 2009. Brazil is the fifth milk producer in the world with about 29 million tons in 2009 [109].

Bovine papillomatosis leads to economic depreciation of animals, deterioration of the animal leather, decrease of milk production through mastitis, and decrease of beef production through weight loss and death by cancer. The lesions may progress to cancer due to the synergistic action of genetic or environmental co-factors [110,111].

Currently, 12 BPVs are known and classified into three genera: *Deltapapillomavirus* (BPV-1 and -2), *Epsilonpapillomavirus* (BPV-5 and -8) and *Xipapillomavirus* (BPV-3, -4, -6, -9, -10, -11 and -12). One BPV type is still unclassified (BPV-7) [12,13, 111]. All BPV types characterized so far are associated with different histopathological lesions. *Xipapillomaviruses* are pure epitheliotropic (causing true papillomas), *Deltapapillomaviruses* induce fibropapillomas and *Epsilonpapillomaviruses* appear to cause both fibropapillomas and true papillomas [110]. Members of *Deltapapillomavirus* genus are also associated with sarcoids in horses and other animals [112,117]. As far as we know, three types of BPV (-1, -2 and -4) are associated with cancer. When connected with environmental factors (e.g. bracken fern ingestion), BPV-1 and -2 cause urinary bladder tumors, and BPV-4 causes upper alimentary tract cancer.

The genome of all PVs are very similar, however some interesting features are noted in BPV genomes. First, unlike HPV, the major transforming protein in BPVs is E5. In *Xipapillomavirus* genus, E6 gene is replaced by an E5-like gene (initially defined as E8). Nevertheless, the knowledge that most functions of this protein are shared with the E5 protein of BPV-1 prompted its redefinition as E5 protein, not E8 [44, 119]. Similarly to most PVs types, BPV types that belong to *Deltapapillomavirus* and *Epsilonpapillomavirus* still retain the E6 gene [110]. In addition, other unusual genes are present in some BPV types: E3 (a small putative gene) in BPV-2 and -4; L3 (a small ORF that overlaps with 3' half of L1 or L2 gene) in BPV-3, -4 and -12; and L4 (a small putative gene) in BPV-4. The significance of these genes is still unclear.

Epidemiology

Papillomas are more often found in cattle than in any other domestic animal. In Brazil, for instance, there is a high prevalence of this disease in cattle in all regions of the country. Bovine of any age could be affected, however calves and yearlings have higher incidence, 46.21% in one study [122]. It is likely that specific factors like nutrition, ticks infestation and age contribute to this incidence. Also, the immune state of the affected cattle population influences the pathology. Among the breeds, Holsteins present a higher occurrence of the disease, although Olson et al. [123] observed that there was not a strong genetic influence in experimentally induced papillomas. Papillomatosis in dairy cattle seems to prefer udder and teats areas in adult animals and heifers. In calves, the most affected sites are: the neck, dewlap, muffle, ears, muzzle, periocular and perilabial regions [124–126].

The transmission between animals is probably made by fomites: contaminated milking equipment, halters, nose leads, grooming and earmarking equipment, rubbing posts and wire fences, and other objects contaminated by contact with affected cattle. Warts are rare in animals that are artificially inseminated, so sexual transmission of these lesions is probable in cattle. The way the cattle are held influences the incidence of bovine papillomatosis. Housed cattle are more often affected with bovine papillomatosis than cattle on pasture. Some BPV hosts could get naturally infected with BPV after exposure to diseased cattle or contaminated fomites in [125,122]. Ticks are probably a major problem in bovine papillomatosis transmission. Willian et al. [1992] observed that all the diseased animals were infested with ticks. A study showed that some tick control measures reduced the incidence of bovine papillomatosis [55,127].

Bovine papillomavirus is widespread in normal cattle skin and dermal tissues. The epidemiological importance of these latent infections is to be elucidated, however warts may grow, at sites of injury, through the reactivation of the latent DNA by physical irritation, immune depression and suppression. The lesion formation may occur in the absence of any previous clinical symptom, indicating a latent infection [122,128].

Until recently, it was believed that BPVs were associated with distinct lesions, in different anatomic sites. BPV-1, -2 and -5 infect both mesenchymal and epithelial cells, causing common cutaneous warts and fibropapillomas [122]. On the other hand, BPV types that are members of *Xipapillomavirus* genus induce non-fibroblast epithelial and cutaneous lesions [122]. According to Borzacchiello and Roperto [110], BPV-1 causes teats and penile fibropapillomas, and urinary bladder tumours; BPV-2 is associated to cutaneous warts (usually on the forehead, neck, upper thorax and back), alimentary fibropapillomas and urinary bladder tumours; BPV-3 causes cutaneous papillomas [129]; BPV-4 is associated with pure epithelial papillomas of the upper gastrointestinal tract; BPV-5 induces rice grain fibropapillomas of the udder; BPV-6, frond papillomas of the teats; BPV-8 causes cutaneous papillomas; BPV-9 and -10 are associated to epithelial squamous papillomas of the udder.

In addition, BPV-4 is a viral type related to mucosal infections and it is responsible for upper digestive tract cancer in cattle when associated with bracken fern (*Pteridium aquilinum*). However, BPV-4 was already detected in skin lesions [52,130], BPV-5 and -6 are usually associated with skin lesions in udder and teats, but they were found in lesions of the shoulder clod and around the eye of some animals, unusual sites for these viral types [130].

Although BPVs are considered an exclusively infecting epithelial cells group of viruses, there is evidence for the presence of BPV DNA, RNA and proteins in peripheral blood mononuclear cells [110,131–137], placenta and amniotic liquid [132,136], gametes [131,138,139], urine [139], uterus [133,136] ovarium [133], semen [133,139,140], also in milk and colostrum [132,139]. These findings could support the hypothesis of the BPV vertical transmission [130–132,141].

BPV Types Distribution

Bovine papillomavirus has been detected in cattle worldwide. Several studies showed the incidence of BPV in all continents [33]. BPV-1 is a viral type frequently found in skin lesions [130,131,132]. In Germany, BPV-1, -6, -8 and -10 were found in bovine warts cowshed [142]. Heifers presented benign teat tumors caused by BPV-6 in Japan [143]. BPV-1, -3, -5, -6, -8, -9 and -10 were detected in papilloma specimens, also in Japan [113,144,145]. In India, bovine cutaneous warts caused by BPV-1 and -2 were reported [119,146]. Recently, [147] identified BPV-10 in teat warts from cattle at a dairy farm also in India. BPV was found in equine sarcoid in western USA and Canada [148,149]. In Africa, BPV was found in skin lesions of zebra, giraffe and antelope [118,120]. In addition, BPV-4 and -2 were detected in cattle in Australia [52,150].

Cattle from Brazil have also been investigated for the presence of BPV. BPV-1, -2, -6 and -8 were identified in skin warts of cattle from southern Brazil [151–153]. In northeastern Brazil, ten different types of BPV were also detected in skin lesions. Despite BPV-3 is rarely reported in the world, since the majority of studies about BPV infection focus on BPV-1 and -2, there is a high incidence of BPV-2 and -3 in northeastern Brazil [130]. BPV was extensively detected in other tissues in Brazil [139,140,154,155].

BPV is also associated with cancer in cattle. BPV-4 infection and associated tumors of upper alimentary tract have been found in Brazil, the Nasampolai Valley of Kenya, Western Highlands of Scotland and in southern Italy [156,157]. BPV-1 and -2 associated field cases of urinary bladder cancer in cattle were reported in continental Europe, Azores Islands, and some regions of Kenya, Brazil, New Zealand, India and China [110].

Some reports describe the occurrence of co-infection with different types of BPV worldwide [119,130,136,139,142,144,158,159]. In Brazil, a high co-infection rate was found. The animals were constantly affected by diverse warts over the body, which could be a consequence of co-infection. This infection with multiple BPV types could prevent the regression of the lesions. Besides that, the presence of infection with multiple viral types could lead to the suppression of the immunological system making the clinical state of the animals worse [130].

Genetic Diversity and Evolution

The PVs is a very diverse group of viruses, which comprises more than 150 genomes of HPV types, more than 80 genomes of non-human Papillomavirus types, but less than 50 host species have at least one PVs associated. Inside this context, only 12 BPV types are known so far, which is a limited number when compared to HPV types. However, 14 new putative types were isolated and probably will increase considerably that number [144,153,160]. That shows the great genetic diversity of these viruses, as PVs types are defined based on L1 gene sequence dissimilarity [12].

Besides that, PVs are also classified into subtypes and variants, which are important to show a more detailed and refined scenario of PV diversification, increasing insights into the representativeness of each PVs type [33]. The differences found in nucleotide sequences of variants of PVs could be responsible for changes in the oncogenic potential, cellular location, host immune response, function of PVs proteins (affecting pathogenesis), or binding capacity of a transcription factor [70,72,73,161]. The role of the nucleotide polymorphisms in BPV genomes is still scarce. However, they clearly have a great biological and evolutionary importance, based on HPV studies.

A few studies related to BPV genetic variability were done. The majority of these studies were associated with equine sarcoid, aiming at understanding differences between the disease in bovine and equine [154,157]. A novel BPV-11 subtype was identified in Brazil, but the role of this genetic variation is still unknown [130]. Seven different BPV-1 E5 variants were detected in equine sarcoids, one of them was present in all isolates from Europe, a possible synonymous polymorphism characteristic of European variants of BPV-1. One polymorphism was found to alter an aliphatic side chain amino acid residue, suggesting that it could have important implications for the structure of E5 protein. An interesting observation is that most of the sequence variation found in this study was associated with an adaptation to the codon usage in mammalian cells [163].

Five different BPV-1 LCR variants and six different BPV-1 E2 variants were detected in Europe. Three non-synonymous sequence changes occurred within the transactivation domain of E2 and two within the 'hinge' region. In this study, specific BPV-1 LCR and E2 variants were infecting horses, suggesting that sarcoids are linked to variant BPV-1 genomes that preferentially infect horses and are kept within equids [164]. In another study, putative BPV-1 variants from Poland were divided into three phylogenetic groups and one separate isolate, showing the great genetic diversity of this virus in equine sarcoids [165]. All those findings suggest that the genetic variability found in BPVs could lead to different pathological processes, as well as it could increase the adaptation to other hosts. So, it is very important to study and understand the evolutionary mechanisms that guide the diversification of PVs.

In the evolutionary point of view, BPV are found in at least three distantly related lineages. First, BPV that belong to *Deltapapillomavirus* and *Epsilonpapillomavirus* form a paraphyletic group with some ovine PVs (OvPV-1 and OvPV-2), which infects a close related host. Other PVs that infect Artiodactyla are also close relatives of those BPVs. However, this group is clustered together with some equine and canine PVs. Second, BPVs that belong to *Xipapillomavirus* are grouped together with caprine PVs (ChPV-1). This group is related with a large cluster that involves some human, canine and rodent PVs. So far, BPV-7, an unclassified PV, has an uncertain phylogenetic position, which makes it difficult to infer its relatedness to other PVs. Therefore, all this diversity found in PVs that infect one host (*Bos taurus*) is a case of evolutionary incongruence between host and PV phylogeny, indicating that co-divergence alone cannot explain the PV diversity [30–32,34,166].

Some conflicting phylogenetic positions within *Xipapillomavirus*, which include BPV-3, -4, -6, -9, -10, -11 and -12, have been shown when analyzing early or late genes phylogenies [166,167]. In general, the topological inconsistency between early and late genes phylogenies have been explained with ancient recombination events [30,31,166]. This also could be the explanation for the contradicting positions of BPV within *Xipapillomavirus*. For BPV, at least three lineages seem to originate the currently known types. These lineages probably passed through a prior divergence process preceding the host divergence. This could also explain the evolutionary proximity of BPV to PVs that infect distantly related hosts [33]. In addition, zoonotic transmission of PVs is a rare event but it occurs in BPV as they were found in zebras, horses, buffaloes and other animals [117,118,166,168]. We could speculate that such a host switch could lower the evolutionary pressures on BPVs in cattle and may allow changes in infection timing and virulence in other hosts [169]. Therefore, other evolutionary mechanisms could be associated with BPV diversification. However sampling is still a limiting factor [33].

Conclusion

PVs are a diverse viral group that can infect humans, mammalian non-human, birds and reptile species. Nowadays, there are more than 200 PVs identified, which are divided into 29 genera. Basically, the genome of PVs presents genes that are expressed in an early stage and late stage. Furthermore, the PVs can present more than one non-coding sequence, named long control region (LCR). All the PVs present E1, E2, L1 and L2 genes. On the other hand, E3, E4, E5, E6, E7, E8 and E9 can be present or absent, according the genus.

To date, there are 151 HPVs and 12 BPV types described. Due to their medical and veterinary importance, HPV and BPV are the most studied viral groups. Concern to HPV, this viral group is the major cofactor involved in anogenital and non-anogenital cancers. HPV-16 is most prevalent specie in cervical, anal, vulvar, vagine and penile cancers worldwide. However, the subsequent most prevalent HPV types depend on several conditions, such as grade cervical lesion, cervical cancer types and the region where the patients live. Furthermore, variation in nucleotide sequence of HPV can result in changes in amino acid sequences, which can be the important to their oncogenic potential. Thus, these differences in prevalence of HPV types and intratype variability can have an important impact in adoption of public policies. Similarly, BPVs are involved with malign tumors in bovines together with environmental cofactors. As mentioned in HPV, changes in nucleotide sequence of BPVs can result in changes in amino acid sequences, which can lead to oncogenic potential differences. Thus, the understanding of the genetic differences among PVs, notably HPV and BPV, could explain why some HPV or BPV types are most oncogenic while other do not.

REFERENCES

[1] Lange CE, Favrot C, Ackermann M, Gull J, Vetsch E, Tobler K. Novel snake papillomavirus does not cluster with other non-mammalian papillomaviruses. Virology Journal 2011;8:436.

[2] Shope RE. Immunization of rabbits to infectious papillomatosis. J Exp Med 1937;65(2):219-231.

[3] Rous P, Beard JW. The progression to carcinoma of virus-induced rabbit papillomas (shope). J Exp Med 1935;62(4):523-548.

[4] zur Hausen H. Papillomaviruses and cancer: from basic studies to clinical application. Nat. Rev. Cancer 2002;2(5):342-350.

[5] Dürst M, Gissmann L, Ikenberg H, zur Hausen H. A papillomavirus dna from a cervical carcinoma and its prevalence in cancer biopsy samples from different geographic regions. Proceedings of the National Academy of Sciences 1983;80(12):3812 -3815.

[6] Who | strengthening cervical cancer prevention and control [s.d.].

[7] Hoory T, Monie A, Gravitt P, Wu T-C. Molecular epidemiology of human papillomavirus. Journal of the Formosan Medical Association 2008;107(3):198-217.

[8] Pfister H. Chapter 8: human papillomavirus and skin cancer. J. Natl. Cancer Inst. Monographs 2003;(31):52-56.

[9] Gottschling M, Bravo IG, Schulz E, Bracho MA, Deaville R, Jepson PD, Van Bressem M-F, Stockfleth E, Nindl I. Modular organizations of novel cetacean papillomaviruses. Mol. Phylogenet. Evol. 2011;59(1):34-42.

[10] Wood CE, Chen Z, Cline JM, Miller BE, Burk RD. Characterization and experimental transmission of an oncogenic papillomavirus in female macaques. J Virol 2007;81(12):6339-6345.

[11] Doorbar J. The papillomavirus life cycle. Journal of Clinical Virology 2005;32:7-15.

[12] de Villiers E-M, Fauquet C, Broker TR, Bernard H-U, zur Hausen H. Classification of papillomaviruses. Virology 2004;324(1):17-27.

[13] Bernard H-U, Burk RD, Chen Z, van Doorslaer K, Hausen H zur, de Villiers E-M. Classification of papillomaviruses (pvs) based on 189 pv types and proposal of taxonomic amendments. Virology 2010;401(1):70-79.

[14] Bravo IG, de Sanjosé S, Gottschling M. The clinical importance of understanding the evolution of papillomaviruses. Trends in Microbiology 2010;18:432-438.

[15] Bodaghi S, Wood LV, Roby G, Ryder C, Steinberg SM, Zheng Z-M. Could human papillomaviruses be spread through blood? J Clin Microbiol 2005;43(11):5428-5434.

[16] Kay P, Allan B, Denny L, Hoffman M, Williamson A-L. Detection of hpv 16 and hpv 18 dna in the blood of patients with cervical cancer. J. Med. Virol. 2005;75(3):435-439.

[17] Pornthanakasem W, Shotelersuk K, Termrungruanglert W, Voravud N, Niruthisard S, Mutirangura A. Human papillomavirus dna in plasma of patients with cervical cancer. BMC Cancer 2001;1(1):2.

[18] Shimada T, Yamaguchi N, Nishida N, Yamasaki K, Miura K, Katamine S, Masuzaki H. Human papillomavirus dna in plasma of patients with hpv16 dna-positive uterine cervical cancer. Japanese Journal of Clinical Oncology 2010;40(5):420 -424.

[19] Widschwendter A, Blassnig A, Wiedemair A, MÃ¼ller-Holzner E, MÃ¼ller HM, Marth C. Human papillomavirus dna in sera of cervical cancer patients as tumor marker. Cancer Letters 2003;202(2):231-239.

[20] Rombaldi RL, Serafini EP, Mandelli J, Zimmermann E, Losquiavo KP. Perinatal transmission of human papilomavirus dna. Virol. J. 2009;6:83.

[21] Weyn C, Thomas D, Jani J, Guizani M, Donner C, Van Rysselberge M, Hans C, Bossens M, Englert Y, Fontaine V. Evidence of human papillomavirus in the placenta. J. Infect. Dis. 2011;203(3):341-343.

[22] Foresta C, Garolla A, Zuccarello D, Pizzol D, Moretti A, Barzon L, Palù G. Human papillomavirus found in sperm head of young adult males affects the progressive motility. Fertil. Steril. 2010;93(3):802-806.

[23] Pittayakhajonwut D, Angeletti PC. Viral trans-factor independent replication of human papillomavirus genomes. Virol. J. 2010;7:123.

[24] Ahola H, Bergman P, Ström AC, Moreno-Lopéz J, Pettersson U. Organization and expression of the transforming region from the european elk papillomavirus (eepv). Gene 1986;50(1-3):195-205.

[25] DiMaio D, Mattoon D. Mechanisms of cell transformation by papillomavirus e5 proteins. Oncogene 2001;20(54):7866-7873.

[26] Schulz E, Gottschling M, Bravo IG, Wittstatt U, Stockfleth E, Nindl I. Genomic characterization of the first insectivoran papillomavirus reveals an unusually long, second non-coding region and indicates a close relationship to betapapillomavirus. Journal of General Virology 2009;90(3):626 -633.

[27] Nonnenmacher M, Salmon J, Jacob Y, Orth G, Breitburd F. Cottontail rabbit papillomavirus e8 protein is essential for wart formation and provides new insights into viral pathogenesis. J. Virol. 2006;80(10):4890-4900.

[28] Eriksson A, Stewart AC, Moreno-Lopéz J, Pettersson U. The genomes of the animal papillomaviruses european elk papillomavirus, deer papillomavirus, and reindeer papillomavirus contain a novel transforming gene (e9) near the early polyadenylation site. J Virol 1994;68(12):8365-8373.

[29] Shah SD, Doorbar J, Goldstein RA. Analysis of host–parasite incongruence in papillomavirus evolution using importance sampling. Molecular Biology and Evolution 2010;27(6):1301 -1314.

[30] Gottschling M, Stamatakis A, Nindl I, Stockfleth E, Alonso Á, Bravo IG. Multiple evolutionary mechanisms drive papillomavirus diversification. Molecular Biology and Evolution 2007;24(5):1242 -1258.

[31] Shah SD, Doorbar J, Goldstein RA. Analysis of host–parasite incongruence in papillomavirus evolution using importance sampling. Molecular Biology and Evolution 2010;27(6):1301 -1314.

[32] Gottschling M, Göker M, Stamatakis A, Bininda-Emonds ORP, Nindl I, Bravo IG. Quantifying the phylodynamic forces driving papillomavirus evolution. Molecular Biology and Evolution 2011.
[33] Freitas AC. Recent insights into bovine papillomavirus. African Journal of Microbiology Research 2011;5(33).
[34] Batista MVA, Ferreira TAE, Freitas AC, Balbino VQ. An entropy-based approach for the identification of phylogenetically informative genomic regions of papillomavirus. Infect. Genet. Evol. 2011;11(8):2026-2033.
[35] Favre M, Ramoz N, Orth G. Human papillomaviruses: general features. Clin. Dermatol. 1997;15(2):181-198.
[36] Fehrmann F, Laimins LA. Human papillomaviruses: targeting differentiating epithelial cells for malignant transformation. Oncogene 2003;22(33):5201-5207.
[37] McCance DJ. Transcriptional regulation by human papillomaviruses. Curr. Opin. Genet. Dev. 2005;15(5):515-519.
[38] Stubenrauch F, Laimins LA. Human papillomavirus life cycle: active and latent phases. Semin. Cancer Biol. 1999;9(6):379-386.
[39] Stanley M. Immune responses to human papillomavirus. Vaccine 2006;24 Suppl 1:S16-22.
[40] Zhang P, Nouri M, Brandsma JL, Iftner T, Steinberg BM. Induction of e6/e7 expression in cottontail rabbit papillomavirus latency following uv activation. Virology 1999;263(2):388-394.
[41] zur Hausen H. Papillomavirus infections--a major cause of human cancers. Biochim. Biophys. Acta 1996;1288(2):F55-78.
[42] Campo MS. Animal models of papillomavirus pathogenesis. Virus Res. 2002;89(2):249-261.
[43] Schiffman M, Rodriguez AC, Chen Z, Wacholder S, Herrero R, Hildesheim A, Desalle R, Befano B, Yu K, Safaeian M, Sherman ME, Morales J, Guillen D, Alfaro M, Hutchinson M, Solomon D, Castle PE, Burk RD. A population-based prospective study of carcinogenic human papillomavirus variant lineages, viral persistence, and cervical neoplasia. Cancer Research 2010;70(8):3159 -3169.
[44] Doorbar J. Papillomavirus life cycle organization and biomarker selection. Dis. Markers 2007;23(4):297-313.
[45] Bruni L, Diaz M, Castellsagué M, Ferrer E, Bosch FX, de Sanjosé S. Cervical human papillomavirus prevalence in 5 continents: meta-analysis of 1 million women with normal cytological findings. Journal of Infectious Diseases 2010;202(12):1789 -1799.
[46] Woodman CBJ, Collins SI, Young LS. The natural history of cervical hpv infection: unresolved issues. Nat Rev Cancer 2007;7(1):11-22.
[47] Li N, Franceschi S, Howell-Jones R, Snijders PJF, Clifford GM. Human papillomavirus type distribution in 30,848 invasive cervical cancers worldwide: variation by geographical region, histological type and year of publication. International Journal of Cancer 2011;128(4):927-935.
[48] Broso PR, Buffetti G. [the papanicolaou classification in the bethesda system (national cancer institute, bethesda, maryland)]. Minerva Ginecol 1993;45(11):557-563.
[49] Trimble CL, Piantadosi S, Gravitt P, Ronnett B, Pizer E, Elko A, Wilgus B, Yutzy W, Daniel R, Shah K, Peng S, Hung C, Roden R, Wu TC, Pardoll D, The Johns Hopkins Medical Institutions B. Spontaneous regression of high-grade cervical dysplasia:

effects of human papillomavirus type and hla phenotype. Clinical Cancer Research 2005;11(13):4717 -4723.

[50] Pett MR, Alazawi WOF, Roberts I, Dowen S, Smith DI, Stanley MA, Coleman N. Acquisition of high-level chromosomal instability is associated with integration of human papillomavirus type 16 in cervical keratinocytes. Cancer Res. 2004;64(4):1359-1368.

[51] Steben M, Duarte-Franco E. Human papillomavirus infection: epidemiology and pathophysiology. Gynecol. Oncol. 2007;107(2 Suppl 1):S2-5.

[52] Bloch N, Sutton RH, Breen M, Spradbrow PB. Bovine papillomavirus type 4 in australia. Aust. Vet. J. 1995;72(7):273-274.

[53] Ziegert C, Wentzensen N, Vinokurova S, Kisseljov F, Einenkel J, Hoeckel M, von Knebel Doeberitz M. A comprehensive analysis of hpv integration loci in anogenital lesions combining transcript and genome-based amplification techniques. Oncogene 2003;22(25):3977-3984.

[54] WHO/ICO Information Centre on HPV and Cervical Cancer (HPV Information Centre). Human papillomavirus and related cancers in brazil. 2010.

[55] Baldez da Silva MFPT, Chagas BS, Guimarães V, Katz LMC, Felix PM, Miranda PM, Lima AA, Arraes LC, Martins DBG, Lima Filho JL, Stocco RC, Crovella S, Freitas AC, Beçak W. Hpv31 and hpv33 incidence in cervical samples from women in recife, brazil. Genet. Mol. Res. 2009;8(4):1437-1443.

[56] Cerqueira DM, Raiol T, Véras NMC, Gal Milanezi N, Amaral FA, Macedo Brígido M, Martins CRF. New variants of human papillomavirus type 18 identified in central brazil. Virus Genes 2008;37:282-287.

[57] Stewart AC, Eriksson AM, Manos MM, Muñoz N, Bosch FX, Peto J, Wheeler CM. Intratype variation in 12 human papillomavirus types: a worldwide perspective. J. Virol. 1996;70(5):3127-3136.

[58] Calleja-Macias IE, Villa LL, Prado JC, Kalantari M, Allan B, Williamson A-L, Chung L-P, Collins RJ, Zuna RE, Dunn ST, Chu T-Y, Cubie HA, Cuschieri K, von Knebel-Doeberitz M, Martins CR, Sanchez GI, Bosch FX, Munoz N, Bernard H-U. Worldwide genomic diversity of the high-risk human papillomavirus types 31, 35, 52, and 58, four close relatives of human papillomavirus type 16. J. Virol. 2005;79(21):13630-13640.

[59] Prado JC, Calleja-Macias IE, Bernard H-U, Kalantari M, Macay SA, Allan B, Williamson A-L, Chung L-P, Collins RJ, Zuna RE, Dunn ST, Ortiz-Lopez R, Barrera-Saldaña HA, Cubie HA, Cuschieri K, von Knebel-Doeberitz M, Sanchez GI, Bosch FX, Villa LL. Worldwide genomic diversity of the human papillomaviruses-53, 56, and 66, a group of high-risk hpvs unrelated to hpv-16 and hpv-18. Virology 2005;340(1):95-104.

[60] Xi LF, Critchlow CW, Wheeler CM, Koutsky LA, Galloway DA, Kuypers J, Hughes JP, Hawes SE, Surawicz C, Goldbaum G, Holmes KK, Kiviat NB. Risk of anal carcinoma in situ in relation to human papillomavirus type 16 variants. Cancer Res. 1998;58(17):3839-3844.

[61] Villa LL, Sichero L, Rahal P, Caballero O, Ferenczy A, Rohan T, Franco EL. Molecular variants of human papillomavirus types 16 and 18 preferentially associated with cervical neoplasia. J. Gen. Virol. 2000;81(Pt 12):2959-2968.

[62] Lizano M, Guido MC, García-Carrancá A, Berumen J, Casas L. Association between human papillomavirus type 18 variants and histopathology of cervical cancer. Journal of the National Cancer Institute 1997;89(16):1227 -1231.

[63] Calleja-Macias IE, Kalantari M, Huh J, Ortiz-Lopez R, Rojas-Martinez A, Gonzalez-Guerrero JF, Williamson A-L, Hagmar B, Wiley DJ, Villarreal L, Bernard H-U, Barrera-Saldaña HA. Genomic diversity of human papillomavirus-16, 18, 31, and 35 isolates in a mexican population and relationship to european, african, and native american variants. Virology 2004;319(2):315-323.

[64] Ho L, Chan SY, Burk RD, Das BC, Fujinaga K, Icenogle JP, Kahn T, Kiviat N, Lancaster W, Mavromara-Nazos P. The genetic drift of human papillomavirus type 16 is a means of reconstructing prehistoric viral spread and the movement of ancient human populations. J Virol 1993;67(11):6413-6423.

[65] Yamada T, Wheeler CM, Halpern AL, Stewart AC, Hildesheim A, Jenison SA. Human papillomavirus type 16 variant lineages in united states populations characterized by nucleotide sequence analysis of the e6, l2, and l1 coding segments. J. Virol. 1995;69(12):7743-7753.

[66] Yamada T, Manos MM, Peto J, Greer CE, Munoz N, Bosch FX, Wheeler CM. Human papillomavirus type 16 sequence variation in cervical cancers: a worldwide perspective. J Virol 1997;71(3):2463-2472.

[67] Ong CK, Chan SY, Campo MS, Fujinaga K, Mavromara-Nazos P, Labropoulou V, Pfister H, Tay SK, ter Meulen J, Villa LL. Evolution of human papillomavirus type 18: an ancient phylogenetic root in africa and intratype diversity reflect coevolution with human ethnic groups. J Virol 1993;67(11):6424-6431.

[68] Arias-Pulido H, Peyton CL, Torrez-Martínez N, Anderson DN, Wheeler CM. Human papillomavirus type 18 variant lineages in united states populations characterized by sequence analysis of lcr-e6, e2, and l1 regions. Virology 2005;338(1):22-34.

[69] Bernard H-U. The clinical importance of the nomenclature, evolution and taxonomy of human papillomaviruses. J. Clin. Virol. 2005;32 Suppl 1:S1-6.

[70] Chagas BS, Batista MVA, Guimarães V, Balbino VQ, Crovella S, Freitas AC. New variants of e6 and e7 oncogenes of human papillomavirus type 31 identified in northeastern brazil. Gynecol. Oncol. 2011;123(2):284-288.

[71] Franco EL, Rohan TE, Villa LL. Epidemiologic evidence and human papillomavirus infection as a necessary cause of cervical cancer. Journal of the National Cancer Institute 1999;91(6):506 -511.

[72] Bernard H-U. Gene expression of genital human papillomaviruses and considerations on potential antiviral approaches. Antivir. Ther. (Lond.) 2002;7(4):219-237.

[73] Sichero L, Villa LL. Epidemiological and functional implications of molecular variants of human papillomavirus. Braz. J. Med. Biol. Res. 2006;39(6):707-717.

[74] Kast W, Brandt R, Sidney J, Drijfhout J, Kubo R, Grey H, Melief C, Sette A. Role of hla-a motifs in identification of potential ctl epitopes in human papillomavirus type 16 e6 and e7 proteins. The Journal of Immunology 1994;152(8):3904 -3912.

[75] Ho L, Chan SY, Burk RD, Das BC, Fujinaga K, Icenogle JP, Kahn T, Kiviat N, Lancaster W, Mavromara-Nazos P. The genetic drift of human papillomavirus type 16 is a means of reconstructing prehistoric viral spread and the movement of ancient human populations. J. Virol. 1993;67(11):6413-6423.

[76] Ong CK, Chan SY, Campo MS, Fujinaga K, Mavromara-Nazos P, Labropoulou V, Pfister H, Tay SK, ter Meulen J, Villa LL. Evolution of human papillomavirus type 18: an ancient phylogenetic root in africa and intratype diversity reflect coevolution with human ethnic groups. J. Virol. 1993;67(11):6424-6431.
[77] Xi LF, Demers GW, Koutsky LA, Kiviat NB, Kuypers J, Watts DH, Holmes KK, Galloway DA. Analysis of human papillomavirus type 16 variants indicates establishment of persistent infection. The Journal of Infectious Diseases 1995;172(3):747-755.
[78] Londesborough P, Ho L, Terry G, Cuzick J, Wheeler C, Singer A. Human papillomavirus genotype as a predictor of persistence and development of high-grade lesions in women with minor cervical abnormalities. Int. J. Cancer 1996;69(5):364-368.
[79] Xi LF, Koutsky LA, Galloway DA, Kiviat NB, Kuypers J, Hughes JP, Wheeler CM, Holmes KK. Genomic variation of human papillomavirus type 16 and risk for high grade cervical intraepithelial neoplasia. Journal of the National Cancer Institute 1997;89(11):796 -802.
[80] Zehbe I, Wilander E, Delius H, Tommasino M. Human papillomavirus 16 e6 variants are more prevalent in invasive cervical carcinoma than the prototype. Cancer Res. 1998;58(4):829-833.
[81] Nindl I, Rindfleisch K, Lotz B, Schneider A, Dürst M. Uniform distribution of hpv 16 e6 and e7 variants in patients with normal histology, cervical intra-epithelial neoplasia and cervical cancer. International Journal of Cancer 1999;82(2):203-207.
[82] Hildesheim A, Schiffman M, Bromley C, Wacholder S, Herrero R, Rodriguez AC, Bratti MC, Sherman ME, Scarpidis U, Lin Q-Q, Terai M, Bromley RL, Buetow K, Apple RJ, Burk RD. Human papillomavirus type 16 variants and risk of cervical cancer. Journal of the National Cancer Institute 2001;93(4):315 -318.
[83] Zehbe I, Voglino G, Wilander E, Delius H, Marongiu A, Edler L, Klimek F, Andersson S, Tommasino M. P53 codon 72 polymorphism and various human papillomavirus 16 e6 genotypes are risk factors for cervical cancer development. Cancer Res. 2001;61(2):608-611.
[84] Villa LL, Sichero L, Rahal P, Caballero O, Ferenczy A, Rohan T, Franco EL. Molecular variants of human papillomavirus types 16 and 18 preferentially associated with cervical neoplasia. J. Gen. Virol. 2000;81(Pt 12):2959-2968.
[85] Xi LF, Kiviat NB, Hildesheim A, Galloway DA, Wheeler CM, Ho J, Koutsky LA. Human papillomavirus type 16 and 18 variants: race-related distribution and persistence. J. Natl. Cancer Inst. 2006;98(15):1045-1052.
[86] Sichero L, Ferreira S, Trottier H, Duarte-Franco E, Ferenczy A, Franco EL, Villa LL. High grade cervical lesions are caused preferentially by non-european variants of hpvs 16 and 18. Int. J. Cancer 2007;120(8):1763-1768.
[87] Kämmer C, Tommasino M, Syrjänen S, Delius H, Hebling U, Warthorst U, Pfister H, Zehbe I. Variants of the long control region and the e6 oncogene in european human papillomavirus type 16 isolates: implications for cervical disease. Br. J. Cancer 2002;86(2):269-273.
[88] de Boer MA, Peters LAW, Aziz MF, Siregar B, Cornain S, Vrede MA, Jordanova ES, Kolkman-Uljee S, Fleuren GJ. Human papillomavirus type 16 e6, e7, and l1 variants in

cervical cancer in indonesia, suriname, and the netherlands. Gynecol. Oncol. 2004;94(2):488-494.
[89] Matsumoto K, Yoshikawa H, Nakagawa S, Tang X, Yasugi T, Kawana K, Sekiya S, Hirai Y, Kukimoto I, Kanda T, Taketani Y. Enhanced oncogenicity of human papillomavirus type 16 (hpv16) variants in japanese population. Cancer Lett. 2000;156(2):159-165.
[90] Hu X, Pang T, Guo Z, Mazurenko N, Kisseljov F, Pontén J, Nistér M. Hpv16 e6 gene variations in invasive cervical squamous cell carcinoma and cancer in situ from russian patients. Br J Cancer 2001;84(6):791-795.
[91] Song YS, Kee SH, Kim JW, Park NH, Kang SB, Chang WH, Lee HP. Major sequence variants in e7 gene of human papillomavirus type 16 from cervical cancerous and noncancerous lesions of korean women. Gynecologic Oncology 1997;66(2):275-281.
[92] De Boer MA, Peters LAW, Aziz MF, Siregar B, Cornain S, Vrede MA, Jordanova ES, Fleuren GJ. Human papillomavirus type 18 variants: histopathology and e6/e7 polymorphisms in three countries. Int. J. Cancer 2005;114(3):422-425.
[93] De la Cruz-Hernández E, García-Carrancá A, Mohar-Betancourt A, Dueñas-González A, Contreras-Paredes A, Pérez-Cardenas E, Herrera-Goepfert R, Lizano-Soberón M. Differential splicing of e6 within human papillomavirus type 18 variants and functional consequences. Journal of General Virology 2005;86(9):2459 -2468.
[94] Rose BR, Thompson CH, Zhang J, Stoeter M, Stephen A, Pfister H, Tattersall MH, Cossart YE. Sequence variation in the upstream regulatory region of hpv 18 isolates from cervical cancers. Gynecol. Oncol. 1997;66(2):282-289.
[95] Combita A-L, Touze A, Bousarghin L, Christensen ND, Coursaget P. Identification of two cross-neutralizing linear epitopes within the l1 major capsid protein of human papillomaviruses. J. Virol. 2002;76(13):6480-6486.
[96] Kirnbauer R, Taub J, Greenstone H, Roden R, Dürst M, Gissmann L, Lowy DR, Schiller JT. Efficient self-assembly of human papillomavirus type 16 l1 and l1-l2 into virus-like particles. J. Virol. 1993;67(12):6929-6936.
[97] Bishop B, Dasgupta J, Chen XS. Structure-based engineering of papillomavirus major capsid l1: controlling particle assembly. Virology Journal 2007;4(1):3.
[98] Xi LF, Koutsky LA, Hildesheim A, Galloway DA, Wheeler CM, Winer RL, Ho J, Kiviat NB. Risk for high-grade cervical intraepithelial neoplasia associated with variants of human papillomavirus types 16 and 18. Cancer Epidemiol. Biomarkers Prev. 2007;16(1):4-10.
[99] De Boer MA, Peters LAW, Aziz MF, Siregar B, Cornain S, Vrede MA, Jordanova ES, Fleuren GJ. Human papillomavirus type 18 variants: histopathology and e6/e7 polymorphisms in three countries. Int. J. Cancer 2005;114(3):422-425.
[100] Raiol T, Wyant PS, de Amorim RMS, Cerqueira DM, Milanezi N von G, Brígido M de M, Sichero L, Martins CRF. Genetic variability and phylogeny of the high-risk hpv-31, -33, -35, -52, and -58 in central brazil. Journal of Medical Virology 2009;81:685-692.
[101] Comerford SA, McCance DJ, Dougan G, Tite JP. Identification of t- and b-cell epitopes of the e7 protein of human papillomavirus type 16. J. Virol. 1991;65(9):4681-4690.
[102] Cento V, Rahmatalla N, Ciccozzi M, Perno CF, Ciotti M. Intratype variations of hpv 31 and 58 in italian women with abnormal cervical cytology. J. Med. Virol. 2011;83(10):1752-1761.

[103] Muñoz N. Human papillomavirus and cancer: the epidemiological evidence. J. Clin. Virol. 2000;19(1-2):1-5.
[104] Forslund O. Genetic diversity of cutaneous human papillomaviruses. J. Gen. Virol. 2007;88(Pt 10):2662-2669.
[105] Feltkamp MCW, de Koning MNC, Bavinck JNB, Ter Schegget J. Betapapillomaviruses: innocent bystanders or causes of skin cancer. J. Clin. Virol. 2008;43(4):353-360.
[106] United States Department of Agriculture (USDA). Livestock and poultry: world market and trades. 2011.
[107] Food and Agricultural Organization (FAO). World cow's milk production. 2010.
[108] Borzacchiello G, Roperto F. Bovine papillomaviruses, papillomas and cancer in cattle. Veterinary Research 2008;39(5):19.
[109] Leal AM, Ferraz OP, Carvalho C, Freitas AC, Beniston RG, Beçak W, Campo MS, Stocco dos Santos RC. Quercetin induces structural chromosomal aberrations and uncommon rearrangements in bovine cells transformed by the e7 protein of bovine papillomavirus type 4. Vet Comp Oncol 2003;1(1):15-21.
[110] Hatama S, Ishihara R, Ueda Y, Kanno T, Uchida I. Detection of a novel bovine papillomavirus type 11 (bpv-11) using xipapillomavirus consensus polymerase chain reaction primers. Arch. Virol. 2011;156(7):1281-1285.
[111] Hatama S, Nobumoto K, Kanno T. Genomic and phylogenetic analysis of two novel bovine papillomaviruses, bpv-9 and bpv-10. Journal of General Virology 2008;89(1):158 -163.
[112] Zhu W, Dong J, Shimizu E, Hatama S, Kadota K, Goto Y, Haga T. Characterization of novel bovine papillomavirus type 12 (bpv-12) causing epithelial papilloma. Arch. Virol. 2012;157(1):85-91.
[113] Löhr CV, Juan-Sallés C, Rosas-Rosas A, García AP, Garner MM, Teifke JP. Sarcoids in captive zebras (equus burchellii): association with bovine papillomavirus type 1 infection. Journal of Zoo and Wildlife Medicine 2005;36(1):74-81.
[114] Kidney BA, Berrocal A. Sarcoids in two captive tapirs (tapirus bairdii): clinical, pathological and molecular study. Vet. Dermatol. 2008;19(6):380-384.
[115] Silvestre O, Borzacchiello G, Nava D, Iovane G, Russo V, Vecchio D, D'Ausilio F, Gault EA, Campo MS, Paciello O. Bovine papillomavirus type 1 dna and e5 oncoprotein expression in water buffalo fibropapillomas. Veterinary Pathology Online 2009;46(4):636 -641.
[116] van Dyk E, Oosthuizen MC, Bosman A-M, Nel PJ, Zimmerman D, Venter EH. Detection of bovine papillomavirus dna in sarcoid-affected and healthy free-roaming zebra (equus zebra) populations in south africa. Journal of Virological Methods 2009;158(1–2):141-151.
[117] Pangty K, Singh S, Goswami R, Saikumar G, Somvanshi R. Detection of bpv-1 and -2 and quantification of bpv-1 by real-time pcr in cutaneous warts in cattle and buffaloes. Transboundary and Emerging Diseases 2010;57(3):185-196.
[118] van Dyk E, Bosman AM, van Wilpe E, Williams JH, Bengis RG, van Heerden J, Venter EH. Detection and characterisation of papillomavirus in skin lesions of giraffe and sable antelope in south africa. J S Afr Vet Assoc 2011;82(2):80-85.
[119] Campo MS. Bovine papillomavirus: old system, new lessons? 2006.

[120] MacLaughlan SD, Lachance JA, Gjelsvik A. Correlation between smoking status and cervical cancer screening: a cross-sectional study. J Low Genit Tract Dis 2011;15(2):114-119.

[121] Olson C, Olson RO, Hubbard-Van Stelle S. Variations of response of cattle to experimentally induced viral papillomatosis. J. Am. Vet. Med. Assoc. 1992;201(1):56-62.

[122] Hama, C, Matsumoto, T, Franceschini, P. H. Papilomatose bovina: avaliação clínica de diferentes produtos utilizados no controle e tratamento., vol. 2. Ciência Veterinária, [s.d.].

[123] Melo. C. B, Leite, R. C. Papilomatose bovina, vol. 6. Ciência Veterinária nos Trópicos, [s.d.].

[124] Muro, L. F. F., Bottura, C. R. P., Piccinin, A. Papilomatose bovina., vol. 1. [s.d.].

[125] Mbuthia, G. P., Ngatia, T. A., & Wamokoya, J. P. O. Occurrence of bovine skin diseases in kenya., vol. 41. Bulletin of Animal Production in Africa, [s.d.].

[126] Campo MS, Jarrett WF, O'Neil W, Barron RJ. Latent papillomavirus infection in cattle. Res. Vet. Sci. 1994;56(2):151-157.

[127] Pfister H, Linz U, Gissmann L, Huchthausen B, Hoffmann D, Zur Hausen H. Partial characterization of a new type of bovine papilloma viruses. Virology 1979;96(1):1-8.

[128] Carvalho CCR, Batista MVA, Silva MAR, Balbino VQ, Freitas AC. Detection of bovine papillomavirus types, co-infection and a putative new bpv11 subtype in cattle. Transboundary and Emerging Diseases 2012.

[129] Freitas AC de, Carvalho C de, Brunner O, Birgel-Junior EH, Dellalibera AMMP, Benesi FJ, Gregory L, Beçak W, Santos R de CS dos. Viral dna sequences in peripheral blood and vertical transmission of the virus: a discussion about bpv-1. Brazilian Journal of Microbiology 2003;34:76-78.

[130] Freitas, A. C., Silva, M. A. R., Carvalho, C. C. R., Birgel Jr, E. H., Santos, J. F., Beçak, W., & Stocco dos Santos, R. C. Papillomavirus dna detection in non epithelial tissues: a discussion about bovine papillomavirus. in a. méndez-vilas (ed.), formatex microbiology book series. Mérida, EX: Formatex, [s.d.].

[131] Stocco R de C. Papillomatosis in cattle: in situ detection of bovine papillomavirus dna sequences in reproductive tissues [s.d.].

[132] Brandt S, Haralambus R, Schoster A, Kirnbauer R, Stanek C. Peripheral blood mononuclear cells represent a reservoir of bovine papillomavirus dna in sarcoid-affected equines. Journal of General Virology 2008;89(6):1390 -1395.

[133] Roperto S, Brun R, Paolini F, Urraro C, Russo V, Borzacchiello G, Pagnini U, Raso C, Rizzo C, Roperto F, Venuti A. Detection of bovine papillomavirus type 2 in the peripheral blood of cattle with urinary bladder tumours: possible biological role. Journal of General Virology 2008;89(12):3027 -3033.

[134] Yaguiu A, Dagli MLZ, Birgel EH Jr, Alves Reis BCA, Ferraz OP, Pituco EM, Freitas AC, Beçak W, Stocco RC. Simultaneous presence of bovine papillomavirus and bovine leukemia virus in different bovine tissues: in situ hybridization and cytogenetic analysis. Genet. Mol. Res. 2008;7(2):487-497.

[135] Roperto S, Comazzi S, Ciusani E, Paolini F, Borzacchiello G, Esposito I, Lucà R, Russo V, Urraro C, Venuti A, Roperto F. Pbmcs are additional sites of productive infection of bovine papillomavirus type 2. Journal of General Virology 2011;92(8):1787 -1794.

[136] Carvalho C de, Freitas AC de, Brunner O, Góes LGB, Cavalcante AY, Beçak W, Santos R de CS dos. Bovine papillomavirus type 2 in reproductive tract and gametes of slaughtered bovine females. Braz. j. microbiol 2003;34(supl.1):82-84.
[137] Lindsey CL, Almeida ME, Vicari CF, Carvalho C, Yaguiu A, Freitas AC, Beçak W, Stocco RC. Bovine papillomavirus dna in milk, blood, urine, semen, and spermatozoa of bovine papillomavirus-infected animals. Genet. Mol. Res. 2009;8(1):310-318.
[138] Silva MAR, Pontes NE, Da Silva KMG, Guerra MMP, Freitas AC. Detection of bovine papillomavirus type 2 dna in commercial frozen semen of bulls (bos taurus). Anim. Reprod. Sci. 2011;129(3-4):146-151.
[139] Stocco dos Santos RC, Lindsey CJ, Ferraz OP, Pinto JR, Mirandola RS, Benesi FJ, Birgel EH, Pereira CA, Be√ßak W. Bovine papillomavirus transmission and chromosomal aberrations: an experimental model. Journal of General Virology 1998;79(9):2127 -2135.
[140] Schmitt M, Fiedler V, Müller M. Prevalence of bpv genotypes in a german cowshed determined by a novel multiplex bpv genotyping assay. Journal of Virological Methods 2010;170(1–2):67-72.
[141] Maeda Y, Shibahara T, Wada Y, Kadota K, Kanno T, Uchida I, Hatama S. An outbreak of teat papillomatosis in cattle caused by bovine papilloma virus (bpv) type 6 and unclassified bpvs. Veterinary Microbiology 2007;121(3–4):242-248.
[142] Ogawa T, Tomita Y, Okada M, Shinozaki K, Kubonoya H, Kaiho I, Shirasawa H. Broad-spectrum detection of papillomaviruses in bovine teat papillomas and healthy teat skin. Journal of General Virology 2004;85(8):2191 -2197.
[143] Tomita Y, Literak I, Ogawa T, Jin Z, Shirasawa H. Complete genomes and phylogenetic positions of bovine papillomavirus type 8 and a variant type from a european bison. Virus Genes 2007;35(2):243-249.
[144] Singh V. Papillomatosis in indian cattle: occurrence and etiopathology. Indian Journal of Veterinary Pathology [s.d.];33(1):52-57.
[145] Rai GK, Saxena M, Singh V, Somvanshi R, Sharma B. Identification of bovine papilloma virus 10 in teat warts of cattle by dnase-sispa. Veterinary Microbiology 2011;147(3–4):416-419.
[146] Carr EA, Théon AP, Madewell BR, Griffey SM, Hitchcock ME. Bovine papillomavirus dna in neoplastic and nonneoplastic tissues obtained from horses with and without sarcoids in the western united states. Am. J. Vet. Res. 2001;62(5):741-744.
[147] Wobeser BK, Davies JL, Hill JE, Jackson ML, Kidney BA, Mayer MN, Townsend HGG, Allen AL. Epidemiology of equine sarcoids in horses in western canada. Can Vet J 2010;51(10):1103-1108.
[148] Spradbrow PB, Ford J. Bovine papillomavirus type 2 from bovine cutaneous papillomas in australia. Australian Veterinary Journal 1983;60(3):78-79.
[149] Silva MS e, Weiss M, Brum MCS, dos Anjos BL, Torres FD, Weiblen R, Flores EF. Molecular identification of bovine papillomaviruses associated with cutaneous warts in southern brazil. Journal of Veterinary Diagnostic Investigation 2010;22(4):603 -606.
[150] Claus MP, Vivian D, Lunardi M, Alfieri AF, Alfieri AA. Phylogenetic analysis of bovine papillomavirus associated with skin warts in cattle herds from the state of paraná. Pesquisa Veterinária Brasileira 2007;27(7):314-318.

[151] Claus MP, Lunardi M, Alfieri AF, Ferracin LM, Fungaro MHP, Alfieri AA. Identification of unreported putative new bovine papillomavirus types in brazilian cattle herds. Vet. Microbiol. 2008;132(3-4):396-401.

[152] Wosiacki SR, Claus MP, Alfieri AF, Alfieri AA. Bovine papillomavirus type 2 detection in the urinary bladder of cattle with chronic enzootic haematuria. Mem. Inst. Oswaldo Cruz 2006;101(6):635-638.

[153] Diniz N, Melo TC, Santos JF, Mori E, Brandão PE, Richtzenhain LJ, Freitas AC, Beçak W, Carvalho RF, Stocco RC. Simultaneous presence of bovine papillomavirus in blood and in short-term lymphocyte cultures from dairy cattle in pernambuco, brazil. Genet. Mol. Res. 2009;8(4):1474-1480.

[154] Jarrett WFH, McNeil PE, Grimshaw WTR, Selman IE, McIntyre WIM. High incidence area of cattle cancer with a possible interaction between an environmental carcinogen and a papilloma virus. Nature 1978;274(5668):215-217.

[155] Borzacchiello G, Ambrosio V, Roperto S, Poggiali F, Tsirimonakis E, Venuti A, Campo MS, Roperto F. Bovine papillomavirus type 4 in oesophageal papillomas of cattle from the south of italy. J. Comp. Pathol. 2003;128(2-3):203-206.

[156] Leishangthem GD. Detection of bovine papillomaviruses in cutaneous warts/papillomas in cattle. Indian Journal of Veterinary Pathology [s.d.];32(1):15-20.

[157] Claus MP, Lunardi M, Alfieri AA, Otonel RAA, Sartori D, Fungaro MHP, Alfieri AF. Multiple bovine papillomavirus infections associated with cutaneous papillomatosis in brazilian cattle herds. Braz. arch. biol. technol 2009;52(spe):93-98.

[158] Antonsson A, Hansson BG. Healthy skin of many animal species harbors papillomaviruses which are closely related to their human counterparts. Journal of Virology 2002;76(24):12537 -12542.

[159] Giannoudis A, Herrington CS. Human papillomavirus variants and squamous neoplasia of the cervix. J. Pathol. 2001;193(3):295-302.

[160] Reid S, Smith K, Jarrett W. Detection, cloning and characterisation of papillomaviral dna present in sarcoid tumours of equus asinus. Veterinary Record 1994;135(18):430 - 432.

[161] Chambers G, Ellsmore VA, O'Brien PM, Reid SWJ, Love S, Campo MS, Nasir L. Sequence variants of bovine papillomavirus e5 detected in equine sarcoids. Virus Research 2003;96(1–2):141-145.

[162] Nasir L, Gault E, Morgan IM, Chambers G, Ellsmore V, Campo MS. Identification and functional analysis of sequence variants in the long control region and the e2 open reading frame of bovine papillomavirus type 1 isolated from equine sarcoids. Virology 2007;364(2):355-361.

[163] Szczerba-Turek A, Siemionek J, Wasowicz K, Szweda W, Raś A, Platt-Samoraj A. Partial sequence analysis of the l1 gene of bovine papillomavirus type 1 detected by pcr with my09/my11 primers in equine sarcoids in poland. Pol J Vet Sci 2010;13(2):241-246.

[164] Freitas AC. Recent insights into bovine papillomavirus. African Journal of Microbiology Research 2011;5(33).

[165] García-Vallvé S, Alonso A, Bravo IG. Papillomaviruses: different genes have different histories. Trends Microbiol. 2005;13(11):514-521.

[166] Bogaert L, Martens A, Kast WM, Van Marck E, De Cock H. Bovine papillomavirus dna can be detected in keratinocytes of equine sarcoid tumors. Vet. Microbiol. 2010;146(3-4):269-275.

[167] Turner PE, Elena SF. Cost of host radiation in an rna virus. Genetics 2000;156(4):1465- 1470.

In: Genetic Diversity: New Research
Editors: J. A. C. Amaya and M. M. F. Jimenez
ISBN: 978-1-62081-598-4
© 2012 Nova Science Publishers, Inc.

Chapter 2

GENETIC DIVERSITY AND PHYLOGENY OF WILD SPECIE TEOSINTE (*ZEA NICARAGUENSE*, ILTIS AND BENZ) WITHIN GENERA *ZEA*

Carlos Henry Loáisiga Caballero
Department of Crop Production,
National Agrarian University, Nicaragua
Alvaro Nicolas Benavides Gonzales
Department of Crop Production,
National Agrarian University, Nicaragua
Pernilla Ellneskog-Staam
Department of Plant Breeding and Biotechnology,
Swedish University of Agricultural Sciences, Sweden
Oscar Rocha
Department of Biological Science,
Kent State University US
Agnese Brantestam
Nordic Genetic Center, Sweden
Mulatu Geleta
Department of Plant Breeding and Biotechnology,
Swedish University of Agricultural Sciences, Sweden
Salomon Björn
Department of Plant Breeding and Biotechnology,
Swedish University of Agricultural Sciences, Sweden
Arnulf Merker
(In Memoriam); Department of Plant Breeding and Biotechnology,
Swedish University of Agricultural Sciences, Sweden

ABSTRACT

Maize (*Zea mays* subsp. *mays*) is a member of the grass family *Poaceae* (*Gramineae*), together with many other important agricultural crops. The wild species commonly known as teosintes are the closest relatives of maize. Teosintes are wild grasses with a native distribution area from Mexico to Nicaragua (Mesoamerican region) and are an important genetic resource for maize improvement. Teosinte from Nicaragua (*Zea nicaraguensis*) is to a large extent an unutilized genetic resource and its properties in terms of desirable traits, such as water-lodging adaptation and disease resistance have to be phenotypically and genetically characterized. This chapter includes results from cytological, genetic, morphological and phylogenetic studies. Cytogenetic studies were based on C-banding techniques. Microsatellite (SSR) markers and DNA sequences of cpDNA regions were used for genetic diversity and phylogenetic studies respectively, whereas morphological studies involved various quantitative morphological traits. The C-banding pattern revealed that *Z. nicaraguensis* is more similar to *Z. luxurians* than other teosintes and cultivated maize. Among the Meso-American teosintes, *Z. diploperennis, Z. perennis* and *Z. nicaraguensis* showed the highest values in the number of rare and unique alleles and the results also indicated that the gene flow between *Z. nicaraguensis* and *Z. luxurians* has been more frequent than between other teosintes in the Mesoamerican region. The morphological characterization revealed that the traits of the number of lateral branch nodes bearing ears, the glume width, the number of tiller nodes bearing ears and the number of tillers per plant were the most important in discriminating between taxa and the principal component analysis grouped many traits around the *Z. nicaraguensis, Z. luxurians* and *Z. mays* subsp. *huehuetenangensis*. Finally, the variation among cpDNA sequences need more analysis to give a definitive phylogenetic resolution among the five *Zea* species, but even so our results support the idea that *Z. nicaraguensis* could be treated as a separate species distinct from *Z. luxurians*.

Keywords: *Zea nicaraguensis*, chromosome, teosinte genetic diversity, microsatellites, gene flow, characterization, chloroplast, taxonomy, phylogenetic

1. INTRODUCION

1.1. History of Genus *Zea*

Zea L. is a genus that comprises a group of annual and perennial grass species native to Mexico and Central America. Meso-America is a region between Nayarit state in Mexico and approximately the middle part of the Pacific coast of Nicaragua (Figure 1). This is recognized as a center of origin of world agriculture and the center of diversity for approximately 225 cultivated plant species (Vavilov, 1931; Hernandez, 1985; Ortega-Pazcka, 2003; Engels et al., 2006). The most significant contribution of this center to the world is the maize crop, *Zea mays* L., Hernandez (1985).

The genus includes both the wild taxa commonly known as "teosintes" and the cultigen maize *Zea mays* subsp. *mays*. Since maize is an American crop, it was unknown to the Europeans until Columbus first encountered it on the island of Cuba. Knowledge about teosintes came to Europe much later, and it was only in 1832 that a German botanist, Schrader (1833), provided the annual form of teosinte with a scientific name, *Euchlaena mexicana* Schrader. In 1910, the American botanist, A. S. Hitchcock, discovered a perennial form of teosinte, which he named *Euchlaena perennis* Hitchcock (Hitchcock, 1922). These two grasses are so unlike maize in the appearance of their ears that their close relationship to maize was not recognized at first, but gradually, through the efforts of several botanists (Beadle, 1932; Collins, 1921; Harshberger, 1896), the close relationship between teosinte and maize was established. Reeves and Mangelsdorf (1942) gave formal recognition to this close relationship by transferring these species into the genus *Zea*, as *Z. mexicana* (Schrader) Kuntze and *Z. perennis* (Hitchcock) Reeves & Mangelsdorf. The teosinte is generally considered to be the progenitor of maize. (*e.g.* Emerson & Beadle, 1932; Galinat, 1974; Darlington, 1963; Wilkes, 1972; Iltis, 1983; Kato, 1984; McClintock et al., 1981; Doebley, 1984, 1990a, 2004).

Since the end of the 19th century, several hypotheses have been postulated to explain how maize originated. The most influential idea was the tripartite hypothesis of the 1930s (Mangelsdorf & Reeves, 1939), although it was discredited around the late 1970s, when the hypothesis of evolution of maize from teosinte had progressively won agreement among scientific communities (Beadle, 1972; Doebley, 1990a; Matsuoka et al., 2002a; Doolitle & Marbry, 2006).

There are two alternative hypothesis regarding the domestication and diversification of maize. The first hypothesis states that maize evolved at several domestication centers from different teosinte populations from 8000 years ago (i.e., multi-centric origin), whereas the second hypothesis states the unique event of maize domestication from teosinte populations of *Z. mays* subsp. *parviglumis* at Raza Balsas near Michoacan, Mexico (i.e., unicentric origin). These hypotheses are associated with the origin and diversification of maize and the different ways it adapted to the environment and to specific cultures (Kato et al., 2009). It is believed that maize evolution has resulted in between 220 to 300 races of maize on the American continents (Brown & Goodman, 1977; Vigouroux et al., 2008).

Teosintes grow in the valleys and mountains of the Meso-American region as wild plants in various habitats along streams and hillsides, and they can also be found as a weed in cultivated fields. For example, *Z. mays* subsp. *parviglumis* grows in the Balsas River drainage of southwest Mexico and for that reason it has been also called Blasas teosinte (Doebley, 2004), and *Z. nicaraguensis* grows around the forest peripheries where its stems and prop roots are covered by 10–40 cm of standing water, which is unique among teosintes (Iltis & Benz, 2000; Loáisiga et al., 2010).

Although maize and teosinte plants share a similar robust growth form, their female inflorescence or ears are strikingly different. The teosinte ear possesses only about 5 to 12 kernels, each sealed tightly in a stony casing. Collectively, the kernel and its stony casing are known as a fruitcase. At maturity, the teosinte ear disarticulates such that the individual fruitcases become the dispersal units. Protected within its casing, the teosinte kernel can survive the digestive tracts of birds and grazing mammals, enabling the seeds to be easily dispersed. Another factor to consider is that found in Nicaraguan teosinte, which grows in slow-moving water where the excess of flow water contributes to the dispersal of the seeds (Loáisiga et al., 2010).

By comparison, the massive maize ear can bear 500 or more kernels, each of which is attached to the central axis of the ear or cob. The kernels are naked without adequate protection from predation and are easily digested by any animal that consumes them. Since the kernels are firmly attached to the cob and the ear does not disarticulate, a maize ear left on the plant will eventually fall to the ground with its full suite of kernels. When hundreds of maize kernels germinate so close to one another in the next season, the emerging plants are unable to develop normally (Sanchez et al., 1998).

Figure 1. Geographical distribution of the accessions of Meso-American teosintes used in this study.

As stated above, the Nicaraguan teosinte is remarkable for its ability to grow in as much as 10-40 cm of standing or slow-moving water. For this reason, Iltis & Benz (2000) suggested that this species could provide maize breeders with a potentially valuable source of germplasm that may lead to the development of maize capable of growing in water-logged soils. Standley, (1950) reported a population of teosinte in Honduras in the mountains of the Department of Francisco Morazán, not far from Tegucigalpa, in the mountains of Copan, and even Nicaragua. However, this finding has not been verified by any other author. Loáisiga et al., (2010), could not find teosinte populations there and suggested that the population reported by Standley might be extinct. Nevertheless, this possible population was an important biological bridge between *Z. luxurians* and *Z. nicaraguensis* both in the same region. Bearing in mind that many civilizations were established in this area (Meso-America), principally the Maya and Chorotega, it is likely that plant material was distributed by humans.

During the 19th century, the origin and evolution of Indian corn became a topic of great interest among botanists. The most perplexing fact that these scientists had to contend with was that maize, unlike the Middle Eastern cereals wheat, barley, and rye, apparently lacked an associated wild form that could be considered its ancestor. This situation provided ample room for speculation, but little hope for resolution of the ancestry of maize. However, the focus of the debate changed with the discovery of teosintes and the subsequent demonstration that some types of teosinte and maize formed fully fertile hybrids (Harshberger, 1896).

There are two hypotheses that have had their supporters over the last 50 years. There has also been a diversity of opinion within each of these two camps. Most researchers working in the field support the "teosinte theory" (Beadle 1972, 1980; Benz, 1987; deWet & Harlan,

1972; Doebley, 1983; Galinat, 1963; Iltis, 1983; Kato, 1984, McClintock et al., 1981). However, Mangelsdorf, (1974, 1986) never swayed in his belief that teosinte is not the ancestor of maize. Many analyses suggest that *Z. mays* subsp. *parviglumis* and maize share a more recent common ancestor than they do with the other teosintes, and the fact that teosinte is wild and maize is fully domesticated suggests that their common ancestor was also a teosinte (Doebley, 1990).

1.2. Systematic of the Genus *Zea*

Iltis and Doebley (1980) proposed a classification for *Zea*, which reflected its presumed evolutionary relationships. This system of classification was based on the morphological and ecological features of the taxa and includes four species, one of which has three subspecies. This was subsequently modified by Doebley (1990) and with the inclusion of the species describe by Iltis & Benz (2000) the genus currently includes five species, one of which has four subspecies. The basic chromosome number for the genus *Zea* is x=10. All species are diploid with the chromosome number 2n=2x=20 except for the tetraploid Z. *perennis* wich has 2n=4x=40.

- *Zea diploperennis* (diploperennis teosinte) as its name signifies, is a diploid (2*n*= 2x=20) perennial species. It has a very narrow geographic distribution, being found only in a small region of the Sierra de Mazatlan in the southwestern part of the state of Jalisco in Mexico. The number of tassel branches ranges from 2 to 10, normally more than *Z. perennis*, and the fruitcase is trapezoidal in shape.
- *Zea perennis* (perennial teosinte) is a tetraploid (2n=4x=40) perennial species and is the only polyploid species in the genus. It has a narrow geographic distribution on the northern slopes of Volcán de Colima in the state of Jalisco in Mexico. This specie is quite similar to Z. *diploperennis*, but it can be distinguished from Z. *diploperennis* by being a more slender plant that lacks tuber-like short shoots and having a different ploidy level.
- *Zea luxurians* (Guatemala or Florida teosinte) is a diploid (2n=2x=20) annual specie native to southeastern Guatemala and Honduras. However, this specie is now extinct in Honduras. It is also known from a single collection made in Oaxaca, Mexico in 1845. No one has been able to recollect it from this locality since that time, but there is now a recent report that it has been rediscovered (Cuevas, 2006), although this information remains to be verified. *Zea luxurians* differ from the two perennial species through its lack of rhizomes, but it shares other similarities with the two perennial species. The outer glumes of its male spikelets have numerous fine veins that distinguish it from all other *Zea* species. In this character it resembles the species of the sister genus, *Tripsacum*.
- *Zea nicaraguensis* (Nicaraguan teosinte) is native to the Pacific coast of Nicaragua. It is a diploid (2n=2x=20) and grows at very low elevations compared to the altitudes for other teosinte species and show a long life cycle, which takes 9-10 months from germination to seed maturation. This species has the ability to

grow normally in almost 0.5 m of standing or slow-moving water. It is a maize-like, erect, candelabra-branched annual species. The outer glumes have two prominent keels that merge at the apex (9-10 mm long). Other characteristics are its transverse rugulose form, especially when young, the presence of stout basal tillers and prominent prop roots.

- *Zea mays* is the specie that comprises both cultivated maize and annual teosintes. Adhering to the biological species concept, Iltis & Doebley (1980) placed three annual teosintes in *Z. mays* emphasizing their close biological relationship to maize, since maize and these teosintes are easily cross-hybridized and their hybrids are fully fertile, it comprises:
  - *Zea Mays* Subsp. H*uehuetenangensis* (Iltis and JF. Doebley) JF. Doebley
    Zea mays subsp. *huehuetenangensis* (Huehuetenango teosinte) is a variant of teosinte found in western Guatemala. This diploid (2n=2x=20) perennial subspecies is distinguished from the other subspecies of *Z. mays* by its longer life cycle, which takes 7-8 months from germination to seed maturation in its native habitat and its height is only comparable with that of *Zea nicaraguensis*. Its floral morphology is similar to *Z. mays* subsp. *parviglumis* (see below) and it is difficult to distinguish these two subspecies on the basis of their tassel and ear morphology alone.
  - *Zea mays* subsp. *mexicana* (Schrad.) Iltis
    Zea mays subsp. *mexicana* (Mexican annual teosinte) is a diploid (2n=2x=20) perennial subspecies that has a rapid life cycle with only 4 months from germination to seed maturation. It is found in central and northern Mexico, ranging from Puebla to the Nobogame Valley in Chihuahua. Its fruitcases and tassel spikelets are relatively large, and distinguish it from *Z. mays* subsp. *parviglumis* and *Z. mays* subsp. *huehuetenangensis*. The populations in the Race Chalco Valley in Mexico commonly have dark red, hairy leaf sheaths. Populations throughout the Mesa Central (Race Central Plateau) typically have green to pale red, glabrous to sparsely hairy leaf sheaths. The northernmost population in the Nobogame Valley of Chihuahua (Race Nobogame) is the shortest and earliest flowering teosinte.
  - *Zea mays* subsp. *parviglumis* Iltis and JF.Doebley
    Zea mays L. subsp. *parviglumis* (Mexican annual teosinte) is a diploid (2n=2x=20) perennial subspecies found in the valleys along the western escarpment of Mexico from Nayarit to Oaxaca. Populations are distributed along an east-west axis from Oaxaca to Jalisco. Populations in the center of this range in the Balsas River drainage are known as Race Balsas or Balsas teosinte. It takes about 6-7 months from seed germination to plant maturity. The plants typically have green to pale red, glabrous leaf sheaths. The epithet "parviglumis" means small glume, referring to the small tassel spikelets.
  - *Zea mays* subsp. mays
    Zea mays subsp. *mays* (maize) differ significantly in plant and ear morphology from the three teosinte subspecies of *Z. mays*.

Iltis & Doebley (1980) did not place maize and three teosintes together in *Z. mays* based on the absence of reproductive isolation alone. They also stressed the similarities in the structure of their tassels. For example, all of them have relatively few widely-spaced veins of the glumes of tassel spikelets compared to *Z. diploperennis*, *Z. perennis*, *Z. luxurians* and *Z. nicaraguensis*. Additionally, molecular characteristics tie maize and the three subspecies together. The general morphological characteristics and geographic distributions of all *Zea* taxa are summarized in Figure 1 and Table1.

1.3. Relative of *Zea* Genera

The closest relatives to *Zea* are found in the sister genus *Tripsacum* (*Gramigrasses*). The *Tripsacum* species are rhizomatous perennials that range from eastern North America to South America (Cuttler & Anderson, 1941; Randolph, 1959; Randolph & Hernandez, 1947). The basic chromosome number is x=18 and most species are diploids (2n=2x=36) or tetrapolid (2n=2x=72) but triploids (2n=2x=54) and hexaploids (2n=2x=108) have also been reported (Berthaud et al., 2004; deWet & Harlam, 1976; Randolph, 1976).

Since *Tripsacum* is the most closely related genus to *Zea* it has held a prominent place in *Zea* research. Despite the fact that several botanists have studied it in varying depth over the past 60 years, *Tripsacum* remains a taxonomically difficult group due to the presence of different ploidy levels even within a species and due to the fact that some species are apparently facultative apomicts, which can give rise to minutely but consistently differentiated forms (deWet et al., 1981). Like maize and teosinte, *Tripsacum* is monoecious and has been suggested as a potential donor of genes for disease and insect resistance that may be used in improving maize.

1.4. Chromosome Set in Teosintes

Generally, the teosintes have larger chromosomes than those found in cultivated maize where they have an average length of 11.2 μm. Among the wild species, *Z. nicaraguensis* has the largest chromosomes, with an average length of 19.6 μm (Ellneskog-Staam et al., 2007) and *Z. perennis* has the smallest average chromosome size and the smallest heterochromatic blocks (Tito et al., 1991).

Cytological studies have dealt mainly with aspects of meiotic behavior, and the information on the chromosome knob constitution of the perennial teosinte is limited as to the presence of knobs in different chromosomes. Information available about *Z. perennis* is that of Longley (1937) who showed pachytene chromosomes. Regarding *Z. diploperennis,* the first pachytene chromosome observations showed the presence of small and medium knobs (Kato, 1984; Pasupuleti & Galinat, 1982). Previous surveys of chromosome knob constitution in populations of the genus *Zea* have led to a new interpretation about the origin and evolution of cultivated maize and its relationship with the diverse teosintes (Kato, 1976, 1984; McClintock, 1978; McClintock et al., 1981).

Chromosome knobs are enlarged structures consisting of condensed heterochromatin on mitotic and meiotic chromosomes. They can be used for taxonomic studies in the genus *Zea* because they exhibit fixed numbers and positions on chromosomes within specific taxa, but vary between taxa (Kato, 1976). In most cases, the main chromosome knobs in maize are internal or subterminal. In contrast, the chromosomes of *Z. nicaraguensis, Z. luxurians, Z. diploperennis,* and *Z. perennis* (Section *Luxuriantes*) have many terminal chromosome knobs and lack internal knobs. The main cytological difference between the *Z. nicaraguensis* and *Z. luxurians* genomes is that chromosome 10 of *Z. nicaraguensis* is knobless (Ellneskog-Staam et al., 2007). On the other hand, all of the subspecies of *Z. mays* (Section *Zea*), in general, have many internal knobs and few or no terminal knobs except *Z. mays* subsp. *huehuetenangensis,* which has many terminal chromosome knobs (Wilkes, 1967). These chromosomal knobs contain thousands to millions of tandem 180 base pairs (bp) and 350 bp repetitive DNAs that have high frequencies of preferential segregation as a result of meiotic drive (Buckler et al., 1999).

Table 1. A summary of the principal morphological characteristics of Meso-American teosintes

Taxa/ trait	*Z. mays* subsp *mexicana*	Z. mays subsp parviglumis	Z: mays subsp *huehuetenan-gensis*	*Zea diploperennis*	*Zea perennis*	*Zea luxurians*	*Zea nicaragu-ensis*
Habitat	Annual	Annual	Annual	Perennial	Perennial	Annual	Annual
Height	1.5-4.0 m	2.0-5.0 m	Up 5.0 m	2.0-2.5 m	1.5-2.0 m	3.0-4.0 m	Up to 6.0 m
Tassel branches	Slender, 10-20 branches	Slender, 20-100 branches	Slender 20 or more branches	Thicker, lax 2-10 branches	Thicker, erect 2-8 branches	Thicker, erect 4-20 branches	Slender, lax 27-38 branches
Male spikelet mm	Paired, longer pedicels 6-10	Paired, 4-7 mm long, small size pedicels 4-7	Paired longer pedicels 5-7	Paired, 10 mm shorter pedicels 1.5-3.5	Shorter pedicels, 1.5-3.8	Shorter pedi-cels 1.3-3.5	Paired, shorter pedicels1.5-3.5
Female spikelet	Slender and distichous	Slender and distich ous	Slender and distichous	Slender and distichous	Slender and distichous	Slender and distichous	Slender and distichous
Flowering date	Septem-ber to October	September to October	Late November to January	September to October	September to October	September to October	Mid. October to November
Fruitcases mg	Large, acute Tri angular 60-90	Small, blunt tri- angular 30-80	Small, triangular 30-60	Trapezoidal 68-75	Trapezoidal 70-83	Trapezoidal 76-99	Trapezoidal 76-99
Rhizomes mm	Absent	Absent	Absent	Slender, cord-like and tuber-like short internodes 2-6	Slender, cord-like with long inter-nodes 1-6	Absent	Absent
Ploidy	Diploid	Diploid	Diploid	Diploid	Tetraploid	Diploid	Diploid
Altitude	1700-2600	500-1800	900-1600	1400-2400	1500-2000	800-1100	10-20
Lat/long	20°05□-101°33	19°49□-104°11	15°40□ -91°45	109°35□ -104°12	19°38□-103°26	14°38□-89°38	12°53□-86°58

Adapted from Iltis et al., (1979); Iltis & Doebley (1980) and Loaisiga et al., (Unpublished).

2. Discussion

2.1. Teosinte *Zea nicaraguensis* versus Wild *Zea* Species

Both cultivated maize and the teosinte species are diploids (2n=20) (except for *Z. perennis* which is a tetraploid specie (Molina and Garcia 1999). The chromosome number of *Z. nicaraguensis* has not yet been reported. This is the first scientific document writing on this specie. Today, this species is found at only two locations in Pacific Coastal Nicaragua, but was much more common and widespread just some 20 years ago. Young plants of *Z. nicaraguensis* are consumed by cattle while mature plants, which can grow as high as 5m, are used to make fences and shelters. The species occurs at 6-15 m above sea level, and has the ability to grow in standing or slowly moving water. Because of this ability, due to a high capacity to form root aerenchyma and adventitious roots at the soil surface, it may be useful in wide hybridizations with maize, in order to improve maize growth in water-logged soils (Iltis and Benz 2000; Mano et al. 2006). Crosses between maize and *Z. nicaraguensis* have already been performed and hybrids have been obtained in an ongoing project by the authors. Mano et al. (2005) have identified Quantitative Trait Loci (QTL) for adventitious root formation and work on transferring these QTLs from teosinte into maize (Mano et al. 2006).

The closest relative to Nicaraguan teosinte, *Z. nicaraguensis*, is the Guatemalan teosinte, *Z. luxurians* (Doebley and Iltis 1980). The two species show great morphological similarity but the developmental behavior differs considerably between the two, supporting a taxonomic segregation. Nicaraguan teosinte has longer and more abundant tassel branches, a larger number of spikelets per branch and a habitat differing from its Guatemalan counterpart (Iltis and Benz 2000 and Loáisiga et al. 2011). When we compare the chromosome numbers of *Z. luxurians, Z. diploperennis, Z. mays* and *Z. nicaraguensis*, the result revealed that the banding patterns among individuals within species were the same except for a few minor bands, which were not seen in all cells. This is probably more due to a more or less successful C-banding procedure used, than to actual differences.

The chromosomes of *Z. mays* are the smallest with an average length of 11.2 µm. The wild species have larger chromosomes with the largest in *Z. nicaraguensis* with an average length of 19.6 µm. The chromosome lengths are not in agreement with the per cent of heterochromatin found. Differences in chromosome contraction or differences in nuclear DNA content not related to heterochromatin could explain the observed differences in chromosome size. Nucleolus Organizing Regions (NOR) could be localized in the two species with the longest chromosomes, *Z. nicaraguensis* and *Z. luxurians* but not in the shorter chromosomes of *Z. diploperennis* and *Z. mays*. The average arm ratios are similar between the four species and individual chromosomes vary from 1.0 to 2.5.

In cultivated maize, *Zea mays* ssp. *mays*, the number of heterochromatic knobs has been found to vary in different maize populations and lines and so the C-banding pattern differs between studies depending on material (Hadlaczky and Kálmán 1975; Molina 1982; Tito et al. 1991). However, in all cases, it has mostly subterminal knobs and only a few terminal ones, which makes it different from the wild teosinte species in section Luxuriantes. The maize cultivar has terminal knobs on both arms of chromosome 1, terminal knobs on one of the arms of chromosomes 5 and 7 and subterminal knobs on one of the arms of chromosomes 2, 4, 6, 8 and 9. Chromosomes 3 and 10 are lacking heterochromatic knobs, Chromosome 1,

2, 3, 4, 6, and 9 have centromeric heterochromatin and chromosomes 5, 7, and 10 regularly show a thin intercalary heterochromatic band. The amount of heterochromatin in a chromosome set, measured as a mean value from the three individuals, is 19.9 % (Table 2).

Also in *Z. diploperennis*, the size and position of heterochromatic knobs have been reported to differ between populations (Kato and Lopez 1990). It has, however, only terminal heterochromatic knobs making it different from cultivated maize. The chromosomes 1 and 2 have terminal knobs on both arms, chromosomes 4, 5, 6, 8 and 9 have terminal knobs on one of the chromosome arms and chromosome 3, 7 and 10 are lacking heterochromatic knobs. Chromosomes 2, 3 and 4 show centromeric heterochromatin and chromosomes 1, 7, 8 and 10 regularly show a thin intercalary heterochromatic band. The C-banding pattern is very similar to the one reported by Molina (1982). The amount of heterochromatin in the chromosome set is lower than in cultivated maize, only 14.0 %.

Zea luxurians has only terminal heterochromatic knobs. Chromosomes 1, 2 and 3 have terminal knobs at both ends while the remaining chromosomes have terminal knobs on one of the arms. Chromosomes 1, 2, 3, 4, 6, 8, and 9 have centromeric heterochromatin and chromosomes 1, 5, 6, 7 and 10 regularly show a thin intercalary heterochromatic band. There is no pair lacking heterochromatic knobs and the mean amount of heterochromatin in a chromosome set is as high as 20,9 %. Tito et al., (1991) reported about a relationship between the number of C-bands and the DNA content where *Z. luxurians* showed a higher DNA content and more C-bands than both *Z. diploperennis* and *Z. mays*. Also we find the number of C-bands on the chromosomes is higher in *Z. luxurians* (13 terminal knobs) than in *Z. diploperennis* (9 terminal knobs) and *Z. mays* (9 terminal or subterminal knobs).

Zea nicaraguensis is, like most other *Zea* species, a diploid with 2n = 20 chromosomes. It has a similar C-banding pattern as *Z. luxurians*, and the two species are therefore probably closely related. Chromosome 1, 2 and 3 have terminal knobs on both arms, chromosome 4-9 have terminal knobs on one of the arms while chromosome 10 is lacking heterochromatic knobs. Chromosomes 1, 2, 3, 4, 6 and 9 have centromeric heterochromatin and chromosomes 1, 5, 7 and 10 regularly show a thin intercalary heterochromatic band (Table 2). The biggest difference compared to *Z. luxurians* is that chromosome 10 is knobless and this is the case in all 10 individuals studied. This leads to a lower heterochromatin content compared to *Z. luxurians*, a mean value of 16.4 %.

From the C-banding pattern it is clear that *Z. nicaraguensis* is more similar to *Z. luxurians* than to *Z. diploperennis* and cultivated maize. The similarity in C-banding pattern most likely reflects a close relationship between the two species. The small differences in banding pattern that exist between them does not speak against regarding them as subspecies as the C-banding pattern can vary considerably even between populations of the same species (as in maize and *Z. diploperennis*).

2.2. Genetic Diversity between Populations of Nicaraguan Teosinte

Currently the maize is the most important crop in the world for production of animal feed. The area under this crop is 140 million of hectares. Yield has been increasing from 325 million in the 80's to 650 million of metric tons in the present decade this represents 4.2 metric tons by hectare (USDA 2004). In Nicaragua it is a basic food with an annual consumption of 70 kg per person and year. The annual sowing area is large but the yield is not enough to supply the national demand (FAO 2000). The production in the country is the 236000 mt annually. The area is 206,000 ha going a mean yield of 1.15 mt/ha. This is far below the world mean value (Somarriba 2000). Nicaragua shows lowest yields of maize in Latin America due to many factors such as; pests, diseases, excess of water, drought or a bad condition of soils and management.

Conservation of plant genetic resource as well as inventory of traits (agronomic, morphological, genetic and biochemical), are necessary in order to describe and differentiate populations of different species and to identify sources useful for genetic crop improvement (FAO 2001). Molecular marker-based studies of genetic diversity in plant species have become important tools in plant breeding, ecology and conservation (Nybom and Bartish 2000). Generally, genetic diversity of a species is strongly associated with phyletic group, life form, geographic range, regional distribution, breeding system, seed dispersal mechanisms, mode of reproduction and successional status (Bhat et al., 1999).

One of the process utilized by plant breeding, consists in looking among wild relatives of the domesticated species for genes or gene complexes that have outstanding characteristics. It can be pest and diseases resistance or tolerance to abiotic stress. One main cause for the low yield in maize in Nicaragua is pest and disease. Recently, teosinte from Nicaragua was classified as new specie, *Zea nicaraguensis*. It is at present in the same geographical groups as *Zea luxurians*.

Lubberstedt et al., (1998) in a study carried out with *Zea luxurians* from Guatemala obtained ranges between 75 and 160 base pairs. When were analyzed the seven teosintes populations (from A to G) from Nicaragua, using 11 microsatellite primers, were find size of pair base from 30 to 194, only in the population G the primer Bnlg 1526 did not amplify anything.

Table 2. Chromosome length, chromosome arm ratio and percent heterochromatin of the chromosomes of species *Z. nicaraguensis, Z. luxurians, Z. diploperennis* and *Z. mays* ssp. *mays*

Chromo	Chromosome length (μm)				Long/short arm ratio				% Heterochromatin			
	Z. *nic*	Z. *lux*	Z. *dipl*	Z. *mays*	Z. *nic*	Z. *lux*	Z. *dipl*	Z. *mays*	Z. *nic*	Z. *lux*	Z. *dipl*	Z. *mays*
1	24.0	20.5	13.9	11.0	1.3	1.3	1.3	1.3	22.9	31.4	33.3	41.6
2	21.3	18.3	12.7	10.7	1.0	1.3	1.1	1.6	24.7	23.3	25.9	23.1
3	22.0	17.5	12.7	13.8	1.3	1.6	1.6	1.2	19.7	23.3	0.0	0.0
4	13.8	13.6	9.7	9.7	2.3	1.9	1.5	2.5	20.1	23.9	20.1	26.2
5	21.3	18.3	13.3	11.9	1.1	1.3	1.2	1.6	12.1	17.9	11.4	22.3
6	17.1	15.7	11.3	11.9	2.4	2.3	1.7	1.4	17.9	20.2	19.5	20
7	18.2	15.7	14.3	10.5	1.3	1.3	1.1	1.1	16.3	21.5	2.2	19.5
8	17.3	16.4	13.6	11.7	1.2	1.1	1.0	1.4	17.3	21.5	16.3	22.2
9	25.3	20.9	19.5	11.7	1.4	1.1	1.2	1.4	11.2	8.6	9.9	21.7
10	16.0	18.1	10.4	9.3	1.1	1.3	1.3	1.0	2.0	17	2.0	2.0
Aver.	19.6	17.5	13.1	11.2	1.4	1.5	1.3	1.5	16.4	20.9	14.1	19.9

The figures are mean values from measurements made from three individuals of each species.

The genetic diversity average between populations was 0.563. This value is high if compared with the values of 0.487 in maize obtained by Sanchez et al., (2000). Matsuoka 2001 in Mexico, working with 42 microsatellites and 4 teosintes, obtained the following results; *Z. diploperennis* (0.49), *Z. luxurians* (0.35), *Z. huehuetenangensis* (0.48) and *Z. parviglumis* (0.68). If we compare the different values in Nicaragua teosinte also is possible to see some similarities among them. For example, according with AMOVA result we can consider each population as a different population, or all populations are below the same specie but with some difference characteristics among them.

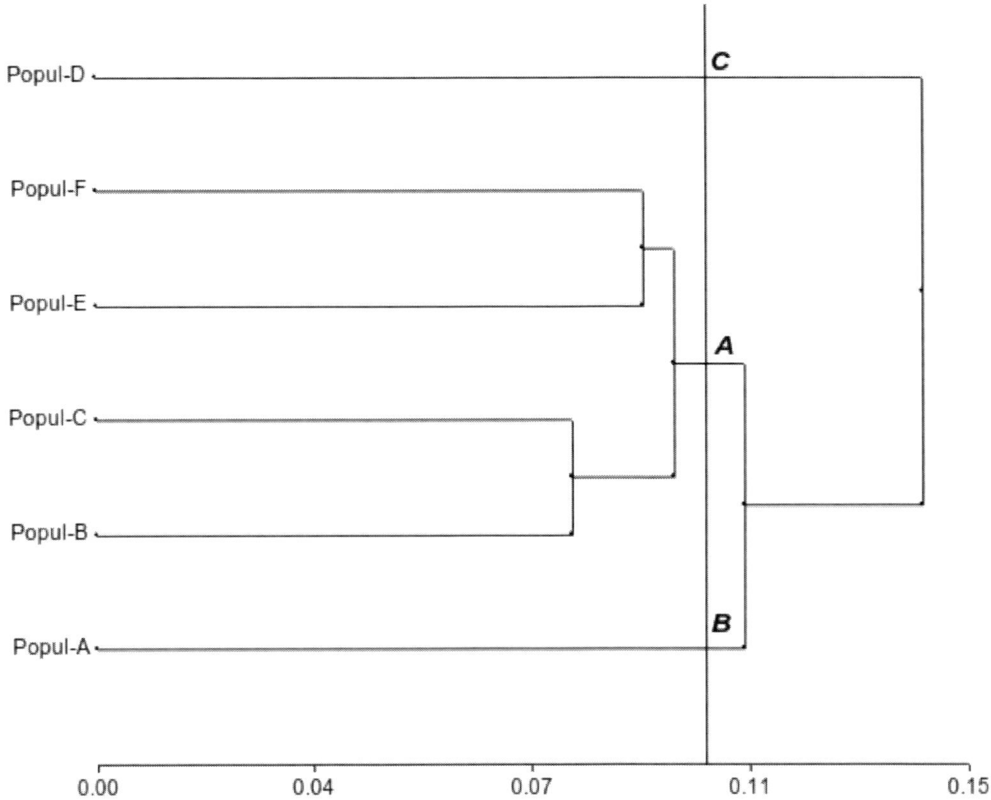

Figure 2. Genetic relationships among seven different populations of Teosinte Zea nicaraguensis using the Rogers Index.

According to Rincon et al., 2005, the size of sample is important to evaluate the genetic diversity, because less variation can be shown when size of population decrease and vice verse. Doebley et al., (1985) find a less variation when he compared the same populations (tree different gramineaes species) but with a different sample size.

The analysis with all populations through the UPGMA clustering based on Roger index, revealed three main clusters; first, the population B-C with 0.08 and populations F and E; second, the population A; third population D with the most the more distanced population from the rest of studied populations. In fact, those two populations (B and C) are closely located and flow between them should be frequent (Figure 2).

Nevertheless, in this figure, the position of the population F it is not expected, because this population is 11 kilometres distance from the principal nucleus. It phenomenon could be possible if we considerer that many years ego (25 – 30 years) all areas around the principal nucleus there was teosintes plants. In addition, has been movement of peoples in this area, it is other factor to consider. In spite of this distance, there are some relations among populations, for example is important to mention that the 35 % of the genetic variation is represented by the two principal coordinates of the principle coordinates analysis (Figure 3).

Doebley et al., (1985), working with 23 different loci in 94 populations of Teosinte (*Zea mays* ssp. *parviglumis*) and 12 individual per population obtained 7.1 alleles. Rincon et al., (2005) obtained different average on number of alleles per locus; *Z. mexicano* (3.08 alleles), *Z. parviglumis* (2.4) and *Z. diploperennis* (2.23). In our case, we find 3.8 as average, its mean that *Z. nicaraguensis* probability has a different genetic diversity in relation to others teosintes in the region.

Also Rincon et al., (2005) working with different populations of Teosinte in Mexico found in each population rare alleles, in this way he suggested that is necessary to keep these populations, because these individuals may possess also other rare characteristics which can be of value for the National Plant Breeding Program. Is important considerer that the crop improvement through breeding process depends in big magnitude if the genetic diversity is utilized, Genet and Belete (2000).

The population E obtained highs values in relation to rare and unique alleles, and in the second place, the population B and C, it could be because there was more plants in the field in those populations that the others. In total, there are 19 and 4 rare and unique alleles among populations. This similar population structure suggest that is convenient to keep all population in especial those with rare and unique alleles.

The knowledge of the phenotypic and genotypic variation, present in the wild species can be useful to prepare or design methods and strategies in plant breeding program in cultivated species. Xiao et al., (1996) demonstrated that the yield of a high-performance rice (*Oryza sativa* L.) cultivar could be increased by introgression of certain genome regions of the low-performance wild species *Oryza rufipogon* Griff., while Marquez (1988) reported similar findings for maize landraces and their relatives. Orr et al. (2004) found that *Z. nicaraguensis* was tolerant to waterlogged environments and, with its stable expression of a reproductive pathway, could be a useful source of germplasm for maize breeding.

Overall, all populations of *Z. nicaraguensis* may possess very valuable traits. Anthropogenic disturbances in the habitat on this specie have had a very strong impact, as mentioned above. During the last 30 years this area has suffered fires and a destruction of the flora and fauna. The hope for the future is that with the appearance of Flora de Nicaragua (Stevens et al., 2000), the growing contingent of well-trained Nicaraguan botanists and environmentalists, assisted by conservationists worldwide, will work to preserve the remaining wild populations of teosintes, *in situ* if possible (Guzman and Iltis 1991). This would have major benefits for Nicaraguan biodiversity (Iltis and Benz 2000). In this way, from an ecological perspective it is necessary to preserve all populations of teosintes in their natural habitats, through a special management plan. Representative samples from all populations should also be preserved in the National Gene Bank (REGEN) at the National Agrarian University (UNA) in Nicaragua.

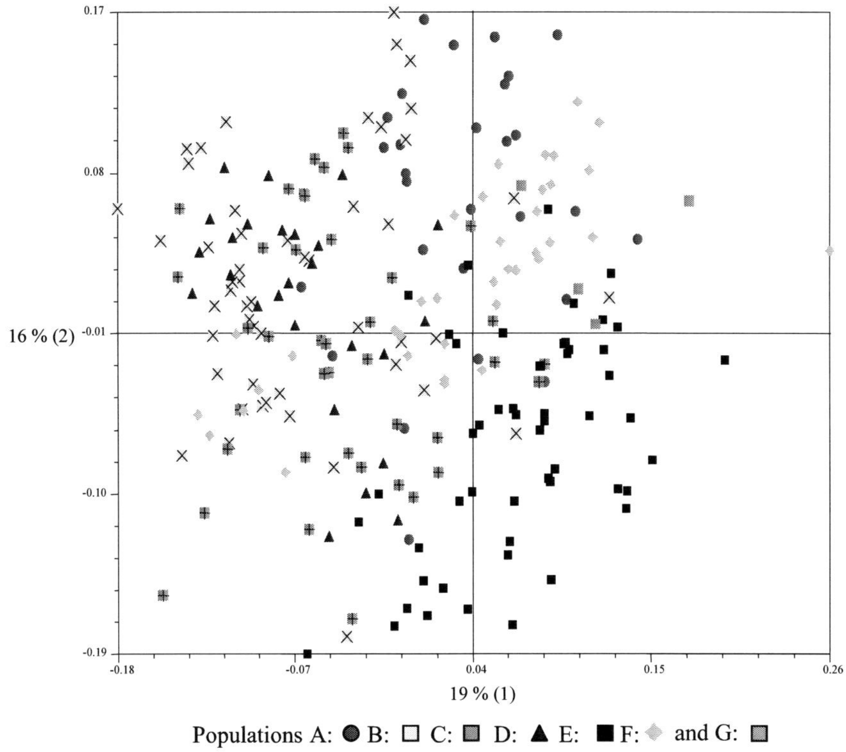

Figure 3. Distribution of seven populations of Teosinte Zea nicaraguensis revealed by the principal coordinate analysis with Roger index.

2.3. Genetic Diversity and Gene Flow among Meso-America Teosintes

The genus *Zea* has a native distribution from Mexico to Nicaragua, a region known as Meso-America, and studies on teosinte species may contribute to our understanding of maize genetics (Lukens and Doebley 1999; Evans and Kermicle 2001), molecular population genetics (Guat et al., 2000), genome evolution (Sanz-Alferez et al., 2003), and crop evolution (Doebley 1990). Moreover, since F1 hybrids between teosinte and maize are fertile, all teosinte species can potentially contribute to maize breeding, acting as sources of traits such as resistance or tolerance to heavy metals in the soil and other adverse environmental conditions (Dale 1994).

Conservation and characterisation of crop genetic resources are necessary to identify and preserve useful genes for genetic improvement. One important strategy consists of screening among wild relatives of crop plants for genes or gene complexes responsible for valuable characteristics for plant breeding, particularly genes associated with pest and disease resistance or tolerance to abiotic stress (Bath et al., 1999). Gene flow between maize (*Zea mays* ssp. *mays*) and teosinte (*Zea* spp.) occurs in Mexico and Central America, and maize x teosinte hybrids have been recorded in archaeological archives (Wilkes 1967; Eubanks 2001).

While gene flow studies have been reported in *Z. mays* ssp. *mexicana* (Baltazar et al., 2002), little information is available on sympatric maize and other teosinte populations from different regions. Such information is essential in order to better understand the potential for gene flow from maize to teosinte. The present work is an effort to characterise the levels of genetic diversity in *Z. nicaraguensis* and its genetic similarity with other species of teosinte.

In general, the allele numbers in wild species tend to be constant over time unless influenced by adverse evolutionary factors (Fontdevila et al., 1999). In many studies with wild *Zea* species from; Doebley et al., (1987), Rincon et al. (2005), Matsuoka et al., (2001) and Loáisiga et al. (2010) has obtained different values of allele's diversity, so has demonstrate a high variability in teosinte species.

Similar findings were made here, when teosintes from Mexico were compared with those from Central America, with a mean of 3.46 and 3.52 alleles per locus, respectively. The higher mean number of alleles in the teosinte population reported by Doebley et al., (1987) 20 years ego compared with more recent studies indicates a possible decrease in genetic diversity over that period for various reasons, principally by antrophonenic process.

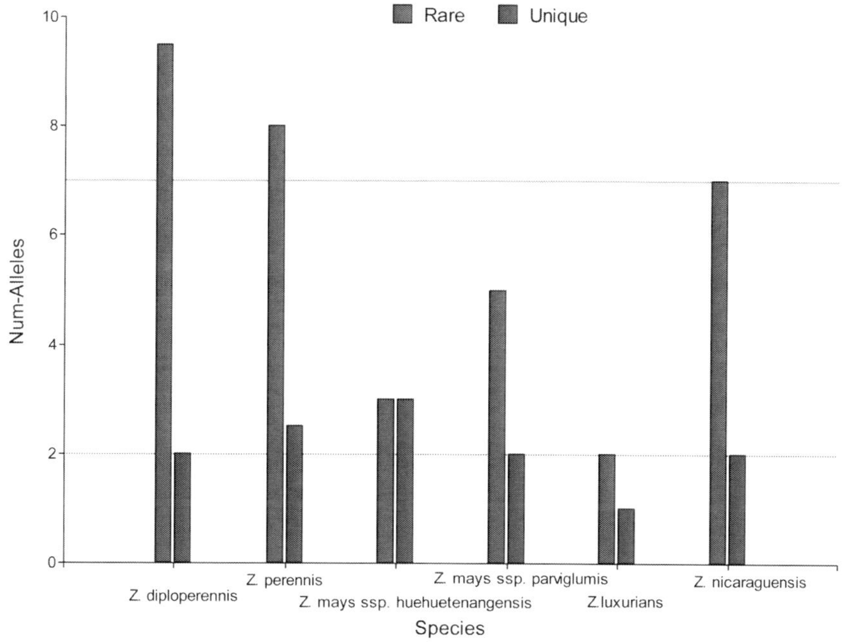

Figure 4. Number of unique and rare alleles in six Meso-America teosintes. The values shown for *Z.. diploperennis* and *Z. perennis* are the mean of two accessions.

There were also differences in allele frequencies between accessions within species. For example, the total numbers of alleles found in the two accessions of *Z. diploperennis* were very similar, but both accessions had unique alleles (4 and 0), indicating differences in their genetic make-up. However, *Z. diploperennis* had almost the same number of rare alleles (8 and 10). Conversely, both accessions of *Z. perennis* had similar numbers of unique alleles (3 and 2) but differed in terms of rare alleles (10 and 7). Comparing data between species only three species had seven or more rare alleles, namely *Z. diploperennis, Z. perennis* and *Z.*

nicaraguensis. As regards unique alleles, almost all species and subspecies had two or more except *Z. luxurians* from Guatemala (Figure 4).

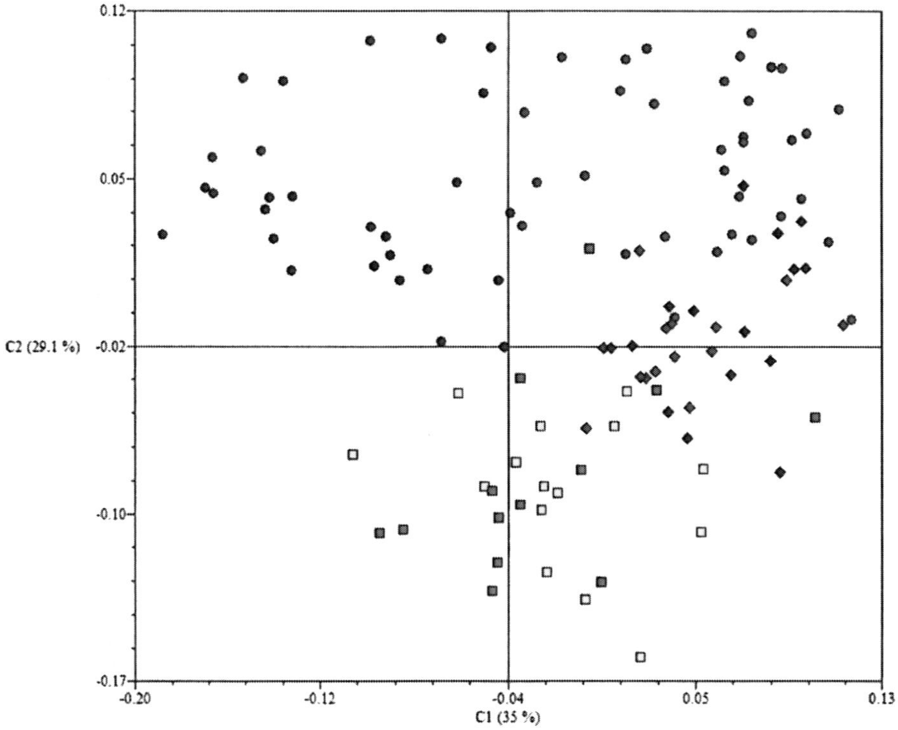

Figure 5. Distribution of six species of Meso-America teosintes with Rogers' genetic distance. *Z. diploperennis*: ● *Z. perennis*: ● *Z. mays* ssp. *huehuetenangensis*: □ *Z. mays* ssp. *parviglumis*: ■ *Z. luxurians*: ◆ and *Z. nicaraguensis* : ◆

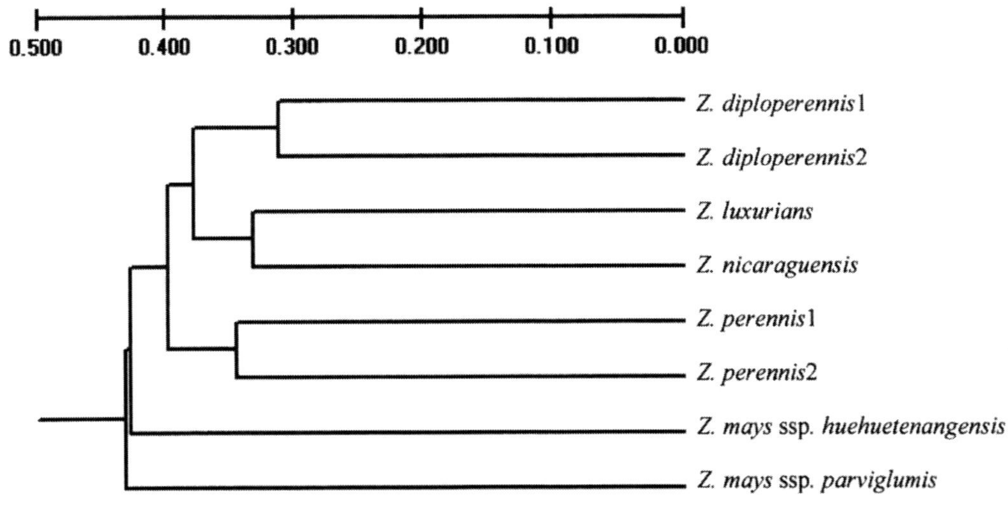

Figure 6. Clustering pattern of eight Meso-America teosinte accessions, generated by UPGMA cluster analysis using the Rogers' genetic distance matrix.

Analyses of molecular variance explained 19.87 % of significant differences between accessions, which means that there was high genetic variation between accessions. This variation can be used to differentiate one accession from another For example, *Z. diploperennis* is spread in quadrant I, *Z. perennis* appears in the quadrant II, *Z. mays* ssp. *huehuetenangensis* and *Z. mays* ssp. *parviglumis* in quadrants III and IV, and *Z. luxurians* and *Z. nicaraguensis* appear together around the line that divides quadrants II and IV. This indicates that each specific accession conserves an intrinsic genetic variation. Doebley et al., (1984) obtained similar results when working with almost the same germplasm (Figure 5).

In spite of this particular genetic structure for each species, there were also similarities between them, as seen in Figure 6. In this case, *Z. luxurians* and *Z. nicaraguensis* showed more similarity to each other than in relation to *Z. mays* ssp. *huehuetenangensis*, despite the fact that the three species are from the same region in Central America. There was higher genetic similarity between *Z. luxurians* and *Z. nicaraguensis* (d=0.771) than between them and *Z. mays* ssp. *huehuetenangensis* (d=0.673). Iltis and Benz (2000) reported great morphological similarity between these species, but also considerable differences in developmental behaviour. Wilkes (1977) and Iltis and Doebley (1980) found the Guatemalan race (*Z. luxurians*) to be distinct from all other teosintes and, accordingly, they treated it as a separate species.

In general, the mean expected heterozygosity in this study was higher (0.567) than observed heterozygosity (0.497) for all accessions. Doebley et al. (1984) obtained the same heterozygosity composition with the same accessions, and also found that the mean expected heterozygosity was higher (0.191) than the observed heterozygosity (0.154). Recombination and gene flow play a major role in the dynamics of genetic diversity in traditional farming systems (Elias et al., 2000a). Other factors, such as low seed production, sterile seeds, seed dormancy, crop management, local ecology, physical distance, flowering synchrony and human selection, influence the level of gene flow (Baltazar et al., 2002).

In Central America, it is evident that the genetic differentiation and gene flow between *Z. nicaraguensis* and *Z. luxurians* is higher (F_{st} = 0.095 and N_m = 2.38) than with *Z. mays* ssp. *huehuetenangensis* (F_{st} = 0.1304 and N_m = 1.667), despite *Z. mays* ssp. *huehuetenangensis* growing closer to *Z. luxurians*. For the other two species in the region (within Guatemala), *Z. luxurians* and *Z. mays* ssp. *huehuetenangensis*, the genetic differentiation and gene flow were low (F_{st} = 0.1333 and Nm =1.629). In other words, gene flow was more frequent between Z. nicaraguensis and Z. luxurians, despite the smaller geographical distance between Z. luxurians and Z. mays ssp. huehuetenangensis (Figure 1).

There are many possible factors behind this phenomenon. First, this area is flatter and there are more communication pathways, leading to more frequent movements by people and, probably, germplasm interchange. Second, Standley (1950) reported a previous teosinte population that is now extinct in San Antonio de Padua, a district of Choluteca Honduras, approximately 90 km north of the *Z. nicaraguensis* populations in Nicaragua. This population could have constituted a biological bridge between the two species. Both these factors could have facilitated gene exchange among the species in the past, leading to the current higher genetic similarity. In general, *Zea nicaraguensis* showed the highest genetic similarity with *Z. luxurians,* but higher genetic diversity values than this species.

2.4. Morphological Similarities between Teosintes (Zea spp.) in Meso-America

For many years, the relationships between *Zea* species were the subject of much controversy. The major difficulty in the taxonomy of maize and the identification of its closest relatives was the absence of a cob-like pistillate inflorescence (or ear) in any other species. Whereas teosintes produce only 6 to 12 kernels in two rows protected by a hard outer covering, the modern cultivated maize has a cob consisting of 20 rows or more, with numerous exposed kernels. In fact, teosintes are morphologically so unlike maize in the structure of the ear that 19th century botanists failed to recognise the close relationship between them, and placed all the teosintes in the genus *Euchlaena* instead (Doebley, 1990).

Despite the profound physical differences, various morphological, cytological and genetic studies eventually delineated the current circumscription of *Zea*, and Wilkes (1967) laid the foundation with the first thorough monograph on teosinte. Wilkes did not attempt to make a formal hierarchy, but instead presented a system of classification using different geographical populations with separate racial designations based on distinguishing morphological features. Later, Doebley and Iltis (1980) and Iltis and Doebley (1980) produced a classification that considered the probable evolutionary relationships among taxa. With the quantitative evaluation of numerous traits and the discovery of many additional populations, Sánchez et al., (1998) provided further characterisation of the genus.

Comparative studies of crops and their wild ancestors are highly relevant to plant breeding. Crops differ from their wild ancestors in their expression of traits of agronomic importance. Knowledge of the mechanism that controls the expression of agronomic traits provides the basis for further improvement of crops. In this regard, identifying the genes for agronomic traits is of great interest, as the expression of agronomic traits is usually controlled genetically (Matsuoka, 2005).

Plant genetic resources are the most fundamental components of all agricultural systems and thus preservation, evaluation and enhancement of germplasm are essential activities. The advances in computing technology and multivariate statistical methods for classifying individuals have provided access to powerful tools for classifying genetic resource conservation and forming core and mini-core subsets. There are several classification methods that can be used in genetic resource conservation, in the formation of core subsets, and in defining possible heterotic groups. In all these methods, the main objective is to select accessions that best represent the entire collection with the minimum loss of genetic diversity. Therefore, the best numerical classification strategy is that which produces the most compact, well-separated groups, i.e., minimum variability within each group and maximum variability among groups (Judez, 1989).

When we compare the Meso-American teosintes using 33 quantitative morphological characteristics, there was significant variation between all accessions, except in spike length, ear length, and grain width (Table 3). When traits placed under *plant characteristics* were considered, *Z. nicaraguensis* showed the highest values in 63% of the traits, whereas *Zea perennis*1 showed the lowest values in 50% of the traits in this category. This means that morphologically, *Z. nicaraguensis* is a more robust species than the other teosinte species. Based on this analysis, *Z. mays* ssp. *huehuetenangensis* is the second more robust species after *Z. nicaraguensis*. However, in the *tassel characteristics*, *Z. mays* ssp. *huehuetenangensis* showed the highest values in 55% of the traits, whereas *Z. diploperennis*1 showed the lowest values in 36% of the traits. *Ear and grain characteristics* did not display any specific pattern among accessions.

In general, 55% of the lowest values were associated with *Z. perennis*1 and *Z. diploperennis*1 and 67% of the highest values were associated with *Z. nicaraguensis* and *Z. mays* ssp. *huehuetenangensis* (Figure 7). This result is in agreement with Bird (1978), who indicated that *Z. mays* ssp. *huehuetenangensis* in north-western Guatemala shares some traits with the Central American teosintes.

Wild *Zea* species are extremely sensitive to even the slightest changes in daylength and moisture, as reported in terms of tillers and branches by Bird (2000); Iltis (1983, 1987), Iltis and Benz (2000) and Guzman (1985). Bird (2000) demonstrated this with *Z. nicaraguensis* in an experiment conducted in Mexico City and North Carolina, in which the plants looked completely different compared with the plants in their natural habitat, it could be explained by their very high phenotypic plasticity.

Wilkes (1977) reported that *Z. mays* ssp. *huehuetenangensis* has a longer vegetative period, the plants do not flower until late November and seeds do not mature until January. Almost the same growth pattern has been recorded for *Z. nicaraguensis,* with both requiring almost nine months from sowing to harvesting, but *Z. nicaraguensis* required 15-20 extra days to complete the vegetative period.

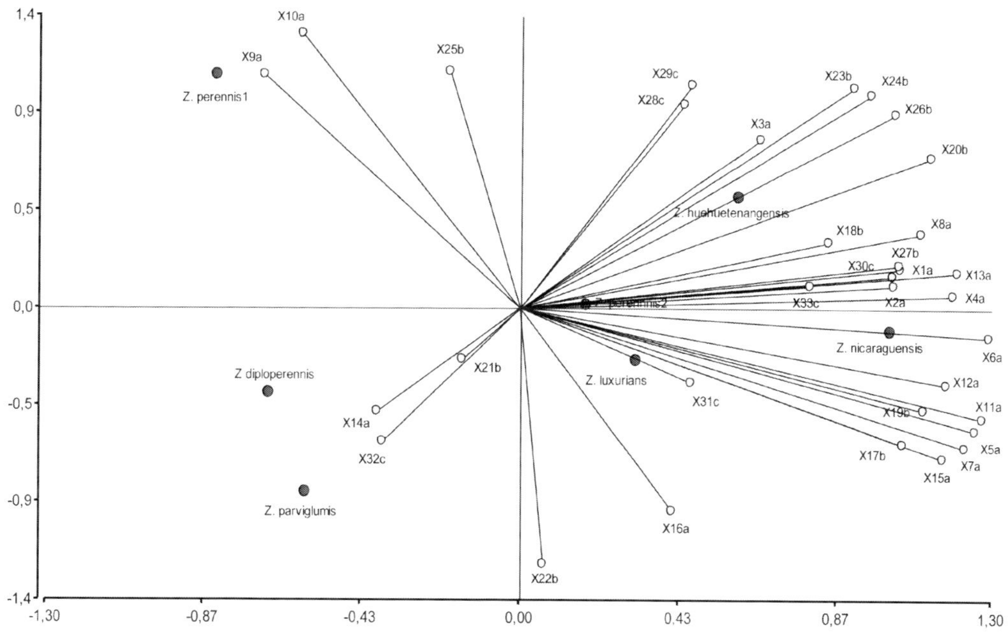

Figure 7. Distribution of the 33 quantitative morphological characteristics of teosintes from Meso-America on first and second principal components.

Plants of *Z. nicaraguensis* grown in a chamber at the University of Minnesota produced mature seeds four months after planting, unlike the common germplasm (Iltis, 1983). These plants were clearly differentiated from the Guatemalan form of the species by their small stature and early maturation. The variation may also be due to the extreme sensitivity of the plants in their response to moisture, daylength, and altitude compared with common teosinte plants (Iltis, 1983, 1987; Ilitis and Benz, 2000; Bird, 2000). Most teosinte populations are spatially isolated by broken topography and high mountains, and consequently the different populations may have acquired certain morphological, ecological, chromosomal and genetic distinctness (Wilkes, 1977).

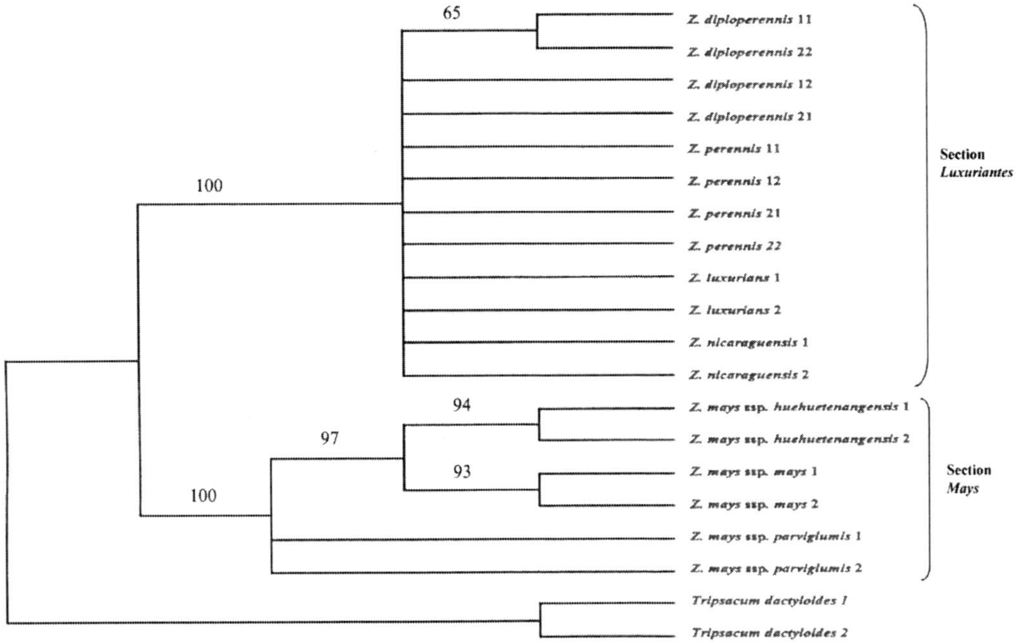

Figure 8. Strict consensus tree generated from sequences data from the eight cpDNA regions in nine *Zea* accessions using two *Tripsacum dactyloides* as outgroup. Indels are treated as binary present/absent characters.

Doebley (1983) found that *Z. mays* ssp. *huehuetenangensis* is very similar to *Z. mays* ssp. *parviglumis*, but differs in that it is adapted to a wetter habitat with a longer growing season. Thus, the plants tend to be much taller (up to 5 m), have larger, glabrous leaves, and are much later in flowering. A corresponding situation is found between *Z. luxurians* and *Z. nicaraguensis*. Iltis and Benz (2000) reported that in the natural habitat of *Z. nicaraguensis*, the plants grow with their stem and prop roots covered by shallow standing water 10-40 cm deep. The plants are erect, annual, unbranched to candelabra-branched, 2-4 (5) m tall, sometimes with one to four stout basal tillers (suckers) up to 2 m long (Iltis and Benz, 2000). Benavides (2003) found *Z. nicaraguensis* plants up 6 m tall in their original habit.

Doebley (1983) also found that *Z. perennis* resembles *Z. diploperennis*, but differs in several key characteristics. Notably, the plants are less robust and about 1.5 m tall, with leaves 1-3 cm wide. They lack the short, thick, tuberous rhizomes of *Z. diploperennis*, but produce an abundance of longer, cordlike rhizomes with much longer internodes. These rhizomes spread aggressively, so that the species grows in thick, dense mats. In contrast, Iltis and Benz (2000) described *Z. mays* ssp. *huehuetenangensis* and *Z. nicaraguensis* as maize-like, erect annual plants, only with *Z. nicaraguensis* being taller (up to 6 m in its natural habitat).

Table 3. Quantitative morphological characteristics evaluated in Meso-American teosintes

X1a	Days to the beginning of flowering	X18b	Spike length (cm)
X2a	Days to complete flowering	X19b	Distance between branches in the tassel (cm)
X3a	Total number of leaves	X20b	Mean branch length in spike (cm)
X4a	Leaf length (cm)	X21b	Internode length in main branches in spike (cm)
X5a	Leaf width (cm)	X22b	Internode length in secondary branches in spike (cm)
X6a	Plant height (cm)	X23b	Spikelet width (mm)
X7a	First ear height (cm)	X24b	Spikelet length (mm)
X8a	Upper ear height (cm)	X25b	Pedicel length (mm)
X9a	Number of lateral branches in main stem	X26b	Glumes width (mm)
X10a	Number of tillers per plant	X27b	Number of veins per glumes
X11a	Number of nodes with ear in main stem	X28c	Ear length (cm)
X12a	Number of nodes with ear in the lateral branches	X29c	Number of grains in ear
X13a	Number of nodes with ear in the tillers	X30c	Hundred seeds weight (g)
X14a	Longitude on the 3rd uppermost last lateral branch	X31c	Grain length (mm)
X15a	Number of nodes on the 3rd uppermost one lateral branch	X32c	Grain width (mm)
X16a	Number of ears on the 3rd uppermost nodes of main stem	X33c	Number of seeds per plant
X17b	Number of branches in tassel		

The closest relative of *Z. nicaraguensis* is presumably *Z. luxurians* of south-eastern Guatemala (Doebley and Iltis, 1980; Doebley, 1983; Iltis et al., 1986), an annual teosinte with similar, strongly two-nerved lower glumes. Iltis and Benz (2000) found that the primary tassel length in *Z. luxurians* was 12-24 cm, compared with 23-32 cm in *Z. nicaraguensis*, the respective number of tassel branches was 4-28 and 27-38, the distance between branches in the tassel was 1-9 cm and 8-12 cm, and the number of spikelet pairs per branch was approx. 20-35 and 36-64. The single 'corpse' of the tassel is borne on a culm fully 25.4 cm long, in comparison with the shorter culm length range (10-23 cm) of the Guatemala *Z. luxurians*.

In our study, the analysis of similarities among accessions using all variables revealed that *Z. nicaraguensis* and *Z. luxurians* were closely related and thus shared many characteristics (Figure 6). In addition, the principal component analysis showed that the first three principal components (explaining 81% of variance) differentiated the accessions into two groups. Hence, this study confirms the close relationship between those species. Close genetic relationships between these two species were also observed by Loáisiga et al., (2010) using molecular markers on the same teosinte populations.

In general, the methods used in this study to analyse the pattern of variation in teosinte proved to be complementary. The correlation analysis revealed a highly significant correlation of physiological traits with green biomass and grain yield. According to component analysis, most of the total variation was explained mainly by eight plant characteristics. Different numerical taxonomic techniques showed that several characteristics were associated, principally between *Z. luxurians* and *Z. nicaraguensis*, but to some degree also between these and *Z. mays* ssp. *huehuetenangensis*.

2.5. A Molecular Phylogeny of *Zea* Specie

Ecological and phylogenetic data provides heterogeneous information about the origin, evolutionary history and present distribution of species and species groups (Wiens 2004). During the last decade an increasing number of studies combined these kinds of data to test biogeographical (Leathwick, 1998), ecological (Anderson et al., 2002a) and evolutionary hypotheses (Graham et al., 2004). Research integrating these fields has revealed new insights in factors influencing the evolution and distribution of species, and species distribution models based on ecological variables have become powerful tools for addressing issues in evolution, biogeography, ecology and conservation biology (Peterson, 2001). DNA analysis has proved to be one of the most effective methods for studies of phylogeny and evolution.

The nuclear genome is inherited in a biparental manner whereas the chloroplast and mitochondrial genome are generally inherited maternally. Biparental inheritance and polyploidy in the nuclear genome may cause difficulties for evolutionary analysis, whereas analyses of chloroplast genomes are effective in the investigation of relationships in plants because of their maternal inheritance (Tomotaro et al., 2002).

Maize (*Zea mays* L.) domestication is inexorably linked to the development of New World cultures and maize remains a dominant food source. Maize is agricultural preeminence has led to its use as a model system for genetics and molecular biology (Doebley 1990a; Kellogg & Birchler 1993). Despite the wealth of information obtained on *Zea mays*, the phylogeny of the genus *Zea* is still unresolved.

Most of the teosintes taxa thus have narrow geographic distributions, often consisting of only a few local populations. The exceptions are *Z. mays* ssp. *mexicana* and *Z. mays* ssp. *parviglumis*, which are widely distributed in Mexico (Fukunaga et al., 2005). The status of *Z. nicaraguensis* has been questioned by Fukunaga et al. (2005). They treat it as a "geographical group" of *Z. luxurians* without ascribing it any rank.

Chloroplast restriction site, isozyme, and cytogenetic analyses (Kato, 1976; Doebley & Goodman 1984; Renfroe & Blanton 1987) have established that maize (*Z. mays* ssp. *mays*) was domesticated from wild *Z. mays* populations in Central Mexico (Doebley, 1990a). The conclusions drawn from those molecular studies were incorporated in the current classification of *Zea* (Doebley, 1990a), but these studies did not resolve the detailed phylogeny of the genus.

The chloroplast genome is useful in providing information on the evolutionary patterns and processes in plants (Raubeson & Jansen, 2005). Chloroplast DNA (cpDNA) sequence variations, either individual or combined with other genomes, have been widely used for inferring phylogenetic relationships of different taxa, including *Hordeum*, *Triticum* and *Aegilops* (Gielle & Taberlet 1994) and *Sorghum* (Dillon et al., 2007). Compared with coding regions, the noncoding chloroplast regions are phylogenetically more informative at lower taxonomic levels because they are under less functional constraints and evolve rapidly (Gielle & Taberlet 1994). Of these regions, *trn*T-*trn*L was reported to be one of the regions that provides the highest number of potentially informative characters across all polygenetic lineages and possesses enough phylogenetic signals for studies at lower taxonomic levels (Shaw et al., 2005).

Comparative DNA sequencing has become a widespread tool for inferring phylogenetic relationships as it is relatively fast and convenient. Phylogenetic inference and elucidation of evolutionary processes that generate biological diversity have been accomplished even at lower taxonomic levels using a non-coding region of chloroplast genome and internal transcribed spacer (ITS) of the nrDNA (Mort et al., 2007; Kårehed et al., 2008). The chloroplast genome is believed to essentially only display maternal inheritance in the majority of angiosperms (Mogensen, 1996; Keeling, 2004; Udall & Wendel 2006). Current systematic treatments of *Zea* are based on evidences from several biological disciplines, e.g., isozyme, chloroplast restriction site, mitochondrial, chromosomal knob and nuclear ribosomal marker studies (Kato, 1976; Smith et al., 1986; Doebley et al., 1984; Doebley et al., 1987a; Allen 1992; Buckler & Holtsford 1996), however, hitherto not by cpDNA sequence variation.

In this study, of the eight cpDNA region primers not all of them successfully amplified the target region in the teosinte accessions. The results revealed a small number of parsimony informative variable sites across all DNA regions; nevertheless the region primers *trn*Y-*trn*D, *trn*Y-*psb*M and *rps*16-*trn*1 may suggest some usefulness for the phylogenetic analysis of the genus *Zea*.

Various phenetic and phylogentic studies have consistently supported a differentiation amongst the taxa of *Zea*. The two perennial species, *Z. perennis* and *Z. diploperennis*, generally form a cluster/clade, with *Z. luxurians* branching out next to them. Although this relationship is clearly supported by cpDNA restriction site variation (Doebley et al., 1987a), it is neither supported nor contradicted by the tree produced by nuclear ribsosomal markers (Buckler & Holtsford 1996; Doebley et al., 1984). Buckler et al., (2006), in research based on isozyme data, found that *Z. nicaraguensis* appears to be allied with *Z. luxurians* but is at the same time consistently distinct from this species, hence supporting Iltis & Benz's (2000) conclusion that "luxurians" from Nicaragua is better treated as a independent species. However, Fukunaga et al., (2005) in a study of 93 microsatellite loci found that *Z. nicaraguensis* appeared nested within the *Z. luxurians* clade. Thus they concluded that "nicaraguensis" could possibly be given the rank of a subspecies of *Z. luxurians* but state that more data is needed.

Our data strongly supports a clade consisting of *Z. perennis, Z. diploperennis, Z. luxurians* and *Z. nicaraguensis* but unfortunately without any resolution within this clade, since all species appear as a basic unresolved polytopy. Thus our data can neither support nor refute any of the previously suggested relationships among the four taxa. It can only be concluded that there is not enough differentiation within the eight cpDNA regions to resolve the phylogeny among these closely related species. Nguni et al., (2010) obtained good resolution using ITS sequences in *S. bicolor* germplasm; for this reason it might be interesting to test ITS sequences in future work with teosinte genera (Figure 8).

The other strongly supported clade in our data consists of *Z. mays* sensu lato and this is in agreement with the current classification of the genus (Doebley, 1990b). However, the relationships within *Z. mays* (i.e., among subspecies) are still unclear. Unfortunately *Z. mays* ssp. *mexicana* was not available for inclusion in this study. Still, our results reveal that *Z. mays* ssp. *huehuetenangensis* is the sister to *Z. mays* ssp. *mays* as judged by cpDNA sequence variation data, with *Z. mays* ssp. *parviglumis* appearing as a weakly supported sister group to these two taxa. However, it is important in this context to remember that this data only represents the maternal inheritance and that when studying infra-specific taxa the phylogenetic relationships are possibly obscured by the presence of gene flow and introgression among populations and subspecies, as indeed have been reported for the *Z. mays* complex by Ross-Ibarra et al., (2009). This could explain the contradictory results regarding affinities within the *Z. mays* complex. For example, isozyme data suggests that *Z. mays* ssp. *huehuetenangensis* is strongly divergent from the other subspecies, which are much more similar to each other (Doebley, 1990a). Unfortunately, there was not enough differentiation within the eight cpDNA regions to resolve the phylogeny among these closely related species in section *Luxuriantes*.

CONCLUSION

Based on the cytological, morphological, molecular and phylogenetic studies of teosintes from Meso-America, the following conclusions can be drawn:

Zea luxurians and *Zea nicaraguensis* are genetically more closely related to each other than to the rest of teosintes and C-banding pattern technique based cytogenetic analysis clearly supports that they are sister taxa.

Teosintes from Central America are more diverse than those in Mexican region, and the genetic variation within accessions is slightly higher than that among accessions in all analyzed taxa.

All populations of *Zea nicaraguensis* belong to the same taxon. The different populations are extremely threatened and the allele richness is high but many alleles are *rare* or *unique*.

Most of the total variation was explained mainly by six vegetative plant part characteristics and two tassel characteristics, according to principal component analysis. Also, several characteristics were associated principally with *Zea luxurians* and *Zea nicaraguensis* and to some degree with *Zea mays* subsp. *huehuetenangensis* as well, using different numerical taxonomic techniques.

Populations of *Zea nicaraguensis* revealed more rare and unique microsatellite alleles than *Zea luxurians.*

FUTURE PROSPECTS

To improve the resolution in the classification among Meso-American teosintes, it is recommended that more cpDNA regions should be used in combination with Internal Transcribed Spacer (ITS) and unlinked single-copy nuclear sequences in order to produce a better resolved phylogeny in genera *Zea*.

The significant genetic variations among teosinte populations from Nicaragua are an important signal for the conservation of all of those populations, with representative samples from each one in their natural habitats, through the implementation of a special management plan to permit the natural process of evolution of this important species.

Further research into *Zea nicaraguensis* is needed in order to find resistant or tolerant genotypes in relation to pests and diseases, and with a particular focus on the ability to cope with water-logged soil conditions as a source of germplasm for breeding.

In order to be prepared for the anticipated climate changes in the world it is urgently necessary to preserve and study all these species in order to find useful traits to incorporate in maize breeding.

REFERENCES

Allien J. (1992). *Teosinte cytoplasmic genomes: interaction with maize nuclear genomes and molecular genetic characterization of the mitochondria*. PhD University of Wisconsin.

Anderson R., Peterson A., Gomez L. (2002a) Using niche-based GIS modeling to test geographic predictions of competitive exclusion and competitive release in South American pocket mice. *Oikos,* 93, 3-16.

Benavides A. (2003) *Caracterización in situ y ex situ del teosinte anual* (Zea nicaraguensis Iltis et Benz) *colectado en la Reserva de Recursos Genéticos de Apacunca.* Chinandega. Nicaragua. Tesis de Maestría. Universidad Nacional Agraria. Universidad Autónoma de Barcelona. 80 p.

Bird R. (1978). *A name change for Central American teosinte*. Taxon 27(4): 361-363.

Bird R. (2000). A Remarkable new teosinte from Nicaragua: Growth and treatment of progeny. *Maize Cooperation Genetics News.* 74: 58-59.

Buckler E., Holtsford H. (1996). Zea systematic: ribosomal ITS evidence. *Mol. Biol. Evol.* 13: 612-622.

Buckler E., Goodman M., Holtsford T., Doebley J. and Sanchez J. (2006). Phylogeography of the wild subspecies of Zea mays. *Maydica* 51: 123-134.

Bath K. Babrekar P. and Lakhanpaul S. (1999). Study of genetic diversity in Indian and exotic sesame (Sesamun indicum L.) germoplasm using random amplified polymorphic DNA (RAPD) marker. *Euphytica* 110:21-33.

Baltazar B. and Schoper J. (2002) Crop-to-crop gee flows: dispersal of transgenes in maize, during field test and commercialization. In: *Proc 7th Int Symp Biosafety Genet Modified Organisms*. Beijing Chinahttp://www.bba.de/gentech /isbgmo.pdf.

Dale P. (1994). The impact of hybrids between genetically modified crop plants and their related species: general considerations. *Molecular Ecology* 3:31-36.

Dillon S. L., Lawrence P. K., Henry R. J. and Price H. J. (2007). Sorghum resolved as a distinct genus based on combined ITS1, ndhF and Adh1 analyses. *Plant Systematics and Evolution*, 268:29-43.

Elias M., Panaud O. and Robert T. (2000a). Assessment of genetic variability in traditional cassava. (Manihot esculenta Crantz). RAPD markers. *Gen Mol Biol* 21:105.

Eubanks M. (2001). The mysterious origin of maize. Econ Bot 55:492-514.

Evans M., and Kermicle L. (2001). Teosinte crossing barrier1, a locus governing hybridization of teosinte with maize. *Theor. Appl. Genet.* 103: 250-265.

Fontdevila A., Moya A. (1999). *Introducción a la genética de poblaciones*. Editorial Síntesis. Madrid España. 350 pp.

Fukunaga K., Hill J., Vigouroux Y., Matsuoka Y., Sanchez J., Lui K., Buckler E. and Doebley J. (2005). Genetic Diversity and Population Structure of Teosinte. Genetics 169: 2241-2254.

Guat B., Le M., Peek S. and Sawkins C. (2000). Maize as a model for the evaluation of plant nuclear genomes. *Proc. Natl. Acad. Sci. USA* 97:7008-7015.

Guzmán M. R. (1985). *Protección e investigación del hábitat del Zea diploperennis.* Universidad de Guadalajara, México. 40 p.

Gielly L. and Taberlet P. (1994). The use of chloroplast DNA to resolve plant phylogenies-noncoding versus RBCL sequences. *Mol Biol Evol*, 11(5):769-777.

Graham C., Ron S., Santos J., Schneider C., and Moritz C. (2004). Integratin phylogenetics and environmental niche model to explore speciation mechanism in dendrobatid frogs. *Evolution,* 58, 1781-1793.

Iltis H. (1983). From teosinte to maize: the catastrophic sexual transmutation. *Science* 222:886-894.

Iltis H. (1987). Maize evolution and agriculture origins, pp. 195-213 in *T.R.* Sodestrom, K. W. Hilo C. S. Cambell M.E. Barkworth (editors). *Grass systematics and evolution.* Smithsonian Institution Press Wasington D.C: USA.

Iltis H., Benz B. (2000). Zea nicaraguensis (Poacea), a new Teosinte from Pacific Coastal Nicaragua. *Novon.* 10: 382-390.

Iltis H., and Doebley J. (1980). Taxonomy of Zea (gramineae). II. Subspecfic categories in the Zea mays complex and a genetic synopsis. *Amer. J. Bot.* 67(6): 994-1004.

Judez L. (1989). *Técnicas de análisis de datos multidimensionales, Bases teóricas y aplicaciones en agricultura.* Ministerio de Agricultura y Pesca. Secretaría General Técnica. Madrid, España.

Hadlaczky, G. Y. and Kálmán, L. (1975). Discrimination of homologous chromosomes of maize with Giemsa staining.; *Heredity* 35(3): 371-374.

Kårehed J., Groeninckx I., Dessein S., Motley T. J., and Bremer B. (2008). The phylogenetic utility of chloroplast anduclear DNA markers and the phylogeny of the Rubiaceae tribe Spermacoceae. *Molecular Phylogenetics and Evolution*, 49:843-866.

Keeling P. J. (2004). Diversity and evolutionary history of plastids and their hosts. *American Journal of Botany*, 91:1481-1493.

Kato Y. 1976 Cytological studies of maize (Zea mays L.) and teosinte (Zea mexicana) in relation to their origin and evolution. *Mass. Agric. Expt. Sta. Bull.* 635:1-185.

Leathwick R. (1998). Are New Zealands Nothofagus species in equilibrium with their environmental? *Journal of Vegetation Science*, 3,157-164.

Lubberstedt T. C Dussle and Melchinger E. (1998). Application of microsatellites from maize to teosinte and other relatives of maize. *Plant Breeding* 117, 447-450.

Matsuoka Y., Mitchel S., Kresovich S., Goodman M., and Doebley J. (2001). Microsatellites in Zea variability, patterns of mutations and use for evolutionary studies. *Theor. Appl Genetic* 104: 436-450.

Matsuoka Y. (2005). Origin matter: Lessons from the search for the wild ancestor of maize. *Breeding Science* 55: 383-390.

Mano, Y., Muraki, M., Fujimori, M., Takamizo, T. and Kindiger, B. (2005). Identification of QTL controlling adventitious root formation during flooding conditions in teosinte (Zea mays ssp. huehuetenangensis) seedlings. *Euphytica* 142: 33-42.

Mano, Y., Omori, F., Takamizo, T., Kindiger, B., McK. Bird, R. and Loáisiga, CH. (2006). Variation for root aerenchyma formation in flooded and non-flooded maize and teosinte seedlings; *Plant and soil* 281: 269-279. 2001.

Molina, M. (1982). Cytogenetic studies on Zea diploperennis. *The nucleus* 26(1): 1-7.

Molina, M. and Garcia, M. D. (1999). Influence of ploidy level on phenotypic and cytogenetic traits in maize and Zea perennis hybrids. *Cytology* 64: 101-109.

Mogensen H. L. (1996). The hows and whys of cytoplasmic inheritance in seed plants. *American Journal of Botany*, 83:383-404.

Mort M. E., Archibald J. K., Randle C. P., Levsen N. D., O'Leary T. R., Topalov K., Wiegand C. M., and Crawford D. J. (2007). Inferring phylogeny at low taxonomy levels: Utility of rapidly evolving cpDNA and nuclear ITS loci. *American Journal of Botany*, 94(2):173-183.

Nybom H. and Bartish I. (2000). Effects on life history traits and samples strategies on genetic diversity estimates obtained with RAPD marker in plants. *Persp. Plant Ecol. Evol. Syst.* 3:93-114.

Nguni D., Mulatu G., Moneim F. and Bryngelsson T. (2010). Molecular phylogeny of Sorghum Moench based on sequence data from non-coding regions of cpDNA and internal transcribed spacer of nuclear rDNA. *Annals of Botany* 105: 471–480.

Peterson A. (2001). Predicting species geographic distributions based on ecological niche modelling. *Condor*, 103, 599-605.

Raubeson L. A., Jansen R. K. 2005. Chloroplast genomes of plants. In: *Plant diversity and evolution*: genotypic and phenotypic variation in higher plants. Wallingford UK: CABI Publishing: 45-68 p.

Rincon G. P., Ramirez J., Sanchez and Kato A. (2005). *Variacion isoenzimatica en poblaciones de Teosinte*. Revista fitotécnica mexicana. Abril – Junio, volumen 28, numero 002. Chapingo México.

Sánchez J., Kato T., Aguilar M., Hernández M., López A. and Ruiz J. (1998). Distribución del teocintle en México. INIFAP. *Libro Técnico Núm*. 2. 149 p.

Sanchez G., M. Goodman and Estuber W. (2000). Isoenzimatic and morphological diversity in the races of maize of Mexico. *Econ. Bot.* 54:43-59.

Somarriba C. 2000. *Texto principal de granos básicos*, Universidad Nacional Agraria (UNA). Managua Nicaragua. 98 p.

Sanz-Alferez S., SanMiguel Y., Jin P., Springer S., Bennetzen L. (2003). Structure and evolution of the Cinful retrotransposon family of maize. *Genome* 46:754-752.

Standley P. (1950). Teosinte in Honduras. *Ceiba* 1:58-61.

Shaw J., Lickey E. B., Beck J. T., Farmer S. B., Liu W. S., Miller J., Siripun K. C., Winder C. T., Schilling E. E. and Small R. L. (2005). The tortoise and the hare II: Relative utility of 21 noncoding chloroplast DNA sequences for phylogenetic analysis. *American Journal of Botany*, 92(1):142-166.

Smith J., Goodman M. and Stuber C. (1984). Variation within teosinte III: Numerical analysis of chromosome knob S data. Eco. Botany 38:97-113.

Tomotaro N., Salomon B, Komatsuda T, von Bathmer R, and Kadowaki K. (2002). Molecular phylogeny of the genus Hordeaum using three chloroplast DNA sequences. *Genome* 45: 1157-1166.

Udall J. A. and Wendel J. F. (2006). Polyploidy and crop improvement. *Crop Science,* 46: S-3-S-14.

Wiens J., and Donghue M. (2004). Historical biography, ecology and species richness. *Trends in Ecology and evolution.* 18, 639-644.

In: Genetic Diversity: New Research
Editors: J. A. C. Amaya and M. M. F. Jimenez
ISBN: 978-1-62081-598-4
© 2012 Nova Science Publishers, Inc.

Chapter 3

CONSERVATION GENETICS OF THE LION: NEW APPROACHES TO SPECIES CONSERVATION

Laura Bertola, Klaas Vrieling and Hans de Iongh
Institute of Environmental Sciences (CML) and
Institute of Biology Leiden (IBL), Leiden University, Leiden, The Netherlands

ABSTRACT

As an ecological and cultural umbrella species, the lion has drawn a lot of attention from researchers from a variety of disciplines. Due to the decrease in suitable habitat and increasing pressure by a growing human population, many lion populations show a declining trend. When it comes to species conservation, initiatives often follow existing taxonomy. This implies that taxonomic distinction reflects the diversity of a taxon sufficiently to preserve it on the long term. In the view of this, it is remarkable that all lion populations in Africa are considered to be the same subspecies, while for other large mammals with a similar geographical distribution, often several subspecies are being distinguished. In the following chapter we will review the published literature dealing with the genetic diversity of the lion and illustrate how upcoming molecular techniques can contribute to the field of species conservation.

Because the lion populations from West and Central Africa were only rarely included in previous studies on genetic diversity, the position of these populations within the species is still unclear. Recent phylogenetic research showed that the West and Central African lion is genetically distinct from their conspecifics in East and southern Africa. Because of geographic barriers and the climatic history of this region, these populations have formed a unique genetic make-up. This provides a strong argument to increase conservation efforts in West and Central Africa. In this region in particular, anthropogenic pressures have led to small, fragmented, and therefore vulnerable populations.

We will also describe how different types of genetic markers can be combined to create a complete picture and come to a well-designed management plan. With the current development in next generation sequencing and genotyping techniques, these approaches come into reach for conservation studies. The resulting data can be used for a wide variety of research questions, for example on taxonomic uncertainties, defining management units, the design of (captive) breeding programs, forensics, and the genetic basis of morphological differences or disease vulnerability.

Despite the fact that we will focus on the case of the lion, the resulting conclusions are also applicable to other species. In species conservation one of the main targets is to halt or to minimize the loss of genetic variation, in order to preserve the species as a dynamic entity with enough evolutionary potential to survive on the long term. Therefore it is of utmost importance to have a detailed picture of the present genetic diversity, both on the phylogenetic level, as on the population genetics level, to properly manage the species under consideration.

Taxonomy and Species Conservation

Species conservation can be defined as the aim to protect and preserve species and their natural habitat for the future. From the term 'species conservation' it can be deduced that taxonomic distinctions play an important role. Obviously, it is easier to protect and conserve something that is officially labeled as being distinctive. Despite the numerous levels on which biodiversity can be studied, entities at the species level are the preferred unit for conservation efforts (Pereira and Cooper, 2006) and the global IUCN Red List takes both species and sub-species into account with its classifications.

Several researchers pointed out that the ultimate goal of conservation is not preserving particular objects (species or ecosystems), but rather preserving the processes of life (Bowen, 1999). Hereby a more sustainable approach of conservation is proposed. Safeguarding the adaptability of both species and entire ecosystems is of exceptional importance in a changing environment, for example as a result of changing climate conditions. Species that are not able to adapt or migrate will perish, which is sadly illustrated by the vulnerability of small and isolated populations. For continued adaptation, genetic variability is a prerequisite. The goal to preserve species as dynamic entities with evolutionary potential, together with the recognition that conservation efforts often strongly follow taxonomy, illustrate why insights in genetic diversity can and should contribute to conservation planning.

This led to the notion that it may be important for conservation practices to also include groups on a level below the distinguished ranks in classical taxonomy. The term Evolutionary Significant Unit (ESU) was first proposed in 1986 by Ryder (1986) and aimed at facilitating conservation efforts on populations that are of particular importance for the long term survival of a species by defining a unit below the species level. However, the concept of an Evolutionary Significant Unit has many, often poorly defined, definitions (Waples, 1991; Avise and Ball, 1990; Dizon *et al.*, 1992; Vogler *et al.*, 1993; Moritz, 1994; Crandall *et al.*, 2000). In general, it is suggested that ESUs ought to be geographically discrete and should display concordant divergence in both molecular and non-molecular traits (Ryder, 1986; Dizon *et al.*, 1992; Vogler *et al.*, 1993). Adaptive variation as well as divergence, as a reflection of the evolutionary history, are typically included in the definition of ESU, but genetic criteria range from significant divergence of allele frequencies (Waples, 1991), to some level of genetic distance (Ryder, 1986), to congruently structured phylogenies of genes (Avise and Ball, 1990). Apart from the ESU there is another entity that can be distinguished which may be of value for conservation practices. Populations that do not show reciprocal monophyly for alleles, yet have diverged in allele frequency, display a pattern which may be important to justify focused conservation efforts. These populations represent groups that are connected with such low levels of gene flow that they are functionally independent. Although they show divergence in their allele frequencies, it is not necessarily the case that there is a phylogenetic distinctiveness of the alleles. Groups which are recognized as such are called Management Units (MU) (Moritz, 1994). Following these definitions, ESUs are concerned with historical population structure, phylogeny and long-term conservation needs. In contrast, MUs address current population structure, allele frequencies and short-term management issues. The risk of extinction can be severely underestimated when all populations from a single species are managed as one homogenous group, without further regard to the distinct genetic lineages within this species.

In the case of the lion, it has become clear that present day taxonomy, only distinguishing the African lion *Panthera leo leo* and the Asiatic lion *Panthera leo persica*, does not sufficiently reflect the different genetic lineages within the species (Barnett *et al.*, 2006a,b; Dubach *et al.*, 2005; Antunes *et al.*, 2008; Bertola *et al.*, 2011). Following taxonomy and considering the African lion as one homogenous group will lead to inadequate management of the species as a whole. Phylogenetic and biogeographic data are of utmost importance to define management strategies for species conservation. This type of data do not only give information on the present day situation and possible (genetic) threats these populations may be facing, but also on the evolutionary history that has shaped the current genetic make up. Thereby it provides a rational basis for prioritising populations for conservation effort, given the fact that it is aimed to minimize the loss of genetic diversity and preserve genetically distinct lineages.

THE AFRICAN LION: NUMBERS AND TRENDS

In the history of lion conservation, there was not much concern for the status of the African lion until 2006, when the lion was proposed for Appendix I at the CITES COP meeting in Bangkok. Although the proposal to uplist the lion was not adopted, the international debate has continued since. The subspecies of the African lion was long considered to be abundant and wide-ranging with no direct threats that would ask for radical conservation measures. However, now it has been recognized that both lion numbers and lion range are at an alarming decline. Most researchers agree on an estimate of less than 40,000 free ranging African lions, representing a decline of almost 50% when comparing this figure to a modelled 75,800 lions in 1980 (Nowell and Jackson, 1996; Ferreras and Cousins, 1996; Chardonnet, 2002; Bauer and Van Der Merwe, 2004). The situation in West and Central Africa looks even more gloomy, where the lion populations have shown the steepest decline. In this region the lion is already qualified as Regionally Endangered, according to IUCN criteria (Bauer and Van Der Merwe, 2004). It has been estimated that 10% of the continental population is located in West and Central Africa, but even this may be a positive estimate. In this part of Africa 20 ecological units important for lion conservation, so called Lion Conservation Units, have been distinguished, whereas East and southern Africa has 66 Lion Conservation Units. According to a recent survey in 15 of the 20 defined Lion Conservation Units in West and Central Africa, researchers were only able to reconfirm the presence of lions in two of the studied areas (Henschel *et al.*, 2010). On top of that, lion populations in West and Central Africa are generally small and isolated and lions almost completely disappeared from non-protected areas (Bauer and Van Der Merwe, 2004). Although the lion's potential area of occurrence might still be relatively large, most of the suitable habitat is fragmented and unprotected (Chardonnet, 2002). In total 18% of the African lion's range is described as currently declining, with again the highest figures in West and Central Africa (39% and 18% respectively) compared to Southern and East Africa (16% and 9% respectively) (Chardonnet, 2002). Also other large mammals populations have shown significant declines in West, Central and East Africa, which will have had a significant impact on large carnivore populations (Craigie *et al.*, 2010). Increasing human population, land cultivation and growing numbers of livestock are the main factors for the decline of the lion's habitat (Chardonnet, 2002; Ferreras and Cousins, 1996). These factors have not only reduced the lion's wild prey base, but little distance to human settlements also leads to human-lion conflicts, like lion predation on livestock. Local pastoralists are known to hunt or poison lions for prevention of or retaliation for stock raiding (Bauer *et al*,. 2001).

Consequently there is an increasing number of lions in peripherally isolated populations or in wildlife parks which are surrounded by human dominated landscapes. Although several parks allow for immigration and have lion populations outside their borders, in many cases there is little or no gene flow (Bauer and Van der Merwe, 2004). The direct surroundings of the national parks can be heavily populated, making it a difficult area to travel through for lions. Genetic erosion and inbreeding are important threats that the lion is facing in these regions (Björklund, 2003).

INBREEDING IN WILD POPULATIONS

Inbreeding is especially likely to result in population decline and ultimately even extinction in species with low reproductive rates, such as large mammals (Mills and Smouse, 1994). There are several studies illustrating the vulnerability of cats to low genetic diversity and inbreeding depression. From captive breeding programmes of lions in zoos, we have learned that lions are particularly sensitive to inbreeding depression (Joep Wensing, personal communication). Few studies are available on inbreeding depression in wild ranging lion populations, but there are some relevant studies, also on other large cat species.

In the case of the Florida panther (*Puma concolor coryi*) the population was reduced to 30-50 individuals as a result of habitat destruction, and the population exhibited an unusual high frequency of traits that are typically linked to inbreeding depression, like a kinked tail and cowlick, low sperm quality and heart defects (Roelke *et al.*, 1993). These defects increased in frequency as the population further declined in numbers. Conservationists made a controversial decision: they argued that the only way to preserve the Flordia panther was by hybridization with individuals from another subspecies, thereby affecting the "genetic integrity" of the Florida panther. In 1995 eight female pumas from Texas (*Puma concolor stanleyana*) were introduced to improve population numbers and genetically enrich the existing population. The population was followed intensively and results of this restoration experiment looked promising; researchers concluded that numbers of individuals had increased threefold, genetic heterozygosity doubled, survival and fitness measures improved and that inbreeding correlates declined significantly (Pimm *et al.*, 2006; Johnson *et al.*, 2010). The authors also warned that, although the introduction of new genetic material seemed to benefit the population, this approach to save small and isolated populations, does not guard the population for persisting pressures like continued habitat loss.

Famous examples of lion populations with low genetic diversity are the lions of the Ngorogoro crater in Tanzania and the lions of the Gir forest in India. The lions of the Ngorogoro crater in Tanzania illustrate well how fast inbreeding can play a role in a natural population. In the early sixties an outbreak of the biting fly *Stomoxys calcitrans* reduced the population from around 70 to 10 individuals (nine females and one male) (Fosbrooke, 1963; Packer *et al.*, 1991). Apparently a few males immigrated into the crater in 1964-195, but apart from this event, no immigration occurred in the following 25 years. The population recovered well and by 1975 there was a number of 75-125 lions estimated to be in the Ngorogoro crater. However, all animals were descendents of only 15 founder individuals and genetic diversity in this population proved to be much lower than in the neighbouring Serengeti ecosystem (O'Brien *et al.*, 1987a; Yuhki and O'Brien, 1990; Packer *et al.*, 1991).

With its extremely low heterozygosity, the isolated lion population in the Gir forest (India) is the only population of the Asiatic subspecies of the lion. Modern lions, represented in the African and Asiatic subspecies, share a common ancestor in the very recent past, estimated between 55,000 and 200,000 years ago (O'Brien *et al.*, 1987a; Janczewski *et al.*, 1995). There long was a continuous lion meta-population from Africa, throughout the Middle East into Asia and even southern Europe (Flower and Lydekker, 1891; Vogt and Specht, 1889). Possibly due to prosecution by humans, the European lion disappeared during the first century AD. Lion populations in North Africa, the Middle East and other parts of Asia all went extinct within the last 150 years (Mazák, 1970). It is hypothesized that gene flow between the African and Asian continent was already hampered much earlier, probably as a result of increasing human densities along the Nile and the Sinai Peninsula. Also the Gir forest lion population may have suffered from isolation even before these extinction events occured. Several markers, including allozymes (O'Brien *et al.*, 1987a), MHC-RFLP (Yuhki and O'Brien, 1990), minisatellites (Gilbert *et al.*, 1991) and microsatellites (Driscoll *et al.*, 2002), confirmed that the Gir forest lions are genetically even further impoverished than the Ngorogoro crater lions. Even unrelated individuals showed identical minisatellite DNA fingerprints (Gilbert *et al.*, 1991). This strong reduction in genetic variability can be explained by a bottleneck event that isolated the Gir forest population from a larger metapopulation as long as 2600 years ago (Driscoll *et al.*, 2002).

Both inbred lion populations, from the Ngorogoro crater and from the Gir forest, and the outbred population from the Serengeti plains were furthermore subjected to a physiological study. It was shown that reduced genetic diversity correlates well with an increased incidence of abnormal sperm and with decreased testosterone levels, which is a hormone crucial for normal spermatogenesis (Wildt *et al.*, 1987a). Small, enclosed game reserves in South Africa, in which lion populations have been introduced by humans, are a good opportunity to further study the effects of small populations and bottleneck events in this species (Trinkel *et al.*, 2008; Trinkel *et al.*, 2010).

Another large cat species, the cheetah, has been covered by extensive genetic studies which may be relevant for lion genetics. In the eighties the cheetah was shown to be genetically monomorphic for 47 allozyme loci, which is a much lower frequency of polymorphism than is known for other cats and mammals in general (O'Brien *et al.*, 1983). An experiment illustrated that skin grafts from one animal were not rejected by another, meaning that even the genetic basis for the very diverse immune system was identical between individuals (O'Brien *et al.*, 1995). This reduction in genetic diversity was explained by one or more historic population bottleneck and consecutive inbreeding. Very high incidences of structurally abnormal sperm were reported for both captive (Wildt *et al.*, 1983) and free-ranging cheetahs (Wildt *et al.*, 1987b). The argument of genetic impoverishment was often used for disappointing breeding results with cheetahs in captivity. However, not all experts agreed on this, as high vulnerability to disease and reduced reproductive success was mainly known from captive individuals and not from wild populations (Caro and Laurenson, 1994). In fact wild populations showed a high level of viability in spite of low genetic variability. It was shown that wild ranging Namibian cheetahs actually respond effectively to pathogens (Thalwitzer *et al.*, 2010). Other recent studies have identified a MHC diversity higher than previously thought (Castro-Prieto *et al.*, 2011; Castro-Prieto *et al.*, 2010).

Despite this, the possible effect of inbreeding should not be underestimated. Purging of deleterious alleles as a result of subsequent bottlenecks can lead to healthy populations in a constant environment. However, when the environment changes (e.g. in captivity) or a stress factor is added, these populations could show to be less resilient than populations with a higher level of genetic diversity.

There are also cases known where populations retain more genetic variability than they theoretically "should", based on the small population size. One of the possible mechanisms is enhanced reproductive success among outbred individuals. Simple models based on random mating, ignoring genetic relations between mates and making no differences between species, seem to be less useful than was previously assumed (Amos and Balmford, 2001). There are several cases that illustrate that species are able to manage themselves, as long as they have certain opportunities, like to disperse to a new area and finding an unrelated mate. In wolves it was shown that the inbreeding coefficient (f), which is the chance that two alleles are identical by descent, and heterozygosity are not as strongly correlated as was previously expected (Bensch *et al.*, 2006). Apparently there is a positive selection for heterozygosity and even in populations with very low immigration, this mechanism may circumvent or decrease inbreeding. Cheetahs in the Serengeti solve the problem in another way, by mating with more than one male within one oestrus cycle. Microsatellite analysis showed that 43% of the studied litters were fathered by more than one male (Gottelli *et al.*, 2007). The low rate of paternity assignment further illustrated that male cheetahs outside the study area contributed substantially to reproduction in this population (Gottelli *et al.*, 2007).

These cases illustrate that in small populations mechanisms may act by which the population stays genetically more healthy than one would initially expect, based on the population size. There is no reason to believe that this is not the case for lions, although the small and isolated populations will continue to be vulnerable.

THE LION: EVOLUTIONARY HISTORY AND PRESENT DAY DIVERSITY

The fact that the lion range once extended from Africa into Asia and even southern Europe is reflected by the phylogenetic pattern. The study described in Bertola *et al.* (2011) is the first phylogenetic study on lions which includes a larger number of West and Central African samples. Although a number of genetic lineages within the African lion was already described in other studies (Barnett *et al.*, 2006a,b; Dubach *et al.*, 2005; Antunes *et al.*, 2008), the phylogenetic position of the West and Central African lion was not solved. In the study by Bertola *et al.* (2011) 53 individuals from 15 countries were sampled and sample size was further increased by adding GenBank data of 28 cytochrome b sequences from five countries (Dubach *et al.*, 2005) and 45 control region sequences from 19 countries (Barnett *et al.*, 2006a). Phylogenetic analysis of cytochrome b sequences only showed a clear structure, but the number of pairwise differences was relatively low. The non-coding control region is much more variable for the studied individuals, however, phylogenetic trees derived from these analysis often lack structure. Combining the two genetic regions led to a clearer picture, that proved to be very robust.

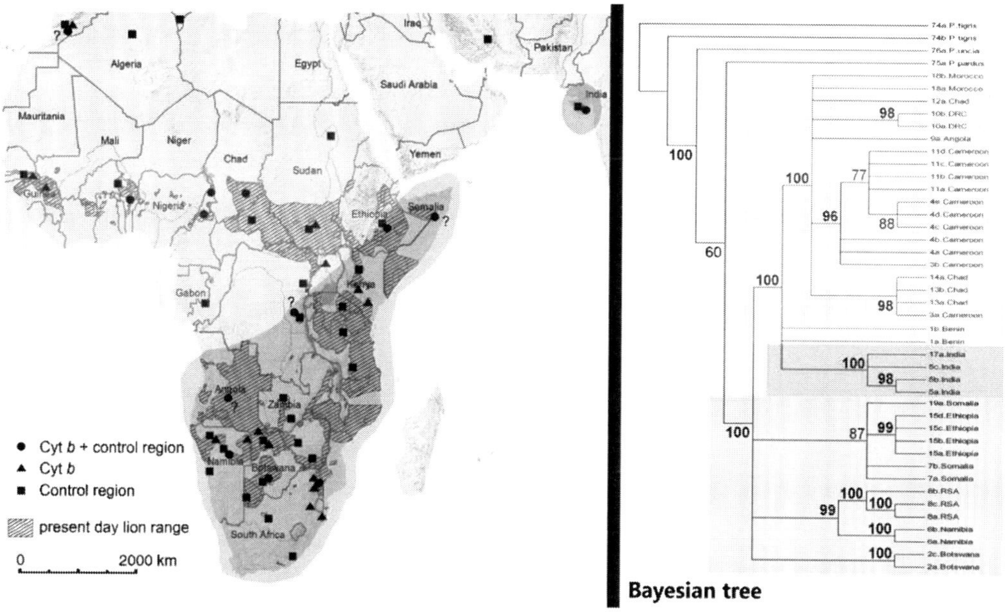

Figure 1. Map and phylogenetic tree indicating the genetic relationship between the analyzed lion sequences. Dots in the map indicate the samples from which cytochrome b (cyt b), tRNAThr, tRNAPro and the left domain of the control region sequences are known, triangles are cytochrome b sequences, and the squares show sample locations from which only a part of the control region was sequenced. For several samples only the country of origin was known, and there was no information available on the exact locality. In these cases the geographical centre of the known lion range within the country is indicated (lion range data are from Bauer et al., 2008). The solid, dashed and dotted lines illustrate the groups that are being distinguished in the phylogenetic analysis. The phylogenetic tree is the result from a Bayesian analysis of cytochrome b + part of the control region lion sequences. The numbers represent the percentages for Bayesian posterior probability (PP). DRC = Democratic Republic of the Congo, RSA = Republic of South Africa.

From a genetics point of view, there is a remarkable difference between lions from West and Central Africa and lions from East and southern Africa (Figure 1, data from Bertola *et al.*, 2011). The data described in Bertola *et al.* (2011) show a strong distinction between samples from Chad, CAR and Sudan, which cluster with West and Central Africa, and on the other hand Ethiopia, Kenya and DRC, which are in the southern and East Africa clade. This might be explained by the geological pattern and the location of the central African rain forest, the Congo river and the Rift Valley, which stretches from Ethiopia to Tanzania and from DRC to Mozambique. This has previously been identified as a possible barrier for lion dispersal

(Dubach *et al.*, 2005). The fact that the Congo river has changed its course several thousand years ago and the fact that there is still evidence of a small lion population in Central Congo, south of the Congo river, needs further investigation in this respect.

Figure 2. Lions from diverse regions of their natural range, showing phenotypical differences. Lion from West and Central Africa often have a skinfold running down the belly, which is also well known from the Asiatic lion. West and Central African lions often share a sparser mane development with their Indian conspecifics. In this case, the Asiatic lion was held in a European zoo and the different climatic conditions may explain why this individual exhibits a relatively strong mane.

The data also show an unexpected close relationship between lions from West and Central Africa and lions from the Asiatic subspecies. Morphological and ecological data also suggest that lions from West and Central Africa are different from the other African lions: individuals tend to be smaller, live in smaller prides and prefer smaller prey than lions in East and southern Africa. The bellyfold, which is very characteristic for Asiatic lions (Pocock, 1930) is also present in an exceptionally high frequency in lions in West and Central Africa (Figure 2). The dominant presence of maneless male lions in the WAP complex (Niger, Burkina Faso and Benin) and the relatively sparser manes in male lions from Cameroon, is another morphological feature which deserves further investigation (Sogbohossou, 2011).

Lion samples from West African countries were included in one previous study focusing on phylogeny, but only part of the control region was analyzed, and samples connecting West to Central Africa were absent (Barnett *et al.*, 2006a,b). In general, a pattern was found of two major clades, one being located west of the Rift Valley, and one confined to eastern and southern Africa (Barnett *et al.*, 2006b). In the same study it was concluded that sub-Saharan lions are basal amongst modern lions, being in line with the high genetic diversity we observe in southern and East Africa.

The data from Barnett *et al.* (2006b) seem to indicate that West African lions are more closely related to lions from southern and East Africa than they are to Central African lions, but this is based on a very small sample size. In the study of Barnett *et al.* (2006b) India falls between West and Central Africa, while one would expect West and Central Africa to be directly related. This pattern is less explicit after incorporation of these sequences to our data set. Antunes *et al.* (2008) do not include any West or Central African countries. In the mitochondrial DNA analysis, samples from Angola, Morocco and Zimbabwe fall in one clade, close to the India clade. But all samples in this group were derived from captive individuals, and the Moroccan samples that were included are likely to contain a Central African haplotype as has previously been described by Barnett *et al.* (2006a). Sequences derived from wild-ranging Zimbabwean lions that were included in our study (control region) cluster with sequences from lions from neighbouring countries, and not with those from West and Central African lions. In line with the pattern described by Dubach *et al.* (2005) we confirm the distinct position of populations west of the Rift Valley, which were represented in the study of Dubach *et al.* (2005) by two sequences from Uganda. Dubach *et al.* (2005) also identifies the Sabi Sands population (Kruger National Park, RSA) as a distinct clade. These samples cluster strongly with the samples from Timbavati Game Reserve (Kruger National Park, RSA) analyzed in the study by Bertola *et al.* (2011). Further analysis showed that at least nine point mutations in the cytochrome b gene make up the difference between Kruger Park lions and lions from other regions in RSA. The findings of Bertola *et al.* (2011) also support the conclusions of Mazák (2010), where one sub-Saharan Africa cluster and one North-Africa/Asia cluster are distinguished based on cranio-morphological data. Due to low sample size for West and Central African lions in the study by Mazák (2010), their taxonomic and phylogenetic position remained largely unresolved. The data of Bertola *et al.* (2011) show that lions from this region should be considered to be part of a cluster that also includes North Africa and India.

In addition to the described relationship between the lion metapopulations, West and Central African lions also show a lower genetic diversity within the region, compared to the other African lions (Bertola *et al.*, 2011). Because the analyzed differences in the DNA are shared by more populations in West and Central Africa, it is not likely that the loss of genetic diversity was a result of local inbreeding. More likely is the scenario in which this clade of lions represents an evolutionary young group, arisen from relatively recent (re-)colonization of the area after a regional extinction event. This part of Africa had a different climatological history compared to the rest of the continent and was characterized by hyper arid condition in the Late Pleistocene (40-18 thousand years ago) (Dupont *et al.*, 2000; Gasse, 2000; Sarntheim, 1978). It is hypothesized that lions went locally extinct, possibly following the extinction of prey species, and that this region was recolonized when conditions changed to a more favourable climate 15-11 thousand years ago (Gasse, 2000). Based on the genetic diversity observed in the southern and East Africa clade, it is assumed that these populations did not go through such a severe bottleneck. In this scenario the recolonization of western Africa took place from areas in close geographic proximity to India, hereby explaining the close genetic relationship to the Asiatic subspecies. Interestingly, this bottleneck in lion populations coincides with the well-known cheetah bottleneck (Driscoll *et al.*, 2002; Menotti-Raymond and O'Brien, 1993) and Late Pleistocene megafaunal extinctions that occurred over much of the globe (Barnosky *et al.*, 2004; Cardillo and Lister, 2002; Lyons *et al.*, 2004). Data from other mammal species show a similar dichotomy between West and Central Africa and the rest of the continent. This pattern has been described for seven African bovids (Arctander *et al.*, 1999; Nersting and Arctander, 2001; O. Flagstad *et al.*, 2001; Pitra *et al.*, 2002; Van Hooft *et al.*, 2002), African elephant (*Loxodonta africana*) (Eggert *et al.*, 2002), cheetah (*Acinonyx jubatus*) (Freeman *et al.*, 2001), black rhinoceros (*Diceros bicornis*) (Brown and Houlden, 2000), roan antilope (*Hippotragus equinus*) (Alpers *et al.*, 2004) and giraffe (*Giraffa camelopardalis*) (Brown *et al.*, 2007). Interestingly, overall genetic diversity in these species is not necessarily higher than what is currently observed in the lion. It seems that easy to spot variations in morphology or coat pattern aid in making taxonomic distinctions, even if this has no sound biological basis. Since lions show extensive variation in size, coat thickness and color, retention of juvenile spots and male secondary characters, categorization based on morphological traits is thought to be not reliable (Hollister, 1917). Therefore genetics can be an important tool in the field of lion conservation, by targeting underlying, measurable genetic variation.

The Next Generation of Molecular Techniques

Phylogenetic studies, including those on lions, often solely focus on mtDNA to get insight into the relationship between different genetic lineages. However, mitochondrial regions rarely include the genetic variation that explain certain phenotypical differences. Moreover, it only shows the evolutionary history following the maternal line. Inclusion of other genetic markers, with another mode of inheritance, can give a more complete picture of genetic relationships between populations. With the rapid development of current genetic techniques Next generation sequencing (NGS) techniques also come into reach for studies on non-model organisms. These techniques and associated bio-informatic tools allow for large scale, high throughput genetic analysis of entire genomes. In conservation studies, often non-model species are being targeted, and therefore there is often only very limited information on genome size and structure. Using these techniques to directly target sequence variation by analyzing single nucleotide polymorphisms (SNPs) has many advantages. Most SNPs are located in the nuclear DNA, making it possible to study the within population genetic diversity by analyzing levels of heterozygosity. At the same time, SNPs can be used to build phylogenies based on genetic variation that distinguishes populations from each other. Because of the high abundance of SNPs in the genome, they are a very powerful tool compared to other genetic markers. A comparison of SNP frequencies in a collection of organisms representing different taxa has shown that, on average, ~500-1000 base pairs of coding DNA or 200-500 base pairs of non-coding DNA need to be sequenced to find a single SNP (Brumfield *et al.*, 2003). SNP frequency may be much higher locally, as was shown in a study on seabirds where they detected SNPs at a frequency of ~1 in 80 nucleotides, based on nine studied loci (Congdon *et al.*, 2000). It was also shown that regions with high recombination are strongly correlated to a high frequency of SNPs, and regions with low recombination are typically characterized by low SNP variation (Nachman *et al.*, 2001; Lercher and Hurst, 2002).

Currently, the most common molecular marker used for answering research questions related to population structure and within population genetic diversity are microsatellites. Because of a very high mutation rate (in the order of 10^{-4}), microsatellites have a much higher allelic diversity than is commonly found with sequencing a genetic region. This mutation rate is higher than in other neutral genetic regions, commonly explained by slipped strand mispairing (slippage) during DNA replication. This often leads to overestimation of the genetic diversity present. The mutation rate is also very variable and may even vary across alleles within the same locus (Macaubas, 1997). A number of drawbacks (e.g. the occurence of allelic drop out) make this a less than ideal marker. By targeting microsatellites, one only gets insight into the genetic diversity of a non-coding, non-functional genetic region. SNPs can be targeted in both coding and non-coding regions. SNPs have a relatively low mutation rate (10^{-8}-10^{-9} in humans), which implies that there is a very limited chance of two mutations arising on the same location. Therefore SNPs are generally considered to be bi-allelic. This aspect makes SNPs less informative for individual assignment or for the detection of a fine-scale geographic structure than f.e. microsatellites. However, this can be compensated by targeting more loci, to achieve the same statistical power in comparison to a much more variable genetic marker, like microsatellites. It is estimated that approximately three times as many SNPs need to be scored in comparison to microsatellites (Kruglyak, 1997).

Apart from targeting questions related to population genetics, SNPs are also a suitable tool for answering questions on the level of phylogenies. In the past this was often done by sequencing a limited number of genetic regions, often closely linked (f.e. regions on the mtDNA). Because of the genome-wide abundance of SNPs and the high number of data points that can be analyzed fast and cost-effectively, this marker provides a good basis for the analysis of genetic parameters for which often large numbers of unlinked loci are needed. This is clearly much harder to achieve when using a targeted gene approach in which a certain genetic region is selected and sequenced to analyze the genetic variation in this particular part of the genome. However, also in the case of SNPs the number of independently segregating loci is crucial for calculation of the population parameter θ, which is a measure for genetic diversity (Watterson, 1975). This implies that a large number of relative short loci are likely to provide more information than a low number of large loci, although in principle one would expect to encounter SNPs at the same rate, using either approach. For studying non-model organisms, chromosome assignment and thus linkage for individual SNPs is often unknown. There are several ways to deal with this, based on computer simulations, f.e. by assigning probabilities to a certain haplotype and to individual sites within this haplotype (Stephens *et al.*, 2000), or by considering the uncertainties of given haplotypes based on estimated population size and recombination rates (Kuhner and Felsenstein, 2001).

For the development on SNPs one may decide to either sequence targeted regions within the genome, or randomly sequence regions throughout the entire genome. In the first approach one could consider to only target protein-coding region. Exon sequencing allows for detection of non-synonymous substitutions which may be functionally important. However, variability is typically lower in coding region because of a stronger functional constraint. The choice for a specific approach will also be influenced by the desired coverage and available budget for the project. If size of the complete genome and the transcriptome differ by a factor 20, this means that one would require 20 times more runs to achieve a similar coverage in the full genome, compared to transcriptome sequencing. For this reason many researchers choose to reduce their library and sequence selections of the genome with enough coverage for the purpose of their project. Focussing on the transcribed part of the genome, or genes that are expressed in a specific tissue or under specific circumstances, is one approach (see f.e. Bers *et al.*, 2010; Hahn *et al.*, 2009; Trick *et al.*, 2009; Novaes *et al.*, 2008; Vera *et al.*, 2008). However, this is only possible with high quality, fresh samples with preserved RNA which can be used to create a cDNA library from. In many conservation studies, where samples have been collected over a longer period of time, often under suboptimal circumstances, it is unlikely that the samples will provide a good quality and quantity of RNA. Another approach to reduce the library is by using a combination of restriction enzymes on the genomic DNA and selecting a specific range of sizes of products (see f.e. Ramos *et al.*, 2009; Kerstens *et al.*, 2009; Sánchez *et al.*, 2009; Amaral *et al.*, 2009; Tassel *et al.*, 2008). Again, this will be difficult in samples that exhibit a different severity of degradation, due to difference in age, storage method and circumstances under which the sample was taken.

With the rapid increase of output from the different NGS platforms, researchers will quickly switch from sequencing a reduced library to the entire genome. This does not only lead to more data, but also circumvents a possible bias introduced by the step of library reduction. This development has led to the choice to sequence the entire genome of a panel of lions. First analyzes show that there is a very high correlation between the individuals for the genetic regions that have high, respectively low, coverage (unpublished). Even if there are not enough reads to cover the entire genome to reliably genotype each individual in the panel, SNPs can still be detected and selected for future genotyping.

The best approach to follow for the development of the SNP markers, as well as selection of samples that will be used, is highly dependent on the research question. Because only a limited number of individuals is being analyzed, automatically an ascertainment bias is introduced. If one is interested in phylogenetic studies, the development of the SNPs should be based on a panel of representatives of the species. These should preferably be collected throughout the entire geographic range to cover maximum variation. However, if one is interested in genomic variations that are linked to a certain phenotype, f.e. a specific morphological character, such as coat pattern or mane development in lions, or vulnerability for a specific disease, a different panel of individuals should be selected (f.e. Ciobanu *et al.*, 2010). By grouping detected DNA variations, it may be possible to detect SNPs that are associated strongly with the trait of interest. A combination of both approaches is possible, although it will reduce the discriminating power as concessions have to be made. For the lion SNP discovery one of the famous maneless males from Tsavo NP (Kenya) was included as a representative for East Africa. As a representative for southern Africa a white lion from Timbavati (RSA) was included. Because the lions from Timbavati and Sabi Sands showed a remarkable genetic distance to other lions from neighbouring regions (Dubach *et al.*, 2005; Bertola *et al.*, 2011) inclusion of one of these individuals is expected to result in a relatively large number of SNPs, specific for this population.

After the initial detection of SNPs and accompanying flanking regions, these positions (or a subset of these) can be genotyped in a broader sample set. This can be done by non-sequencing techniques, f.e. probes or primers specific to the locus (f.e. Fan *et al.*, 2006), high-res melting (Wittwer *et al.*, 2003) or chiparrays. This allows for a very cost-effective analysis of a much larger number of variable positions compared to f.e. microsatellites. In the process of analysis of the SNPs it is needed to correct for the ascertainment bias that is introduced during detection of loci (Nielsen, 2000). It is obvious that rare SNPs, i.e. alleles with very low frequencies, can easily be missed in the process of SNP detection. This is also directly influenced by the choice whether every variable site is regarded as a SNP or if a cut-off frequency is used to correct for possible sequencing errors. Because of this bias the studied population seems to be more uniform than it actually is, which can lead to underestimation of population structure, or overestimation of migration. In phylogenetic studies, this ascertainment bias leads to a reduced resolution towards the tips of the phylogenetic tree.

Another advantage of SNPs is that it also detects the background variability, which makes it easier to compare different datasets (f.e. from different species). This is of particular interest for questions related to phylogeographic patterns of co-distributed species. Proper comparisons can only be made if there is no global difference in mutation rates between species, which can be assessed by employing relative rate tests on a common set of loci. As nuclear SNPs are measured on the same mutational scale as mitochondrial SNPs, the use of this marker also allows for intergenomic comparison. In species with sex-biased gene flow this allows for direct comparison of maternally (mtDNA) and paternally (Y-chromosome) inherited genetic regions. Sequencing the whole genome of two lions already yielded the complete mitochondrial sequence of both individuals and Y-chromosomal SNPs (unpublished). For lions it is known that they exhibit sex-biased dispersal, in which male lions tend to migrate earlier and longer distances than females and most females never leave their natal pride (Pusey and Packer, 1987). Moreover, it may also be possible to detect an effect of sex-biased selection by humans. It is known that lions have been translocated by humans for centuries (in northern Africa during Roman times, but also from Africa to India). Nuclear data can be used to complement the phylogenetic pattern described based on mitochondrial sequences.

Apart from using NGS methods to detect and use SNPs from the nuclear and Y-chromosomal DNA, the massive parallel sequencing approaches can solve other problems. In the light of the evolutionary history of the lion and the presumed recolonization event in western Africa, genetic data of nowadays extinct lion populations from northern Africa and the Middle East may form the missing link between the two subspecies. For ancient DNA studies, f.e. based on museum samples, typically each PCR product is cloned and a number of clones is being sequenced. With this approach it is possible to distinguish point mutations in DNA as a result of DNA damage from point mutations that contain information about the genetic background of the sequenced individual. With 454 sequencing individual DNA molecules are the starting material for a PCR step, after which the molecules are being sequenced. This results in a population of reads in which degradation of the DNA can be distinguished from actual SNPs. It has been shown that this is a useful alternative to cloning individual PCR products (Gilbert *et al.*, 2007). Museum collections have shown to be a good source of samples, in the case that it is not possible to collect new material in the region of interest. Apart from extinction of the populations of interest, this may also be the case in regions that are difficult to access, f.e. as a result of political instability.

As each of the available Next Generation Sequencing platforms has its own specificities, a combination of techniques can be employed to complement each other, depending on the research question. Platforms that produce long read length are particularly useful for creating a reference genome (in the case there is no reference genome available yet). Typically these platforms have limited throughput and coverage will be very low. Platforms that produce a much higher number of shorter reads can be used to provide the required sequencing depth to detect nucleotide variation. Illumina/Solexa produces reads of about 100 bp long, which can be paired and has a relatively low error rate. Sequencing error typically are substitution errors and deletion/insertion errors are much less common. This is different with the pyrosequencing approach used by 454/Roche, where repeats of one nucleotide often lead to uncertainty about the length this mononucleotide repeat. However, 454/Roche produces a lower number of longer reads (about four times the length of Illumina/Solexa), which may improve the results of a de novo assembly. Other platforms detect genetic variation more directly. Single Molecule Sequencers (f.e. Helicos) detect genetic variation based on single molecules, avoiding biases introduced by additional steps like PCR. Also instead of measuring an enzymatic reaction, incorporation of nucleotides can by measured directly by detecting small shifts in the pH as the result of a free hydrogen atom after nucleotide incorporation (Ion torrent).

Application for Conservation Genetics

The new data on the phylogenetic relationship between and the diversity within different lion populations can have a number of applications. One of the main threats for free-ranging lions is poaching. Many parts of lions are used for traditional medicine, both for the local market, as for the market in the far East. Analyzing genetic data from illegally exported lion material, may give insight into where poaching activities are highest. With a similar approach source population of illegal ivory shipments could be traced (Wasser *et al.*, 2008).

The described pattern of distinct genetic lineages within the subspecies of the African lion should not only have implications for management of wild populations, but also for breeding of captive stocks. In the past decades, most zoos have inadvertently crossed lions originating from different regions in Africa. Despite the lack of a studbook for the African lion, within this group at least seven "subspecies" are being distinguished in the International Species Information System (www.isis.org), in which information on species holdings is collected for zoos, aquariums and related organizations. The "subspecies" within the African lion (*P. l. azandica*, *P. l. bleyenberghi*, *P. l. krugeri*, *P. l. leo*, *P. l. massaica*, P. *l. senegalensis*, *P. l. somaliensis*), account for less then 20% of the almost 2000 registered lions. 15% of all registered lions belong to the Asiatic subspecies, however, the origin of the majority of registered individuals remains unknown, according to ISIS. It is not clear how to deal with so called subspecies, especially when numbers within subspecies are already that low and inbreeding depression is already a known phenomenon in captive held lion populations (Joep Wensing, personal communication). Information on the origin of these lions, or their free-ranging ancestors, is often questionable or incomplete and it would not be sensible to interpret the proposed subspecies name as a reflection of genetic heritage without doing further research. Several cases are known in which individuals have been assigned to a certain subspecies, but details on the breeding history revealed that it was not possible to be sure of the pure line within these individuals because of past hybridization events. One example is illustrated by the phylogenetic position of analyzed Angola lions. To our knowledge, only two studies included Angola lions and in both cases the derived sequence grouped closely to sequences from West and Central Africa (and in one case to a sequence from a captive Zimbabwe lion) (Bertola *et al.*, 2011; Antunes *et al.*, 2008). This is unexpected because of the presumed barrier function of the central African rainforest, although it has also been hypothesized that there may have been corridors in certain climatic periods. Closer investigation showed that in both cases the Angola lions were held in captivity and derived from the same stock. Pedigrees published by Steinmetz *et al.* (2005), in a study focussing on lens abnormalities in captive lions, showed that it is impossible to be certain of the origin of the mitochondrial DNA based on hybridization with (female) lions of unknown origin. The only way of confirming the true phylogenetic position of the Angola lion is to include samples from "true" Angola lions. To date it proved to be impossible to obtain new samples from Angola, partly due to the large number of still uncovered land mines. Analysis of an Angola sample from the collection of NCB Naturalis (Leiden, The Netherlands) is still in progress.

Another point of debate is the breeding programme of the presumed "Barbary" lions. It was already shown by Barnett *et al.* (2006a,b) that the zoo lions that are supposedly pure-blooded descendents of the nowadays extinct Barbary lions, do not match the Barbary lion haplotype derived from museum specimen. All analyzed zoo "Barbary" lions contained a haplotype also found in Central Africa. It may be possible that after analysis of nuclear SNPs these individuals prove to contain some Barbary lion DNA and that these lions are descendants of Barbary lions that were hybridized with female lions from Central Africa. Recently, an article was published that illustrates how genetic material from extinct species can "live on" in the genomes of extant individuals of mixed ancestry (Garrick *et al.*, 2011), as is also the case for the Florida panther (Pimm *et al.*, 2006; Johnson et al, 2010). Given the low number of lions from West and Central Africa held in captivity, this stock of lions may be important for lion conservation, even if these individuals prove to contain no Barbary DNA at all. A similar argument can be used for the Asiatic lions. Although this subspecies does have an official studbook, it is thought that hybridization events with African lions took place in the past (O'Brien *et al.*, 1987b). In this case, it is even more complex, because it is unclear whether during the course of history African lions may have introgressed into Asiatic lions as a result of translocation by humans. It is known that Indian maharajas travelled the African continent and imported wild animals from there (Shekar Kolipaka, personal communication). Although previous genetic analyzes of the mitochondrial DNA have not detected a haplotype shared by both the African and the Asiatic subspecies, it seems to be impossible to exclude that African lions have hybridized with the Asiatic subspecies in recent history, even in the wild.

A new (genetic) inventory of lions in all zoos associated with the World Association of Zoos and Aquariums (WAZA) could help clarifying the picture on the (genetic) diversity of lions currently held in captivity. Application of these data will make it possible to construct a management plan with a balance of genetic diversity on two levels: a greater genetic diversity is maintained when distinct genetic lineages are managed as such, but these units should still be large enough to avoid inbreeding. Giving insight into the genetic diversity within a particular population, conservation genetics will guard us from making serious mistakes in the development of a conservation and management strategy. For example, valuable resources might be wasted when conserving a population which is not endangered, or one might recognize too late that a species needs special attention when extinction is to be avoided. Insight in the genetic make-up of the populations we consider to conserve is a prerequisite for the establishment of effective and efficient conservation practices.

REFERENCES

Alpers, D. L., Van Vuuren, B. J., Arctander, P., Robinson, T. J. (2004). Population genetics of the roan antelope (Hippotragus equinus) with suggestions for conservation. *Molecular ecology*, *13*(7), 1771-84.

Amaral, A. J., Megens, H. -J., Kerstens, H. H. D., Heuven, H. C. M., Dibbits, B., Crooijmans, R. P. M. A., Dunnen, J. T. den, Groenen, M. A. M. (2009). Application of massive parallel sequencing to whole genome SNP discovery in the porcine genome. *BMC genomics*, *10*(1), 374.

Antunes, A., Troyer, J. L., Roelke, M. E., Pecon-Slattery, J., Packer, C., Winterbach, C., Hemson, G., Frank, L., Stander, P., Siefert, L., Driciru, M., Funston, P. J., Alexander, K. A., Prager, K. C., Mills, G., Wildt, D., Bush, M., O'Brien, S. J., Johnson, W. E. (2008). The evolutionary dynamics of the lion Panthera leo revealed by host and viral population genomics. *PLoS genetics*, *4*(11), e1000251.

Arctander, P., Johansen, C., Coutellec-Vreto, M. A. (1999). Phylogeography of three closely related African bovids (tribe Alcelaphini). *Molecular biology and evolution*, *16*(12), 1724-39.

Avise, J. C. Ball, R. M. J. (1990). Principles of genealogical concordance in species concepts and biological taxonomy. *Oxf. Surv. Evol. Biol*, *7*, 45-67.

Barnett, R., Yamaguchi, N., Barnes, I., Cooper, A. (2006a). The origin, current diversity and future conservation of the modern lion (Panthera leo). *Proceedings. Biological sciences / The Royal Society*, *273*(1598), 2119-2125.

Barnett, R., Yamaguchi, N., Barnes, I., Cooper, A. (2006b). Lost populations and preserving genetic diversity in the lion Panthera leo: Implications for its ex situ conservation. *Conservation Genetics*, *7*(4), 507-514.

Barnosky, A.D., Koch, P.L., Feranec, R.S., Wing, S.L. and Shabel, A.B. (2004) Assessing the causes of late Pleistocene extinctions on the continents. *Science*, *306*, 70–75.

Bensch, S., Andrén, H., Hansson, B., Pedersen, H. C., Sand, H., Sejberg, D., Wabakken, P., Akesson, M., Liberg, O. (2006). Selection for heterozygosity gives hope to a wild population of inbred wolves. *PloS one*, *1*(1), e72.

Bers, N. E. M. van, Oers, K. van, Kerstens, H. H. D., Dibbits, B. W., Crooijmans, R. P. M. A., Visser, M. E., Groenen, M. A. M. (2010). Genome-wide SNP detection in the great tit Parus major using high throughput sequencing. *Molecular ecology*, *19*(Suppl 1), 89-99.

Bertola, L. D., Hooft, W. F. van, Vrieling, K., Uit de Weerd, D. R., York, D. S., Bauer, H., Prins, H. H. T., Funston, P. J., Udo de Haes, H. A., Leirs, H., Haeringen, W. A. van, Sogbohossou, E., Tumenta, P. N., de Iongh, H. H. (2011). Genetic diversity, evolutionary history and implications for conservation of the lion (Panthera leo) in West and Central Africa. *Journal of Biogeography*, *38*(7), 1356-1367.

Bowen, B. W. (1999). Preserving genes, species, or ecosystems? Healing the fractured foundations of conservation policy. *Molecular ecology*, *8*(12 Suppl 1), S5-10.

Brown, D. M., Brenneman, R. a, Koepfli, K.-P., Pollinger, J. P., Milá, B., Georgiadis, N. J., Louis, E. E., Grether, G. F., Jacobs, D. K., Wayne, R. K. (2007). Extensive population genetic structure in the giraffe. *BMC biology*, *5*, 57.

Brown, S. M., and Houlden, B. A. (2001). Conservation genetics of the black rhinoceros (Diceros bicornis). *Conservation Genetics*, 365-370.

Brumfield, R. (2003). The utility of single nucleotide polymorphisms in inferences of population history. *Trends in Ecology Evolution*, *18*(5), 249-56.

Cardillo, M. and Lister, A. (2002) Death in slow lane. *Nature*, *419*, 440–441.

Caro T.M., Laurenson M.K. (1994). Ecological and genetic factors in conservation: a cautionary tale. *Science*, *263*, 485-486.

Castro-Prieto, A., Wachter, B., Melzheimer, J., Thalwitzer, S., Sommer, S. (2011). Journal of Heredity, *102*(6), 653-65.

Castro-Prieto, A., Wachter, B., Sommer, S. (2010). Cheetah paradigm revisited: MHC diversity in the world's largest free-ranging population. *Molecular biology and evolution*, *28*(4), 1455-68.

Ciobanu, D. C., Bastiaansen, J. W. M., Magrin, J., Rocha, J. L., Jiang, D. -H., Yu, N., Geiger, B., Deeb, N., Rocha, D. (2010). A major SNP resource for dissection of phenotypic and genetic variation in Pacific white shrimp (Litopenaeus vannamei). *Animal genetics*, *41*(1), 39-47.

Congdon, B. C., Piatt, J. F., Martin, K., Friesen, V. L. (2000). Mechanisms of population differentiation in marbled murrelets: historical versus contemporary processes. *Evolution; international journal of organic evolution*, *54*(3), 974-86.

Craigie, I. D., Baillie, J. E. M., Balmford, A., Carbone, C., Collen, B., Green, R. E., Hutton, J. M. (2010). Large mammal population declines in Africa's protected areas. *Biological Conservation*, *143*(9), 2221-2228.

Crandall, K. A., Bininda-emonds, O. R. P., Mace, G. M., Wayne, R. K. (2000). Considering evolutionary processes in conservation biology. *Science*, *15*(7), 290-295.

Dizon, A. E., Lockyer, C., Perrin, W. F., Demaster, D. P., Sisson, J. (1992). Rethinking the stock Concept—a phylogeographic approach. *Conservation Biology*, *6*, 24–36.

Driscoll, C. A., Menotti-Raymond, M., Nelson, G., Goldstein, D. O'Brien, S. J. (2002). Genomic microsatellites as evolutionary chronometers: a test in wild cats. *Genome research*, *12*(3), 414-423.

Dubach, J., Patterson, B. D., Briggs, M. B., Venzke, K., Flamand, J., Stander, P., Scheepers, L., Kays, R. W. (2005). Molecular genetic variation across the southern and eastern geographic ranges of the African lion, Panthera leo. *Conservation Genetics*, *6*(1), 15-24.

Dupont, L. M., Jahns, S., Marret, F., and Ning, S. (2000). Vegetation change in equatorial West Africa: time-slices for the last 150 ka. *Palaeogeography, Palaeoclimatology, Palaeoecology*, *155*, 95-122.

Eggert, L. S., Rasner, C. A, and Woodruff, D. S. (2002). The evolution and phylogeography of the African elephant inferred from mitochondrial DNA sequence and nuclear microsatellite markers. *Proceedings. Biological sciences / The Royal Society*, *269*(1504), 1993-2006.

Fan, J. -B., Chee, M. S., Gunderson, K. L. (2006). Highly parallel genomic assays. *Nature reviews. Genetics*, *7*(8), 632-44.

Flagstad, O., Syvertsen, P. O., Stenseth, N. C., and Jakobsen, K. S. (2001). Environmental change and rates of evolution: the phylogeographic pattern within the hartebeest complex as related to climatic variation. *Proceedings. Biological sciences / The Royal Society*, *268*(1468), 667-77.

Flower, W.H. and Lydekker, R. (1891). An Introduction to the Study of Mammals Living and Extinct. Adam and Charles Black, London, U.K.

Fosbrooke, H. (1963) The *Stomoxys* plague in Ngorogoro, 1962. *East African Wildlife Journal*, *1*, 124-126.

Freeman, a R., MacHugh, D. E., McKeown, S., Walzer, C., McConnell, D. J., and Bradley, D. G. (2001). Sequence variation in the mitochondrial DNA control region of wild African cheetahs (Acinonyx jubatus). *Heredity*, *86*(Pt 3), 355-62.

Garrick, R. C., Benavides, E., Russello, M. a, Gibbs, J. P., Poulakakis, N., Dion, K. B., Hyseni, C., Kajdacsi, B., Márquez, L., Bahan, S., Ciofi, C., Tapia, W., Caccone, A. (2012). Genetic rediscovery of an "extinct" Galápagos giant tortoise species. *Current Biology*, *22*(1), R10-R11.

Gasse, F. (2000). Hydrological changes in the African tropics since the Last Glacial Maximum. *Quaternary Science Reviews*, *19*, 189-211.

Gilbert, D. A., Packer, C., Pusey, A. E., Stephens, J. C., O'Brien, S. J. (1990). Analytical DNA fingerprinting in lions: parentage, genetic diversity, and kinship. *The Journal of heredity*, *82*(5), 378-86.

Gilbert, M. T. P., Drautz, D. I., Lesk, A. M., Ho, S.Y. W., Qi, J., Ratan, A., Hsu, C.-H, Sher, A., Dalen, L., Gotherstrom, A., Tomsho, L. P., Rendulic, S., Packard, M., Campos, P. F., Kuznetsova, T. V., Shidlovskiy, F., Tikhonov, A., Willerslev, E., Iacumin, P., Buigues, B., Ericson, P. G. P., Germonpre, M., Kosintsev, P., Nikolaev, V., Nowak-Kemp, M., Knight, J. R., Irzyk, G. P., Perbost, C. S., Fredrikson, K. M., Harkins, T. T., Sheridan, S., Miller, W., Schuster, S. C. (2008). Intraspecific phylogenetic analysis of Siberian woolly mammoths using complete mitochondrial genomes. *PNAS*, *105*(24).

Hahn, D. A., Ragland, G. J., Shoemaker, D. D., Denlinger, D. L. (2009). Gene discovery using massively parallel pyrosequencing to develop ESTs for the flesh fly Sarcophaga crassipalpis. *BMC genomics*, *10*(1), 234.

Henschel, P., Azani, D., Burton, C., Malanda, G. U. Y., Saidu, Y., Sam, M., Hunter, L. (2010). Lion status updates from five range countries in West and Central Africa. *CATnews*, *52*, 34-39.

Hollister, N. (1917). Some effects of environment and habit on captive lions. *Proceedings U.S. National Museum*, *53*, 177-193.

Janczewski, D. N., Modi, W. S., Stephens, J. C., O'Brien, S. J. (1995). Mitochondrial 12S Molecular Evolution of Mitochondrial 12S RNA and and Cytochrome b Sequences in Pantherine lineage of Felidae. *Molecular biology and evolution*, *12*(4), 690-707.

Johnson, W. E., Onorato, D. P., Roelke, M. E., Land, E. D., Cunningham, M., Belden, R. C., McBride, R., Jansen, D., Lotz, M., Shindle, D., Howard, J., Wildt, D. E., Penfold, L. M.,

Hostetler, J. A., Oli, M. K., O'Brien, S. J. (2010). Genetic Restoration of the Florida Panther. *Science*, *329*(5999), 1641-1645.

Kerstens, H. H. D., Crooijmans, R. P. M. A., Veenendaal, A., Dibbits, B. W., Chin-A-Woeng, T. F. C., Dunnen, J. T. den, Groenen, M. A. M. (2009). Large scale single nucleotide polymorphism discovery in unsequenced genomes using second generation high throughput sequencing technology: applied to turkey. *BMC genomics*, *10*(1), 479.

Kruglyak, L. (1997). The use of a genetic map of biallelic markers in linkage studies. *Nature genetics*, *17*, 21-24.

Kuhner, M. K., Felsenstein, J. (2000). Sampling among haplotype resolutions in a coalescent-based genealogy sampler. *Genetic epidemiology*, *19 Suppl 1*, S15-21.

Lercher, M. J., Hurst, L. D. (2002). Human SNP variability and mutation rate are higher in regions of high recombination. *Trends in genetics : TIG*, *18*(7), 337-40.

Lyons, S.K., Smith, F.A. and Brown, J.H. (2004) Of mice, mastodons and men: human-mediated extinctions on four continents. *Evolutionary Ecology Research*, *6*, 339–358.

Macaubas, C., Jin, L., Hallmayer, J., Kimura, a, Mignot, E. (1997). The complex mutation pattern of a microsatellite. *Genome research*, *7*(6), 635-41.

Mazák, J. H. (2010). Geographical variation and phylogenetics of modern lions based on craniometric data. *Journal of Zoology*, *281*, 194-209.

Mazák, V. (1970). The Barbary lion, Panthera leo leo, (Linnaeus, 1758): some systematic notes, and an interim list of the specimens preserved in European museums. *Zeitschrift für Säugetierkunde, 35*, 34–45.

Menotti-Raymond, M., and O'Brien, S. J. (1993). Dating the genetic bottleneck of the African cheetah. *Proceedings of the National Academy of Sciences of the United States of America*, *90*(8), 3172-6.

Mills L.S., Smouse P.E. (1994). Demographic consequences of inbreeding in remnant populations. *Amer. Nat.*, *144*, 412–431.

Nachman, M. W. (2001). Single nucleotide polymorphisms and recombination rate in humans. *Trends in genetics : TIG*, *17*(9), 481-5.

Nersting, L. G., and Arctander, P. (2001). Phylogeography and conservation of impala and greater kudu. *Molecular ecology*, *10*(3), 711-9.

Nielsen, R. (2000). Estimation of population parameters and recombination rates from single nucleotide polymorphisms. *Genetics*, *154*(2), 931-42.

Novaes, E., Drost, D. R., Farmerie, W. G., Pappas, G. J., Grattapaglia, D., Sederoff, R. R., Kirst, M. (2008). High-throughput gene and SNP discovery in Eucalyptus grandis, an uncharacterized genome. *BMC genomics*, *9*(1), 312.

O'Brien S.J., Roelke M.E., Marker L., Newman A., Winkler C.A., Meltzer D., Colly L., Evermann J.F., Bush M., Wildt D.E. (1985). Genetic basis for species vulnerability in the cheetah. *Science*, *227*, 1428-1434.

O'Brien, S.J., Martenson, J.S.., Packer, C. Herbst, L., Vos, V. De, Joslin, P., Ott-Joslin, J., Wildt, D. E., Bush, M. (1987a). Biochemical genetic variation in geographic isolation of African and Asiatic lion. *National Geographic Research*, *3*, 114-124.

O'Brien. S. J., Wildt, D. E., Goldmam, D., Merril, C. R., Bush, M. (1983). The cheetah is depauperate in genetic variation *Science*, *22*, 459-462.

O'Brien, S.J., Joslin, P., Smith III, G.L., Wolfe, R., Schaffer, N., Heath, N., Ott-Joslin, J., Rawal, P. P., Bhattacharjee, K. K., Martenson, J. S. (1987b). Evidence for African origins of founders of the asiatic lion species survival plan. *Zoo Biology*, *6*(2), 99–116.

Pereira, H. M., David Cooper, H. (2006). Towards the global monitoring of biodiversity change. *Trends in ecology evolution*, *21*(3), 123-9.

Pimm, S. L., Dollar, L., Bass, O. L. (2006). The genetic rescue of the Florida panther. *Animal Conservation*, *9*(2), 115-122.

Pitra, C., Hansen, A., Lieckfeldt, D., Arctander, P. (2002). An exceptional case of historical outbreeding in African sable antelope populations. *Molecular Ecology*, *11*, 1197-1208.

Pusey, A. and Packer, C. (1987). The evolution of sex-biased dispersal in lions. *Behaviour, 101* (4), 275-310.

Ramos, A. M., Crooijmans, R. P. M. A., Affara, N. A., Amaral, A. J., Archibald, A. L., Beever, J. E., Bendixen, C., Churcher, C., Clark, R., Dehais, P., Hansen, M. S., Hedegaard, J., Hu, Z., Kerstens, H. H., Law, A. S., Megens, H., Milan, D., Nonneman, D. J., Rohrer, G. A., Rothschild, M. F., Smith, T. P. L., Schnabel, R. D., Tassell, C. P. van, Taylor, J. F., Wiedmann, R. T., Schook, L. B., Groenen, M. A. M. (2009). Design of a high density SNP genotyping assay in the pig using SNPs identified and characterized by next generation sequencing technology. *PloS one*, *4*(8), e6524.

Roelke, M. E., Martenson, J. S., O'Brien, S. J. (1993). The consequences of demographic genetic depletion in the endangered reduction and Florida panther. *Current Biology*.

Ryder O. A. (1986). Species conservation and systematics: the dilemma of the subspecies. *Trends Ecol Evol, (1)*:9–10.

Sánchez, C. C., Smith, T. P. L., Wiedmann, R. T., Vallejo, R. L., Salem, M., Yao, J., Rexroad, C. E. (2009). Single nucleotide polymorphism discovery in rainbow trout by deep sequencing of a reduced representation library. *BMC genomics*, *10*(1), 559.

Sarntheim, M. (1978). Sand deserts during glacial maximum and climatic optimum. *Nature*, *272*, 43-46.

Sogbohossou, E.A. (2011). Lions of West Africa : ecology of lion (Panthera leo Linnaeus 1975) populations and human-lion conflicts in Pendjari Biosphere Reserve, North Benin. (Dissertation).

Steinmetz, A., Eulenberger, K., Thielebein, J., Buschatz, S., Bernhard, A., Wilsdorf, A., Ofri, R. (2006). Lens-anomalies and other ophthalmic findings in a group of closely-related angola lions (Panthera leo bleyenberghi). *Zoo Biology*, *25*(5), 433-439.

Stephens, M., Smith, N. J., Donnelly, P. (2001). A new statistical method for haplotype reconstruction from population data. *American journal of human genetics*, *68*(4), 978-89.

Tassell, C. P. van, Smith, T. P. L., Matukumalli, L. K., Taylor, J. F., Schnabel, R. D., Lawley, C. T., Haudenschild, C. D., Moore, S. S., Warren, W. C., Sonstegard, T. S. (2008). SNP discovery and allele frequency estimation by deep sequencing of reduced representation libraries. *Nature Methods*, *5*(3), 247-252.

Thalwitzer, S., Wachter, B., Robert, N., Wibbelt, G., Muller, T., Lonzer, J., Meli, M.L., Bay, G., Hofer, H., Lutz, H. (2010). Seroprevalences to Viral Pathogens in Free-Ranging and Captive Cheetahs (Acinonyx jubatus) on Namibian Farmland. Clin. Vaccine Immunol, *17*, 232-238.

Trick, M., Long, Y., Meng, J., Bancroft, I. (2009). Single nucleotide polymorphism (SNP) discovery in the polyploid Brassica napus using Solexa transcriptome sequencing. *Plant biotechnology journal*, *7*(4), 334-346.

Trinkel, M., Ferguson, N., Reid, a, Reid, C., Somers, M., Turelli, L., Graf, J., Szykman, M., Cooper, D., Haverman, P., Kastberger, G., Packer, C., Slotow, R. (2008). Translocating lions into an inbred lion population in the Hluhluwe-iMfolozi Park, South Africa. *Animal Conservation*, *11*(2), 138-142.

Trinkel, M., Funston, P., Hofmeyr, M., Hofmeyr, D., Dell, S., Packer, C., and Slotow, R. (2010). Inbreeding and density-dependent population growth in a small, isolated lion population. *Animal Conservation*, *13*(4), 374-382.

Van Hooft, W. F., Groen, A. F., and Prins, H. H. T. (2002). Phylogeography of the African buffalo based on mitochondrial and Y-chromosomal loci: Pleistocene origin and population expansion of the Cape buffalo subspecies. *Molecular ecology*, *11*(2), 267-79.

Vera, J. C., Wheat, C. W., Fescemyer, H. W., Frilander, M. J., Crawford, D. L., Hanski, I., Marden, J. H. (2008). Rapid transcriptome characterization for a nonmodel organism using 454 pyrosequencing. *Molecular ecology*, *17*(7), 1636-47.

Vogler, A. P., Knisley, C. B., Glueck, S. B., Hill, J. M., DeSalle, R. (1993). Using molecular and ecological data to diagnose endangered populations of the puritan tiger beetle *Cicindela puritana*. *Molecular Ecology*, *2*, 375-383.

Vogt, C. and Specht, F. (1889). The Natural History of Animals. Blackie and Son, London, U.K.

Waples, R. S. (1991). Pacific salmon, Oncorhynchus spp., and the definition of "species" under the Endangered Species Act. *Mar. Fish. Rev.*, *53*(3), 11-22.

Wasser, S. K., Joseph Clark, W., Drori, O., Stephen Kisamo, E., Mailand, C., Mutayoba, B., Stephens, M. (2008). Combating the illegal trade in African elephant ivory with DNA forensics. *Conservation biology : the journal of the Society for Conservation Biology*, *22*(4), 1065-1071.

Watterson, G.A. (1975). On the number of segregating sites in genetical models without recombination. *Theoretical Population Biology*, *7* (2), 256–276.

Wildt, D. E., Bush, M., Goodrowe, K. L., Packer, C., Pusea, A. E., Brown, J. L., Joslin, P., O'Brien, S. J. (1987a). Reproductive and genetic consequences of founding isolated lion populations. *Nature*, *329*(4), 328-331.

Wildt, D.E., Bush, M., Howard, J.G., O'Brien, S.J., Meltzer, D., Dyk, A. van, Ebedes, H., Brand, D.J. (1983). Unique seminal quality in the South African cheetah and a comparative evaluation in the domestic cat. *Biol Reprod.*,*29*(4), 1019–1025.

Wildt, D.E., O'Brien, S.J., Howard, J.G., Caro, T.M., Roelke, M.E., Brown, J.L., Bush, M. (1987b) Similarity in ejaculate-endocrine characteristics in captive versus free-ranging cheetahs of two subspecies. *Biol Reprod.*, *36*(2), 351–360.

Wittwer, C. T., Reed, G. H., Gundry, C. N., Vandersteen, J. G., Pryor, R. J. (2003). High-resolution genotyping by amplicon melting analysis using LCGreen. *Clinical chemistry*, *49*(6 Pt 1), 853-60.

Yuhki, N. O'Brien, S.J. (1990). DNA variation of the mammalian major histocompatibility complex reflects genomic diversity and population history. *Proceedings of the National Academy of Science USA*, *87*, 836-840.

In: Genetic Diversity: New Research
Editors: J. A. C. Amaya and M. M. F. Jimenez

ISBN: 978-1-62081-598-4
© 2012 Nova Science Publishers, Inc.

Chapter 4

Genetic Diversity for Quality Traits in Wild and Ancient Wheats

Juan B. Alvarez **and Carlos Guzmán***

Departamento de Genética, Escuela Técnica Superior de
Ingeniería Agronómica y de Montes, Edificio Gregor Mendel,
Campus de Rabanales, Universidad de Córdoba, Córdoba, Spain

Abstract

Global Change, a result of human action on ecosystems, is one of the major problems to face humanity at the present time. For crops, this means that there will be a need to find or create new cultivars able to adapt to a changing environment, without a reduction in nutritional standards or industrial quality. The quality characteristics of wheat depend on many factors, including gluten levels, starch levels and hardness, which are affected by seed storage proteins, starch synthases and puroindolines, respectively. There is little diversity among modern cultivars for the abovementioned characters, mainly due to the strong pressure exerted by wheat breeding programmes to adapt cultivars to different climatic conditions. Consequently, the search for allelic variants of these genes in local cultivars or closely related species could serve as a basis to increase the genetic pool of these components in modern wheat. In this context, a large programme has been developed to firstly, evaluate and characterize ancient wheats and some close relatives that could be used in wheat breeding and secondly, to improve wheats in general within the context of sustainable agriculture.

Introduction

Today, one of the major problems facing humanity is so-called Global Change. This is the result of human action on ecosystems and Climate Change is one of its most dramatic impacts. For crops, this will bring about a need to find or create new cultivars able to adapt to

* E-mail: jb.alvarez@uco.es.

a changing environment without a reduction in nutritional standards or industrial quality and that fulfil the demands of a population that has grown used to high quality foodstuffs.

In this context, one fundamental key is the biodiversity present in these crops. Biodiversity is a term that was introduced in the late 1980's by Edward O. Wilson as a shortening of the term biological diversity (Wilson 1988). This is defined as the variation in all living organisms at all levels and is divided into three categories: genetic diversity, species diversity and ecosystem diversity. However, there is a widespread tendency of not integrating these in studies on domesticated plants and animals. This has meant that studies into biodiversity have concentrated on wild plants and animals. In response to this, a new concept arose, called agricultural biodiversity or agrobiodiversity, which is defined as: «*...that part of biodiversity which nurtures people and which are nurtured by people...*» (FAO 1995). Consequently, agrobiodiversity refers mainly to genetic variability in cultivated plants and domesticated animals, together with their progenitors and closely related wild species, which are maintained within agroecosystems and surrounding natural environments (Qualset et al. 1995; Hammer et al. 2003). It is important to emphasise that crop genetic diversity is a biological patrimony, which is an important tool for genetic improvement, for example, to avoid vulnerability to biotic and abiotic stresses. The maintenance of this genetic diversity is a prerequisite for sustainable agriculture, which is the use of natural resources without their exploitation (World Commission on Environment and Development 1987). Consequently, agriculture can not be considered sustainable if it does not include a suitable programme for the conservation of plant genetic resources, together with clear and decisive programmes for the evaluation and characterization of the genetic diversity that they contain.

According to FAO data (FAO 1996), 30 crops constitute the world food base; although just three of the major crops (wheat, rice and maize) supply more than 55% of the global food energy derived from crops. However, the great underlying problem is not only that few crops are used, but also that in many cases their genetic base is very narrow, which presents a real danger for the crops and makes them more vulnerable (Esquinas-Alcázar 2005). It is noteworthy that due to their importance, 25% of the approximately six million accessions stored in Germplasm Banks are of the three major crops, whereas minor crops and wild relatives are poorly represented (Padulosi et al. 2002).

In this chapter, we outline the different studies carried out by our research group into the genetic diversity and quality traits in several ancient wheats and some close relatives, which could be used as new genetic resources in wheat breeding or as new crops in a more diverse agriculture system

Wheat

One of the most important crops worldwide is wheat (*Triticum* sp.) with a cultivated area of approximately 217 million hectares (FAO 2010). In terms of production, ranks as the third most important crop after maize and rice, with 651 million tonnes. Although throughout history, wheat has been used for the manufacture of various products, mainly bread, pasta and beer; nowadays, this crop is associated to the pasta manufacture (durum wheat) or to the baking industry (common wheat) into the Western countries. However, it has also begun to be used to the development of non-food products such as cosmetics and even as an energy crop for bioethanol and biomass production (Bell et al. 1995).

The wheat origin and domestication are placed in the Near East, in the well-known zone as Fertile Crescent (Zohary and Hopf 1988). The wheat is a complex polyploid formed by multiple species of different ploidic level, consequence of the merge of genomes that come from different species of the *Triticeae* tribe, *Poaceae* family. Thus, it can be found diploid (2n = 2× = 14, AA), tetraploid (2n = 4 ×= 28, AABB) and hexaploid species (2n = 6× = 42, AABBDD) -Figure 1.

Among the diploid species, the one that has been cultivated is *Triticum monococcum* L. ssp. *monococcum* or einkorn (A^mA^m), which comes from the domestication of *T. monococcum* ssp. *aegilopoides* Link em. Thell. (syn. *T. boeoticum* Boiss.) in the Fertile Crescent. Another diploid species, *T. urartu* Thum ex. Gandil (A^uA^u), in this case wild, has been proposed as a donor of the A genome in polyploid species of wheat (Dvořák et al. 1988).

The currently accepted hypothesis suggests that *T.urartu* could have generated the wild tetraploid wheats mainly in two different events. On the one hand, crossing it with an *Aegilops* species (section *Sitopsis*), probably *Aegilops speltoides* Tausch. (2n = 2× = 14, SS), generating wild emmer (*T. turgidum* ssp. *dicoccoides* Korn. ex Asch. and Graebner em. Thell., 2n = 4× = 28, A^uA^uBB), from which was obtained cultivated emmer (*T. turgidum* spp. *dicoccum* Schrank em. Thell.) by domestication. The rest of the tetraploid wheat, including durum wheat (*T. turgidum* ssp. *durum* Desf. em. Husn.), derive from this species, as well as hexaploid wheat (Dvořák and Zhang 1990, 1992). Similarly, the crossing of some other species of the section *Sitopsis* with *T. urartu* originated *T. timophevii* ssp. *armeniacum* Jakubz. em. Slageren (2n = 4× = 28, A^uA^uGG), whose domesticated form (*T. timophevii* ssp. *timophevii*) is restricted to western Georgia.

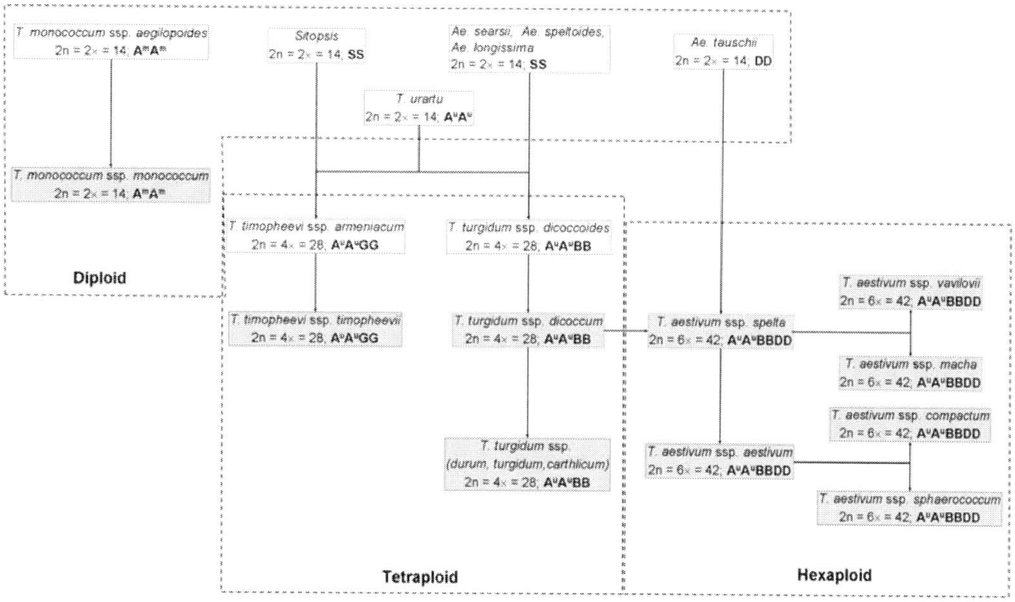

Figure 1. Diagram of the synthesis of different wheats. The shadow species were or are cultivated in some time.

With respect to the hexaploid wheats, several studies (McFadden and Sears 1946, Kerber and Rowland 1974) suggest that the D genome donor is *Ae. tauschii* Coss. (2n = 2× = 14, DD), that crossing with cultivated emmer and subsequent chromosome doubling led to spelt (*T. aestivum* ssp. *spelta* L. em. Thell., 2n = 6× = 42, A^uA^uBBDD), putative ancestor of common or bread wheat (*T. aestivum* ssp. *aestivum* L. em. Thell.) that is the species of the *Triticum* genus more important today. The absence of wild hexaploid wheat species supports this theory of the origin of cultivated hexaploid wheat and also shows that this crossing should happened in a crop field, likely in the Caspian Sea area, natural habitat of *Ae. tauschii*.

WHEAT FOOD QUALITY

Wheat quality can be defined as the ability of a variety to produce flour suitable for a specific product. Consequently, this parameter is a variable that will depend on consumer preferences, the product to be developed and the process to be used in the preparation of the product. This trait is strongly associated with the grain components and their physico-chemical properties.

Grain wheat is mainly composed of proteins (7 - 18%), lipids (1.5 - 2%) and carbohydrates (60 - 75%), together with other minor components such as certain vitamins and minerals (Aykroyd and Doughty 1970). Proteins and carbohydrates, especially starches, have considerable influence on three grain characteristics closely linked to the technical wheat qualities required for baking or pasta manufacture. These are gluten strength, starch properties and grain hardness or texture.

Gluten

Gluten is defined as «*the viscoelastic mass that remains after thoroughly washing out the starch from a dough*» (Miflin et al. 1983). This artificial structure is mainly constituted by endosperm storage proteins or prolamins, named gliadins and glutenins in wheat (Figure 2). These proteins make up a complex mixture that can range between 50 components for the hexaploid wheat and about 20 for the diploid species (Shewry et al. 1988). Because their variations have not repercussion on any physiological function, being cool against the evolution, become these proteins a good marker to evaluate the genetic diversity.

Gliadins are monomeric proteins with molecular weights between 30 kDa and 80 kDa, joined by hydrogen bonds and hydrophobic interactions. According to the mobility with respect to the cathode, gliadins have been classified into α-, β-, γ-and ω-gliadins, being synthesized by genes on the short arm of chromosome 1 and 6. The *Gli-1* loci (*Gli-A1*, *Gli-B1* and *Gli-D1*) present in the group-1 homoeologous chromosomes synthesised the γ- and ω-gliadins, and the *Gli-2* loci (*Gli-A2*, *Gli-B2* and *Gli-D2*) located at the group-6 homoeologous chromosomes codes the α- and β-gliadins (Metakovsky et al. 1984; Wrigley et al. 2006). These proteins have been studied, not only for their importance in viscoelastic properties of gluten as they are responsible for the cohesiveness and extensibility of the dough, but for their high level of polymorphism that become them in an interesting marker to analyze the genetic variation (Lafiandra et al. 1990), that can be used to identify wheat cultivars (Bushuk and Zillman 1978).

The glutenins are polymeric proteins with a wide range in terms of molecular weight (100 kDa and several million), being the different subunits covalently linked by disulfide bonds (SS). These proteins that give elasticity to dough, are classified into high molecular weight subunits (HMWGs) and low molecular weight subunits (LMWGs) (Payne et al. 1984). The HMWGs, with molecular weights ranging from 80-140 kDa, are encoded by the *Glu-1* complex loci located on the long arm of each chromosome of the homoeologous group-1, called *Glu-A1*, *Glu-B1* and *Glu-D1*, respectively (Payne 1987). Each locus consists of two closely linked genes which encode for two types of subunits with different molecular weight and hence different mobility in SDS-PAGE, namely: *x*-type subunit with higher molecular weight and lower mobility in SDS-PAGE, and *y*-type subunit, smaller and with greater electrophoretic mobility (Harberd et al. 1986).

The LMWGs, meanwhile, have molecular weights between 30-50 kDa, being encoded by the *Glu-A3*, *Glu-B3* and *Glu-D3* loci located on the short arm of homoeologous group-1 chromosome (Singh and Shepherd 1988; Liu 1995). Based on their mobility in SDS-PAGE and in their isoelectric point, Jackson et al. (1983) divided the LMWGs in B-, C-and D-LMWGs. Of all these, the B-LMWGs are those comprising the largest number of subunits, and the molecular studies have led to catalogue these subunits in three different types (LMW-i, LMW-m and LMW-s) according to the first amino acid residue of the mature protein: isoleucine, methionine and serine, respectively (D'Ovidio and Masci 2004).

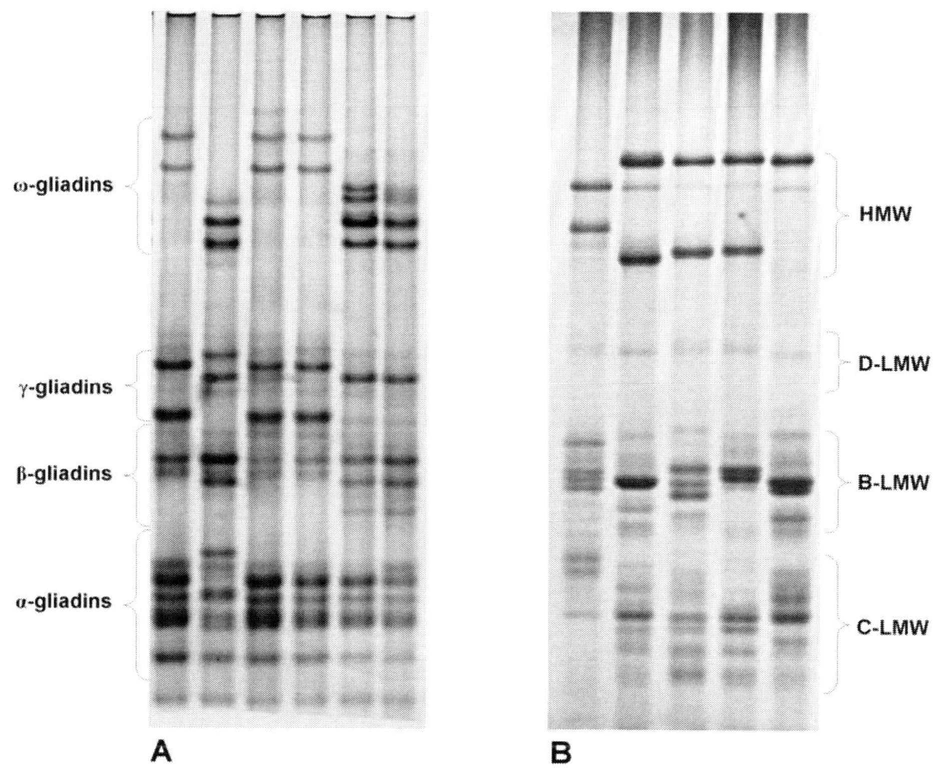

Figure 2. Electrophoretic separation of gliadins (A) and glutenins (B) in polyacrylamide gels.

Different studies have indicated that the HMWGs are the major group of gluten proteins that determine the bread-making characteristics of dough, whereas the variation of LMWGs would be more important in the pasta manufacture (see Wrigley et al. 2006 for review). With respect to these latter, the most qualitative evaluations of LMWGs have been performed on B-type subunits, as they are the most abundant and easiest to detect, being very little is known about the role of C- and D-type subunits (D'Ovidio and Masci 2004).

On the other hand, although certain degree of variability for the different prolamins has been detected in modern wheat, it is important to emphasize the low frequency of some alleles, which suggests that part of the diversity has been lost, perhaps forever. This process has been sponsored by the indiscriminate replacement of landraces with improved varieties, which has been particularly intense in the last century (Frankel, 1970, Frankel and Hawkes, 1975).

Starch

Starch is the major constituent of the endosperm of the wheat grain, with approximately 65% of its dry weight, and has an important role in the appearance, structure and characteristics of food products. It consists of two different structural components: amylose and amylopectin. Amylose is a relatively less-branched α-(1→4)-linked glucose polymer, whereas amylopectin is a branched glucose polymer in which α-(1→4)-linked polymers are connected by α-(1→6) linkages. In normal wheat, starch contains 20-30% amylose and 70-80% amylopectin (Kuroda et al. 1989).

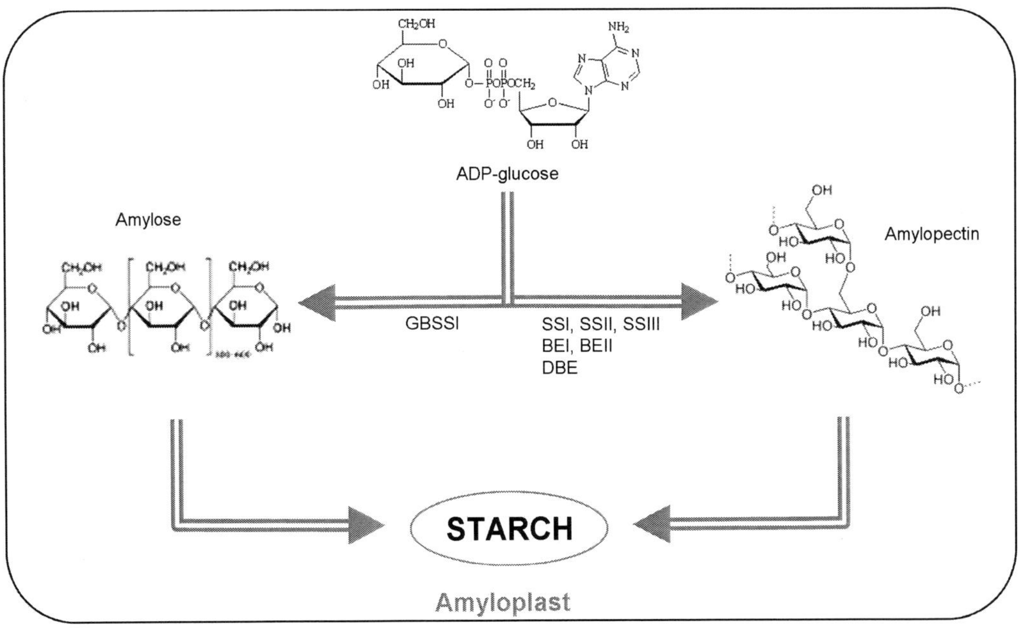

Figure 3. Starch synthesis in the amyloplast.

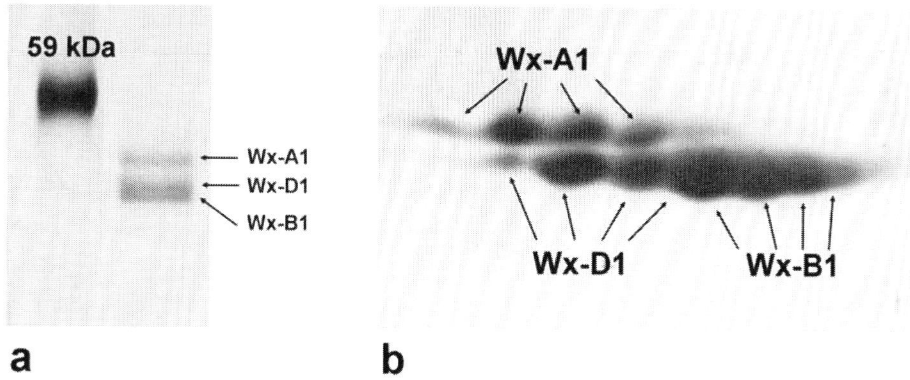

Figure 4. Uni-dimensional (a) and bi-dimensional (b) electrophoretic separation of waxy proteins.

The synthesis of starch in the endosperm occurs in the amyloplast (Figure 3), a plastid no photosynthetic apparatus dedicated to this function. For the synthesis of starch, glucose-1 phosphate is transported into the amyloplast and used as a substrate for several enzymes that act sequentially to form the synthetic route. These enzymes are ADP-glucose pyrophosphorylase, starch synthases (SS) and branching enzymes (BE). These enzymes have been studied in recent years given the importance of starch composition for many of its technological properties. Amylose and amylopectin are synthesized by two different pathways having a common substrate (ADP-glucose). The GBSSI is the key enzyme in amylose synthesis, whereas the soluble starch synthases with starch branching enzymes are believed to have unique functions in amylopectin synthesis (James et al. 2003).

The most studied of the enzymes involved in starch synthesis are waxy proteins (Figure 4). These proteins are synthesized by the *waxy* gene, normally located on chromosome 7 group of each genome (Chao et al. 1989; Ainsworth et al. 1993). Depending on the level of ploidy of each species of wheat have one, two or three waxy proteins. The latter is the case of common wheat where we find three waxy proteins encoded by the *Wx-A1* gene (chromosome 7AS), *Wx-B1* (chromosome 4 AL by a translocation of a segment of chromosome 7BS) and *Wx-D1* (7DS chromosome) (Yamamori et al. 1994). Other enzyme that has been studied is the SGP-1 (SGP-A1, SGP-B1, and SGP-D1), which is coded by the *SSIIa-A*, *SSIIa-B* and *SSIIa-D* genes located on chromosome arms 7AS, 7BS and 7DS, respectively (Li et al. 1999, 2003; Shimbata et al. 2005).

In recent years, numerous studies to find polymorphism in waxy proteins have been carried out (Yamamori et al. 1994, 1995; Rodriguez-Quijano et al. 1998; Nieto-Taladriz et al. 2000). They showed that the level of variability in these proteins is lower, even more when compared with other proteins such as seed storage proteins. However, different alleles were detected, including variants with different electrophoretic mobility in SDS-PAGE, with different isoelectric point and even null alleles. The latter are important because they allow the creation of waxy wheat, which is starch 100% amylose-free. This starch shows different technological properties from normal, affecting its gelatinisation, pasting and gelation, and therefore its quality (Zeng et al. 1997). However, for the development of healthy functional foods appears to be more interesting the use of starches with higher amylose content due to their slow digestibility (Behall and Scholfield 2005). Some studies have been developed on this aspect, using mainly the inactivation or suppression of the *SSII* genes (Morita et al. 2002;

Shimbata et al. 2005; Yamamori et al. 2000, 2006). Starch quality also may be important for processes external to the food industry, such as the generation of bioethanol from grain, which has showed better performance using waxy wheat (Wu et al. 2006).

Grain Hardness

Grain hardness is an important quality parameter of common wheat that determines the end use properties and market classification of wheat crops (Pomeranz and Williams 1990). Based on this trait, common wheat is classified into soft and hard. Soft wheat kernels fracture more easily, producing fine-textured flours with low levels of damaged starch granules, which make this type of flour suitable for cookies, cakes, pastries, and some types of noodles (Morris and Rose 1996). Hard wheat produce coarser-textured flours that exhibit fracture planes and considerable amounts of broken starch granules. The high levels of starch damage in hard wheat flour allows the high water absorption, which makes hard wheat better suited for bread and other foods prepared with viscoelastic dough (Morris and Rose 1996).

The molecular-genetic basis of wheat grain hardness is now well established, being related with "friabilin", a Mr 15-kDa protein associated with the starch membrane, which is abundant in soft wheat starch, scarce in hard wheat starch, and absent in durum wheat (Greenwell and Schofield 1986). The main component of "friabilin" are puroindolines a and b (PINA and PINB); and a third minor protein called the grain softness protein (GSP). These proteins are encoded by the *Pina*, *Pinb* and *Gsp* genes located in the locus *Ha* (*Hardness*), which is present in the short arm of the chromosome 5D of common wheat (Morris 2002). It is a complex locus formed by 10 closely linked genes -Figure 5- (Li et al. 2008), although only two (*Pina* and *Pinb*) have been so far associated to grain hardness (Gautier et al. 1994; Rahman et al. 1994; Giroux and Morris 1997, 1998).

When both puroindolines are present in its functional form, grain texture is soft; while the absence of any of them, either by gene deletion or inactivation, causes the grain texture becomes harder (Gautier et al. 1994, Giroux and Morris 1998; Lillemo and Morris 2000, Morris 2002, Chen et al. 2006). The extreme of this circumstance is durum wheat, which grain texture is very hard as a result of the deletion of both genes (Gautier et al. 2000), occurred during the generation process of tetraploid wheat by partial removal of *Ha* locus from the A and B genomes (Li et al. 2008). Subsequent hybridization with *Ae. tauschii* allowed the recovery of soft character in hexaploid wheat by the contribution of *Pin* genes from the D genome.

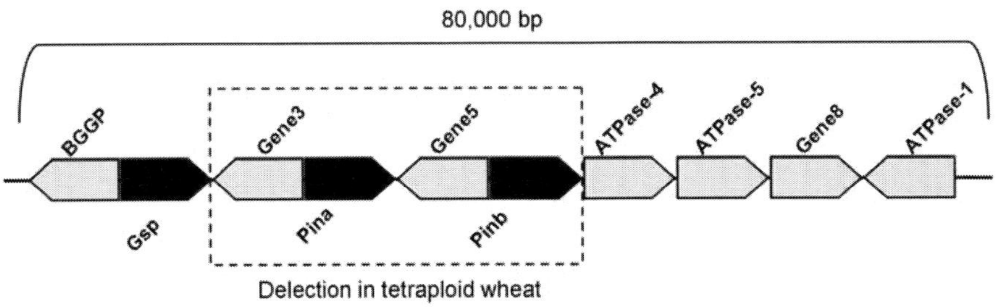

Figure 5. Structure of the *Ha* locus. The genes marked in black have one known function in the grain hardness.

In recent years there have been many studies focused on finding variability in puroindolines that cause differences in grain hardness. Several mutations have been described in both genes involving amino acid changes of the mature protein that alter the hardness of the grain, usually making it harder (Giroux and Morris 1998; Lillemo and Morris 2000, Chen et al. 2006), but some mutations have also been described that cause the grain to be softer (Gedye et al. 2004). The search for new variants continues today, in order to increase the range of textures available, and as in the case of waxy proteins, has been extended to other species (Massa et al. 2004).

Genetic Diversity Analyses

In this context, our group began the current research with the aim of evaluating different types of neglected or underutilized wheats, such as *Triticum* and *Aegilops* species, together with their close relatives. This research began in 1999 and has focused on the species indicated in the following table, although recently other species, mainly of the *Aegilops* genus, have been included in this research line.

Species analysed by our group			
Ploidy level	Common name	Species	Accessions
2×	Einkorn	*T. monococcum* ssp. *monococcum*	45
	Wild diploid wheat	*T. urartu*	169
4×	Persian wheat	*T. turgidum* ssp. *carthlicum*	140
	Emmer	*T. turgidum* ssp. *dicoccum*	102
	Polish wheat	*T. turgidum* ssp. *polonicum*	159
	Khorasan wheat	*T. turgidum* ssp. *turanicum*	77
	Rivet wheat	*T. turgidum* ssp. *turgidum*	313
6×	Mexican Creole wheat	*T. aestivum* ssp. *aestivum*	100
	Spelt	*T. aestivum* ssp. *spelta*	489
Total			1594

The choice of each species was based on the objectives of the present research: the evaluation and characterization of ancient wheats as a source of new allelic variants for enlarging the genetic background of modern wheat and an evaluation of the potential of these ancient crops to be used to create new products as part of the revival in traditional foods that has recently been seen in developed countries.

The first crop types to be incorporated into the programme were represented by three broad collections of hulled wheats (einkorn, emmer and spelt) that were cultivated in Spain in the past and could now be catalogued as neglected or underutilized crops. Today, both einkorn and emmer have practically disappeared in Spain, whereas spelt survives in the marginal farming areas of Asturias (Northern Spain), where the farmers grow it using traditional farming systems mainly for home consumption (Peña-Chocarro 1996; Alvarez et al. 2010). Currently, these crops are becoming a fashionable food, particularly if they are grown using traditional or organic farming methods. They are also said to have superior pharmaceutical and nutritional qualities, not necessarily confirmed scientifically. They are said to have a higher nutrient content and to be useful in the treatment of ulcerous colitis, rheumatoid arthritis, neurodermitis and high blood cholesterol (Strehlow et al. 1994). From the point of view of wheat breeding, they are a potential source of genes that could be used to improve cultivated wheat today. This increased genetic variability has been highlighted in numerous studies which have investigated the variability of storage proteins (Rodriguez-Quijano et al. 1990) and starch (Rodriguez-Quijano et al. 1998, 2003; Nieto-Taladriz et al. 2000).

Other species that in the past had been cultivated in Spain were progressively incorporated to the research collection, although their origins were broader, being accessions from other countries as well as Spain. In this group is included different subspecies of tetraploid wheat as khorasan wheat (*T. turgidum* ssp. *turanicum* Jakubz em. A. Löve and D. Löve), Persian wheat (*T. turgidum* ssp. *carthlicum* Nevski em. A. Löve et D. Löve), Polish wheat (*T. turgidum* ssp. *polonicum* L. em. Thell.) and rivet wheat (*T. turgidum* L. ssp. *turgidum*).

Recently, one collection of ancient bread wheat named Mexican Creole wheat was obtained from the CIMMYT Germplasm Bank. These old wheats are the last trace of the common wheat carried by the Spanish to America during the 16th and the 17th centuries. These wheats are derived from old Spanish cultivars and landraces and their cultivation survives on small-scale traditional farms. They could represent the nexus between the old bread wheats cultivated in Spain and the traditional varieties stored in the Germplasm Banks and could contain genetic variability that is now missing from modern wheats.

To date, only one wild species (*T. urartu*) has been analysed. This species was selected as it contained one characteristic absent from modern cultivated wheats. It contains active *Ay* subunits from the HMWGs, which have been associated with increased gluten strength (Ciaffi et al. 1995).

Endosperm Storage Proteins

Our studies have mainly centred on evaluating the composition of the seed storage proteins present in the three Spanish hulled species taken from a number of Germplasm Banks (32 accessions of einkorn, 102 of emmer and 489 of spelt). They showed wide polymorphism that could be used to transfer new, quality genes to wheat and widen the gene pool available to breeders (Figure 6).

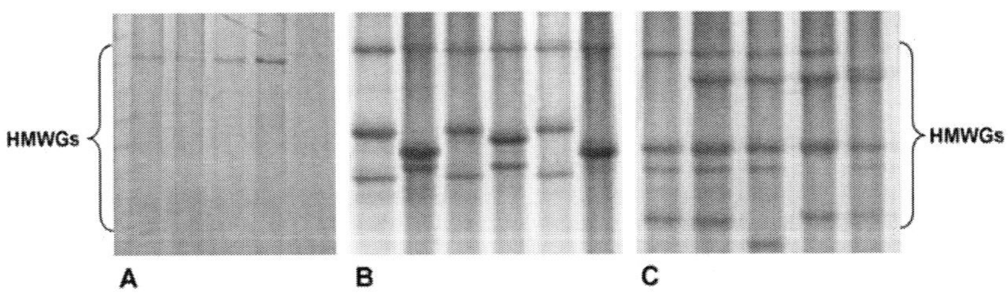

Figure 6. Sample of the variation for high-molecular-weight glutenin subunits in hulled wheats. A, einkorn; B, emmer; and C, spelt.

The variation in seed storage proteins in Spanish einkorn was high, with three allelic variants for the *Glu-A^{m}1* locus, six for the *Glu-A^{m}3* locus, seven for the *Gli-A^{m}1* locus and fourteen for the *Gli-A^{m}2* locus being found among the evaluated accessions. Internal variability was detected in some of these materials, which could be related to the fact that they are landraces. Up to 48 different genotypes were identified based on their origin and seed storage protein composition (Alvarez et al. 2006).

When the emmer collection was evaluated for HMWGs, four allelic variants were detected for the *Glu-A1* locus; one of which had not been previously described. For the *Glu-B1* locus, three of the nine alleles detected had not been previously detected (Pflüger et al. 2001). Similarly, a high degree of variation was found for the LMWGs, where up to 23 different patterns were detected for B-LMWGs. When considering HMWGs and LMWGs collectively, 30 combinations were found between the evaluated lines. In one latter study, we selected 31 lines that presented these allelic variations and evaluated them for morphological and quality traits (Alvarez et al. 2007a; Caballero et al. 2008a). Up to seven different botanical varieties were detected, which suggested a wide diversity between the varieties, although lower than that for the 10 botanical varieties that had been historically described (Tellez-Molina and Alonso-Peña 1952). Furthermore, the allele data obtained for gluten strength also showed some of the alleles detected for HMWGs and LMWGs (Caballero et al. 2008a).

In spelt, three allelic variants were detected for the *Glu-A1* locus. For the *Glu-B1* locus, two of seven alleles detected had not been found before while four of the nine alleles detected for the *Glu-D1* had not been described previously (Caballero et al. 2001). When the three loci were considered together, 25 combinations were found among the evaluated lines. For the LMWGs, extensive variation was found, 46 different patterns for the B-LMWGs and 16 for the C-LMWGs (Caballero et al. 2004a). The B-LMWG patterns exhibited between two to six bands and the C-LMWG patterns had between four to six bands in SDS-PAGE gels. When urea was added to the gel, according to Goldsbrough et al. (1989), a larger number of bands were observed. The patterns had between six to eleven bands and 14 new patterns were found in the B-LMWGs. For the C-LMWGs, up to 10 new patterns, possessing between five to nine bands, were detected. Altogether, 86 patterns were found for the LMWGs. With regards to gliadins, 61 different patterns were found for the ω-gliadins, 44 for the γ-gliadins, 19 for the β-gliadins and 15 for the α-gliadins (Caballero et al. 2004b). The variability detected in spelt was larger than that detected in other hulled wheats. One core collection of 25 lines was evaluated for morphological and quality traits and the results were similar to the results obtained for the emmer lines (Alvarez et al. 2007b; Caballero et al. 2008a).

Within the spelt collection, the diversity of HMWGs was studied electrophoretically in 333 accessions, grouped in 50 populations, originally collected in Asturias (North Spain) during the 1930s. The inter and intra-population distribution of HMWGs alleles at the *Glu-A1*, *Glu-B1* and *Glu-D1* loci, together with the gliadin patterns, were investigated (Caballero et al. 2004b,c). The results showed that the genetic variation is mainly present within populations with the variation between populations being only 21%. The material analysed showed a wide polymorphism for all seed storage proteins, although some allelic variants were clearly hegemonic and other ones appeared less frequently. This suggested the possibility that there was a loss of variability, even before the collecting mission, which could have deteriorated further with the subsequent reduction in the cultivated area of this species in Asturias.

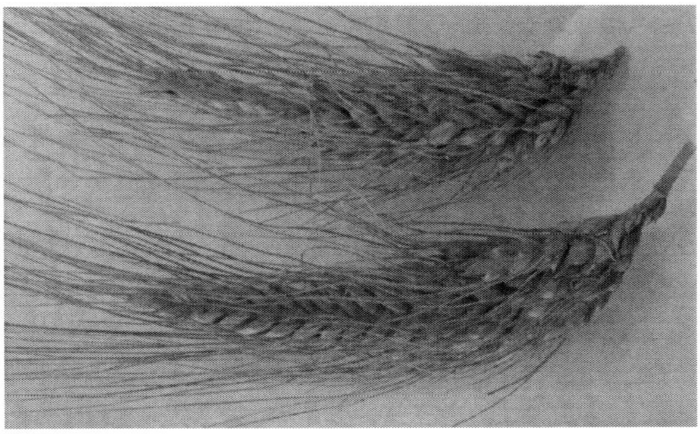

Figure 7. Rivet wheats accession with ramified spikes.

In order to evaluate the current status of this crop, we carried out one collecting mission to Asturias in 2004 (Caballero et al. 2007). Thirty two spelt populations, containing a high genetic diversity, were collected, although it was lower than that described in the past (Caballero et al. 2008b), with the loss of two botanical varieties described by Lagasca and Rojas Clemente in the 19th century (Tellez-Molina and Alonso-Peña 1952), together with nine HMWG alleles that were detected in the old collection (Caballero et al. 2001).

Other neglected wheats that were cultivated in the past in Spain are Persian wheat and Polish wheat. Nowadays, most of the genetic resources for these species are conserved in Germplasm Banks. When the composition of the HMWGs was evaluated, up to eighteen allelic variants (seven alleles at the *Glu-A1* locus and eleven at the *Glu-B1* locus) were found in both species. Ten out of these allelic variants had not been previously identified in wheat, four at the *Glu-A1* locus and six at the *Glu-B1* locus (Caballero et al. 2008c). When both loci were considered, six and seventeen different combinations were found in Persian and Polish wheat, respectively, most of them rare or very rare (frequency under 5%).

For the B-LMWGs, there were up to 12 different patterns in Persian wheat and 17 for Polish wheat (Caballero et al. 2005). Although the variability detected in the material was greater than that detected in other hulled wheats, the frequency distribution was clearly asymmetric in both species. Three out of the 12 patterns represented 90.7% of the variation in Persian wheat, while 86.2% of the variation in Polish wheat was constituted by four patterns. The low frequency of some patterns confirmed the need to protect and conserve these accessions because the possibility of finding the same alleles in other material is very low.

These results suggested that the number of neglected tetraploid wheats in our collection should be increased. Two new species were therefore included: rivet wheat and khorasan wheat. Only rivet wheat has been cultivated in Spain in the past. High genetic diversity was detected in these two wheats, mainly in rivet wheat where lines with ramified spikes were found (Figure 7).

Sixty accessions of rivet wheat originated from Spain. The analysis of these accessions showed a high genetic diversity with 16 botanical varieties and five additional morphological types that could not be associated with any botanical variety described by Lagasca and Rojas Clemente (Tellez-Molina and Alonso-Peña 1952). Furthermore, up to 13 allelic variants (four alleles for the *Glu-A1* locus and nine alleles for the *Glu-B1* locus) and 34 B-LMWG patterns were found in the evaluated lines. Five of them were novel (one for *Glu-A1* locus and four for the *Glu-B1* locus). There were also 34 B-LMWG patterns. (Carmona et al. 2010a). In the worldwide collection, the variation in HMWGs and LMWGs was higher than in the Spanish collection (Carmona et al. 2010b). Twenty alleles (five at the *Glu-A1* locus and fifteen at the *Glu-B1* locus) were identified. Two novel alleles were detected at the *Glu-A1* locus and five at the *Glu-B1* locus. Sixty seven B-LMWGs patterns were also seen, although at very low frequencies.

In khorasan wheat, 11 allelic variants were found, three alleles at the *Glu-A1* locus and eight alleles at the *Glu-B1* locus (Carmona et al. 2010b). Three of these were novel, one at the *Glu-A1* locus (found also in rivet wheat) and two at the *Glu-B1* locus (one of them was also found in rivet wheat). However, the B-LMWGs showed low variability, with only six different patterns being detected. Two of them were detected in 81.2% of cases (67.5% and 14.3%, respectively). Sixteen khorasan wheat lines, which differed in their composition of HMWGs (null, 7+15 *vs.* null, 6+8), were evaluated for gluten strength using SDS-sedimentation volume (Carmona et al. 2009). The data suggested that the null, 7+15 pair had higher SDS-sedimentation values than the other pair evaluated. However, in general, this species had weak gluten and could be used for other baking applications rather than in the pasta industry.

In 2004 in collaboration with Dr. Roberto J. Peña from CIMMYT, a collection of traditional common wheat landraces (Mexican Creole wheat) was received. Different morphological traits were measured with the purpose of evaluating how many of the 22 botanical varieties that Lagasca and Rojas Clemente established for common wheat in Spain throughout the 19th century still existed (Tellez-Molina and Alonso-Peña 1952). These Mexican Creole wheats were grouped into seven botanical varieties (Caballero et al. 2010). However, because there is no data on the botanical varieties taken to the Americas during the 16th and 17th centuries, the level of erosion undergone by these wheats could not be determined.

Up to 21 HMWG allelic variants (four, eight and nine alleles for the *Glu-A1*, *Glu-B1* and *Glu-D1* loci, respectively) were detected. The frequency distribution was very asymmetric: 90% of the lines had only two *Glu-A1* alleles, 88% had two *Glu-B1* alleles and only one *Glu-D1* allele was present in 81% of the lines. This suggested the possibility of erosion, mainly by genetic drift (Caballero et al. 2010).

For *T. urartu*, one broad collection of 169 accessions was analysed for HMWGs, B-LMWGs and gliadins (Caballero et al. 2008d; Martín et al. 2008). This collection was constituted by 70 accessions from Turkey, 91 from Lebanon, 2 from Armenia, 3 from the former USSR, 2 from Iraq and 1 from Iran. Genetic diversity was high within this collection. Seventeen alleles were detected for the *Glu-A^{u}1* locus, 24 for the *Glu-A^{u}3* locus, 26 alleles for the *Gli-A^{u}1* locus, and 26 alleles for *Gli-A^{u}2* locus. However, the geographic distribution of these accessions was very unequal. Furthermore, some locations were scarcely represented while others were represented in five or more accessions (Lebanon and Turkey). The analysis of these populations showed that the Turkish populations showed greater diversity than the Lebanese populations with diversity distributed mainly within populations, whereas the opposite was true for the Lebanese populations (Caballero et al. 2009).

Some of the 17 *Glu-A^{u}1* alleles showed active *Ay* subunits. As these alleles had a potential future use, these *Ay* genes were molecularly characterized by PCR amplification and subsequent digestion with endonucleases (Gutiérrez et al. 2011). The combination of two restriction patterns revealed variations between the active and inactive alleles, and within each type. This variation, especially the one detected among the active alleles, could enlarge the high-quality genetic pool of modern wheat and be used for bread-making quality improvement in durum and common wheat.

With the aim of evaluating the effect of *T. urartu* germplasm in cultivated wheat, an amphiploid derived from the cross between one accession of *T. urartu* (MG26952), carrying active *Ay* subunits, and durum wheat cv. *Yavaros* was synthesised by chromosome doubling with colchicine and the resulting amphiploid was backcrossed with the cv. *Yavaros* (Alvarez et al. 2009). Twenty lines with 28 chromosomes plus *Ax+Ay* subunits from *T. urartu* were selected. These lines were heterogeneous for the *Glu-B3* locus, whose variation has been associated with gluten strength (see Wrigley et al. 2006 for revision). Our data did not show the significant differences in gluten strength that was shown by the SDS-sedimentation test between the lines with and without the *Glu-A^{u}3* allele (Alvarez et al. 2009). However, two different groups could be identified between these lines: 10 had amber grain and soft gluten whereas the other 10 had red grain and strong gluten. Furthermore, the red lines showed higher pigment content than the amber lines. The data suggested that grain colour could have pleiotropic effects on gluten strength and that, although the seed storage proteins composition must be controlled when new interspecific crosses with *T. urartu* are carried out, grain colour is an important trait for selecting the resulting lines.

Waxy Proteins

Since 2006, 536 accessions of Spanish hulled wheat (29 lines of einkorn, 87 of emmer and 420 of spelt) have been evaluated for variation in waxy proteins (Figure 8). Up to 10 *waxy* alleles (two for the *Wx-A^{m}1* locus, two for the *Wx-A1* locus, three for the *Wx-B1* locus and three for *Wx-D1* locus) were detected in the accessions. In all cases, one of these alleles was clearly predominant while the others were rare or very rare.

In einkorn, waxy protein polymorphism was very low in the current collection, so the electrophoretic analyses revealed only two alleles (Guzmán et al. 2009), although this was larger than that found by other researchers (Urbano et al. 2002; Rodriguez-Quijano et al. 2004), who did not find any polymorphism in collections of wild and cultivated einkorn and *T. urartu*. One of the alleles (*Wx-A^{m}1a*) was previously described by Rodríguez-Quijano et al. (2004). The novel allele (*Wx-A^{m}1a'*) showed less mobility and was only found in one accession. Both alleles showed higher mobility than that of the *Wx-A1a* allele detected in common wheat (cv. *Chinese Spring*).

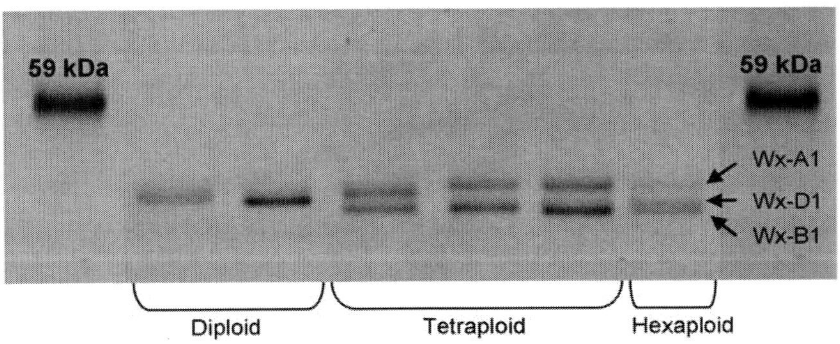

Figure 8. Sample of the variation found for waxy proteins in Spanish hulled wheats.

In contrast, emmer revealed no polymorphism for the *Wx-A1* locus. All the accessions evaluated (87 accessions) showed the *Wx-A1a* allele, which could be similar to the wild allele (Guzman et al. 2011a). For the *Wx-B1* locus, three alleles were detected: *Wx-B1b* (*null* type) that lacks the protein; *Wx-B1c**, which had the same mobility as *Wx-B1c'* detected in the durum wheat cultivar "Mexicali" but lower electrophoretic mobility than the *Wx-B1a* allele, and a new allele, called *Wx-B1g*. This new allele showed a slightly higher mobility than the *Wx-B1a* allele, but lower than the *Wx-B1d* allele found by Yamamori et al. (1995). The *Wx-B1g* allele was the most abundant, being detected in 85 accessions. The *null* type (*Wx-B1b* allele) and the *Wx-B1c** allele were only detected in one line.

Because of the reduced variability for size among the different *waxy* alleles, the variability found was revaluated by PCR-RFLP and DNA sequence analyses (Guzman et al. 2011a). These analyses showed that the *Wx-A1a* allele of emmer wheat is slightly different to the *Wx-A1a* allele of durum wheat, presenting two synonymous changes (silent mutations) in the eleventh exon that did not modify the predicted protein sequence.

For the *Wx-B1* gene, the molecular data confirmed the three alleles previously described by SDS-PAGE separation (Guzmán et al. 2011a). The deduced protein for two of them (*Wx-B1b* and *Wx-B1g*) was the same, although no protein could be detected for the *Wx-B1b* (*null* allele). Both alleles had 15 and 14 differences respectively, with the *Wx-B1a* and *Wx-B1c'* alleles detected in the durum wheat cultivars (Langdon and Mexicali) used as standards. The other *Wx-B1* allele of emmer wheat (*Wx-B1c**) was also different from the standard alleles (*Wx-B1a* and *Wx-B1c'*) with 13 and one amino acid changes, respectively, and two amino acid changes with the *Wx-B1b* and *Wx-B1g* alleles.

The effect of these alleles on the amylose content was also evaluated (Guzmán et al. 2011b), although these data must be considered as preliminary because certain compensation effects have been observed among these genes. All alleles evaluated showed smaller amylose content than the line containing the *Wx-B1a* allele of cv. Langdon. The lowest amylose content was detected in the line that contained the *Wx-B1b* allele, which lacked the *Wx-B1* protein, In contrast, the line containing the *Wx-B1c** allele showed slightly higher amylose content than the line containing the *Wx-B1g* allele, although both lines were not significantly different to the *Wx-B1c'* allele of cv. Mexicali. (Guzmán et al. 2011b).

The electrophoretic analysis of the waxy proteins revealed polymorphism for the three *waxy* loci in the Spanish spelt lines that were evaluated (Guzmán et al. 2010). For the *Wx-A1* locus, two alleles were found: *Wx-A1a,* similar to the allele present in cv. Chinese Spring, and *Wx-A1b* (*null* type), which lacks the protein. The *Wx-A1b* allele was present in 36 accessions (8.57%). However, subsequent studies by Caballero et al. (2008e) showed that one of the *Wx-A1b* (*null* type) alleles did show low activity. This allele has been reassigned as *Wx-A1g*. This result was confirmed with 2D-PAGE, which showed three faint spots in the expected position of the Wx-A1 protein but with less intensity than that shown in the spelt reference sample. The analysis of its sequence indicated the presence of a 160-bp insertion in the fourth intron which showed some characteristics typical of transposable-like elements such as target site duplications and inverted repeats (Guzmán et al. 2012a). Our data revealed that this 160-bp insertion appeared to prevent the complete splicing of all the RNA molecules, thus reducing the yield of correctly processed RNA leading to a significant decrease in the final concentration of Wx-A1 protein. Although further studies, similar to that made by Yamamori (2009) and Yamamori and Yamamoto (2011) with other *Wx* alleles, should be carried out, these data suggest that the allele described in the current survey (*Wx-A1g*) could show different behaviour when in combination with the null alleles for the *Wx-B1* and *Wx-D1* loci, which opens the possibility of being able to obtain quasi-waxy wheats with amylose content above 1% but less than 20%.

For the *Wx-B1* locus, three different alleles were found: *Wx-B1a*, the *null* protein *Wx-B1b* and the *Wx-B1c'* allele, which showed less mobility than *Wx-B1a* allele. Both *Wx-B1b* and *Wx-B1c'* alleles were less common (11.67% and 12.86%) than *Wx-B1a*. The lines evaluated showed a large homogeneity for the *Wx-D1* locus, where 99.52% of the accessions contained the *Wx-D1a* allele. Two additional alleles were detected: the *null* allele (*Wx-D1b*), which was only present in one accession, and a novel *waxy* allele, not previously described, with a slightly lower electrophoretic mobility than the *Wx-D1a* allele. This new allele was provisionally named *Wx-D1g*.

When considering the variation found for the *Wx-A1*, *Wx-B1* and *Wx-D1* loci, eight combinations were detected, where there was a clear dominance by the combination *Wx-A1a*, *Wx-B1a*, *Wx-D1a*; appearing in 69.52% of the samples. In the other loci, three combinations appeared with a relatively high frequency The combination *Wx-A1a, Wx-B1c', Wx-D1a* appeared in 47 accessions, *Wx-A1a, Wx-B1b, Wx-D1a* in 44 and *Wx-A1g, Wx-B1a, Wx-D1a* in 23. The other four combinations were rare with frequencies of less than 5%

Puroindolines

Recently, our studies have included evaluation of the puroindolines found in the ancient and wild wheats. Obviously, the current information on these quality aspects is less than for the previous traits. Furthermore, because the data indicated the absence by deletion of puroindoline genes in the A and B genome of durum wheat (Li et al. 2008), these studies could only be carried out on diploid and hexaploid wheats.

The main difficulty in evaluating grain hardness in diploid wheat (cultivated or wild) is the small size of the grain, together with its low grain yield. If the variability found is used for breeding modern wheat varieties, then the agronomical valuation must be carried out after the transference of this trait because of the clear differences in grain size and yield among the cultivated polyploid wheats and the cultivated or wild diploid wheats. Consequently, the molecular markers that permit detection of novel allelic variants for these puroindoline genes could have been useful tools in the previous evaluation of possible gene sources outlined above.

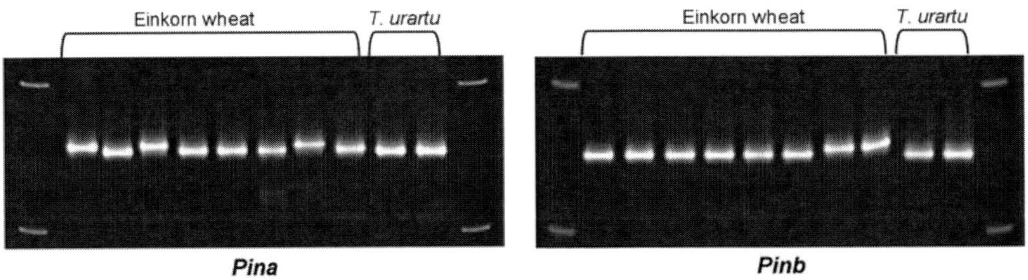

Figure 9. Allelic variation for puroindolines in diploid wheat.

Pina and *Pinb* gene variability was evaluated in the einkorn and *T. urartu* wheat collections (Guzmán et al. 2012b). Four *Pina* and seven *Pinb* alleles were detected in the two collections (Figure 9).

The variability was lower in *T. urartu* than in einkorn. The *T. urartu* lines evaluated were homogeneous for the *Pina* gene, showing two *Pinb* alleles, although one of them ($Pinb\text{-}A^{u}1a$) was only detected in one out of the seven lines evaluated. However, three *Pina* and five *Pinb* alleles were detected in the einkorn lines, including three novel alleles for the *Pinb* locus: $Pinb\text{-}A^{m}1i$, $Pinb\text{-}A^{m}1j$ and $Pinb\text{-}A^{m}1k$.

The effect of these alleles on grain hardness in einkorn was also analysed. Although the data must be considered as preliminary, some differences were found in grain hardness between the different genotypes depending on which *Pin* alleles they showed. Our data suggest that, although further studies must be carried out, these variants could be used to expand the range of available grain textures in durum and common wheat, which would permit the development of new flours adapted to novel uses in the baking and pasta industry.

CONCLUSION

Wheat is an important crop that has been associated with human food for many centuries (Harlan 1981). It is the basis for a diverse range of products, mainly bread, pasta and beer, which are present in most worldwide diets. In some cases, the same wheat type has been used for all of these three different products depending on the geographical or cultural area. For example, durum wheat was used to make bread in the Nile Valley, beer in the Euphrates Valley and pasta in the Yang-tsé Valley (Harlan 1981). In the Mediterranean region, wheat is linked to flour and its consumption (Harlan 1981).

Up until the Industrial Revolution, all baking processes were carried out by hand; this permitted the use of wheat varieties with rheological properties that were very different to the current wheat varieties. Nevertheless, the use of machinery in the baking processes forced people to look for varieties with very specific qualities (Matz 1960). The dough made with these flours must have a certain tolerance to mechanical mixing and over-mixing, a process that is very different from the manual process. At the same time, the world population began to increase considerably, which required the production of larger quantities of food. Consequently, many of the traditional wheats have been neglected, due mainly to their smaller yields and, in many cases, their flour's unsuitability for mechanised production.

This demand for varieties mainly characterized by a high yield and tolerance to mechanical over-mixing remains to this day and has led to the replacement of the numerous local varieties by a few highly productive modern varieties, adapted to the baking and pasta industry. However, these processes and the prevalence of increasingly intensive farming have led to a drastic reduction in the genetic variability found in modern wheats.

From the mid 20th century, plant breeding, based on high-yielding cultivars, has contributed to the narrowing of the genetic base of crops, causing many modern cultivars to be closely related (Esquinas-Alcázar 2005). Subsequently, it is necessary to search for and conserve plant genetic resources to enlarge the gene pool and avoid genetic uniformity, which makes crops vulnerable to biotic and abiotic stresses. In order to face the new challenges arising as a consequence of climate change, sustainable models of agriculture need to be found. Plant genetic resources can be used to enlarge the genetic background of the modern cultivars, although, in many cases, these ancient varieties are underutilized or neglected crops that could be reintroduced into agriculture.

In this chapter, we have evaluated the variability of several collections of ancient and wild wheats for three traits related to technological quality. In general, genetic diversity was high, although a large part of this diversity is at risk of erosion, given that the distribution of the combinations among the evaluated accessions was not random. There was an assumption that, in general, species distribution is conditioned by seed transference among farmers. This produces the homogenization of a crop or cultivar in any one region after several generations, along with their irretrievable loss when these farmers stop farming the crop or cultivar.

The alleles found for these loci were different to those detected in cultivated wheats. The low frequencies of the new alleles confirmed the need to protect and conserve these accessions because it is unlikely that these alleles will be found in other crops. So the loss of these materials would be equivalent to the definitive loss of these alleles. Although the revival of these old crops in modern agriculture is possible, it is more probable that the variation found in these species can be used by the plant-breeders for enlarging the genetic background of the modern wheats.

Acknowledgments

This research was supported since 2001 by several grants from the Spanish Ministry of Science and Innovation, co-financed with European Regional Development Fund (FEDER) from the European Union (AGL2001-2419-C02-02, AGL2004-03361-C02-01, AGL2007-65685-C02-02 and AGL2010-19643-C02-01).

References

Ainsworth, C. C., Clark, J. and Balsdon, J. (1993). Expression, organization and structure of the genes encoding the waxy protein (granule-bound starch synthase) in wheat. *Plant Molecular Biology*, *22*, 67-82.

Alvarez, J. B, Caballero, L. and Martín, L. M. (2007b). Variability for morphological traits and high-molecular-weight glutenin subunits in Spanish spelt lines. *Plant Genetic Resources: Characterization and Utilization*, *5,* 128-130.

Alvarez, J. B, Caballero, L., Nadal, S., Ramírez, M. C. and Martín, A. (2009). Development and gluten strength evaluation of introgression lines of *Triticum urartu in* durum wheat. *Cereal Research Communications*, *37*, 243-248

Alvarez, J. B., Caballero, L., Ureña, P., Vacas, M. and Martín, L. M. (2007a). Characterization and variation of morphological traits and storage proteins in Spanish emmer wheat germplasm (*Triticum dicoccon*). *Genetic Resources and Crop Evolution*, *54*, 241-248.

Alvarez, J. B., Martín, M. A., Caballero, L. and Martín, L. M. (2010). The role of plant genetic resources in the Sustainable Agriculture. In: A. Salazar and I. Rios, (Eds.), *Sustainable Agriculture: Technology, Planning and Management.* New York, USA: Nova Science Publishers, Inc. 145-176.

Alvarez, J. B., Moral, A. and Martín, L. M. (2006). Polymorphism and genetic diversity for the seed storage proteins in Spanish cultivated einkorn wheat (*Triticum monococcum* L. ssp. *monococcum*). *Genetic Resources and Crop Evolution*, *53*, 1061-1067.

Aykroyd, W. R. and Doughty, J. (1970). Wheat in human nutrition. Nutr. Ser. No. 23. Roma, Italy: FAO.

Behall, K. M. and Scholfield, D. J. (2005). Food amylose content affects postprandial glucose and insulin responses. *Cereal Chemistry*, *82*, 654-659.

Bushuk, W. and Zillman, R. R. (1978). Wheat cultivar identification by gliadin electrophoregram. I. Apparatus, method and nomenclature. *Canadian Journal of Plant Science*, *58*, 505-515.

Caballero, L., Alvarez, J. B. and Martín, L. M. (2005). LMW glutenin subunits variation in Persian and Polish wheat. *Czech Journal of Genetics and Plant Breeding*, *41(s.i.)*, 171-174.

Caballero, L., Bancel, E., Debiton, C. and Branlard, G. (2008e). Granule-bound starch synthase (GBSS) diversity of ancient wheat and related species. *Plant Breeding*, *127*, 548-553.

Caballero, L., Martín, L. M. and Alvarez, J. B. (2001). Allelic variation of the HMW glutenin subunits in Spain accessions of spelt wheat. *Theoretical and Applied Genetics*, *103*, 124-128.

Caballero, L., Martín, L. M. and Alvarez, J. B. (2004a). Genetic variability of the low-molecular-weight glutenin subunits in spelt wheat (*Triticum aestivum* ssp. *spelta* L. em Thell.). *Theoretical and Applied Genetics*, *108*, 914-919.

Caballero, L., Martín, L. M. and Alvarez, J. B. (2004b). Variation and genetic diversity for gliadins in Spanish spelt wheat accessions. *Genetic Resources and Crop Evolution*, *51*, 679-686.

Caballero, L., Martín, L. M. and Alvarez, J. B. (2004c). Intra- and interpopulation diversity for HMW glutenin subunits in Spanish spelt wheat. *Genetic Resources and Crop Evolution*, *51*, 175-181.

Caballero, L., Martín, L. M. and Alvarez, J. B. (2007). Agrobiodiversity of hulled wheats in Asturias (North of Spain). *Genetic Resources and Crop Evolution*, *54*, 267-277.

Caballero, L., Martín, L. M. and Alvarez, J. B. (2008a). Relationships between the HMW- and LMW-glutenin subunits and SDS-sedimentation volume in Spanish hulled wheat lines. *Czech Journal of Genetics and Plant Breeding*, *44*, 114-117.

Caballero, L., Martín, L. M. and Alvarez, J. B. (2008b). Genetic diversity in Spanish populations of *Triticum spelta* L. (*escanda*): example of an endangered genetic resource. *Genetic Resources and Crop Evolution*, *55*, 675-682.

Caballero, L., Martín, L. M. and Alvarez, J. B. (2008c). Variation of high molecular weight glutenin subunits in two neglected tetraploid wheat subspecies. *Czech Journal of Genetics and Plant Breeding*, *44*, 140-146.

Caballero, L., Martín, M. A. and Alvarez, J. B. (2008d). Allelic variation for the high- and low-molecular-weight glutenin subunits in wild diploid wheat (*Triticum urartu*) and its comparison with durum wheats. *Australian Journal of Agricultural Research*, *59*, 906-910.

Caballero, L., Martín, M. A. and Alvarez, J. B. (2009). Genetic diversity for seed storage proteins in Lebanon and Turkey populations of wild diploid wheat (*Triticum urartu* Thum. *ex* Gandil.). *Genetic Resources and Crop Evolution*, *56*, 1117-1124.

Caballero, L., Peña, R. J., Martín, L. M. and Alvarez, J. B. (2010). Characterization of Mexican Creole wheat landraces in relation to morphological characteristics and HMW glutenin subunit composition. *Genetic Resources and Crop Evolution*, *57*, 657-665.

Carmona, S., Alvarez, J. B. and Caballero, L. (2010a). Genetic diversity for morphological traits and seed storage proteins in Spanish rivet wheat. *Biologia Plantarum*, *54*, 69-75.

Carmona, S., Caballero, L. and Alvarez, J. B. (2009). Association between the HMW-glutenin subunits and gluten strength characteristics in khorassan wheat lines. *Czech Journal of Genetics and Plant Breeding*, *45*, 169-172.

Carmona, S., Caballero, L., Martín, L. M. and Alvarez, J. B. (2010b). Genetic diversity in khorasan and rivet wheat by assessment of morphological traits and seed storage proteins. *Crop and Pasture Science*, *61*, 938-944.

Chao, S., Sharp, P. J., Worland, A. J., Warham, E. J., Koebner, R. M. D. and Gale, M. D. (1989). RFLP-based genetic maps of wheat homoeologous group 7 chromosomes. *Theoretical and Applied Genetics*, *78*, 495-504.

Chen, F., He, Z. H., Xia, X. C., Xia, L. Q., Zhang, X. Y., Lillemo, M. and Morris, C. (2006). Molecular and biochemical characterization of puroindoline a and b alleles in Chinese landraces and historical cultivars. *Theoretical and Applied Genetics*, *112*, 400-409.

Ciaffi, M., Lafiandra, D., Turchetta, T., Ravaglia, S., Bariana, H., Gupta, R. B. and MacRitchie, F. (1995). Breadbaking potential of durum wheat lines expressing both *x*-and *y*-type subunits at the *Glu-A1* locus. *Cereal Chemistry, 72,* 465-469.

D'Ovidio, R. and Masci, S. (2004). The low-molecular-weight glutenin subunits of wheat gluten. *Journal of Cereal Science*, *39*, 321-339.

Dvořák, J. and Zhang H. B. (1990). Variation in repeated nucleotide sequences sheds light on the phylogeny of the wheat B and G genome. *Proceedings of the National Academic of Sciences USA*, *87*, 9640-9644.

Dvořák, J. and Zhang H. B. (1992). Reconstruction of the phylogeny of the genus *Triticum* from variation in repeated nucleotide sequences. *Theoretical and Applied Genetics*, *84*, 419-429.

Dvořák, J., McGuire, P. E. and Cassidy, B. (1988). Apparent source of the A genomes of wheats inferred from polymorphism in abundance and restriction fragment length of repeated nucleotide sequences. *Genome*, *30*, 680-689.

Esquinas-Alcázar, J. (2005). Protecting crop genetic diversity for food security: political, ethical and technical challenges. *Nature Reviews Genetics*, *6*, 946-953.

FAO. (1995). *Report of the Sixth Session of the Commission on Plant Genetic Resources*, 19-30 June 1995, Document CPGR-6/95 REP. Rome, Italy: Food and Agriculture Organization.

FAO. (1996). The state of the World's plant genetic resources for food and agriculture. Rome, Italy: *Food and Agriculture Organization.*

FAO. (2010). [http://faostat.fao.org/site/291/default.aspx].

Frankel, O. H. (1970). Genetic conservation in perpective. In: O.H. Frankel, and E. Bennett, (Eds.), *Genetic resources in plants: their exploration and conservation.* Oxford and Edinburgh, UK: Blackwell Publishers. 469-489.

Frankel, O. H. and Hawkes, J. G. (1975). Genetic resources: the past ten years and the next. In: O.H. Frankel and J.G. Hawkes, (Eds.), *Crop genetic resources for today and tomorrow*. Cambridge, UK: Cambridge University Press. 1-11.

Gautier, M. F., Aleman, M. E., Guirao, A., Marion, D. and Joudrier, P. (1994). *Triticum aestivum* puroindolines, two basic cystine-rich seed proteins: cDNA sequence analysis and developmental gene expression. *Plant Molecular Biology*, *25*, 43-57.

Gautier, M. F., Cosson, P., Guirao, A., Alary, R. and Joudrier, P. (2000). Puroindoline genes are highly conserved in diploid ancestor wheats and related species but absent in tetraploid *Triticum* species. *Plant Science*, *153*, 81-91.

Gedye, K. R., Morris, C. F. and Bettge, A. D. (2004). Determination and evaluation of the sequence and textural effects of the puroindoline a and puroindoline b genes in a population of synthetic hexaploid wheat. *Theoretical and Applied Genetics*, *109*, 1603–1597.

Giroux, M. J. and Morris, C. F. (1997). A glycine to serine change in puroindoline b is associated with wheat grain hardness and low levels of starch-surface friabilin. *Theoretical and Applied Genetics*, *95*, 857-864.

Giroux, M. J. and Morris, C. F. (1998). Wheat grain hardness results from highly conserved mutations in the friabilin components puroindoline a and b *Proceedings of National Academic of Science USA*, *95*, 6262-6266.

Goldsbrough, A. P., Bulleid, N. J., Freedman, R. B. and Flavell, R. B. (1989). Conformational differences between two wheat (*Triticum aestivum*) 'high-molecular-weight' glutenin subunits are due to a short region containing six amino acid differences. *Biochemical Journal*, *263*, 837-842.

Greenwell, P. and Schofield, J. D. (1986). A starch granule protein associated with endosperm softness in wheat. *Cereal Chemistry*, *63*, 379-380.

Gutiérrez, M. V., Guzmán, C., Martín, L. M. and Alvarez, J B. (2011). Molecular characterization of the *Glu-Ay* gene from *Triticum urartu* for its potential use in quality wheat breeding. *Plant Genetic Resources: Characterization and Utilization, 9*, 334-337.

Guzmán, C., Caballero, L. and Alvarez J. B. (2009). Variation in Spanish cultivated einkorn wheat as determined by morphological traits and waxy proteins. *Genetic Resources and Crop Evolution*, *56*, 601-604.

Guzmán, C., Caballero, L. and Alvarez, J. B. (2011a). Molecular characterization of the *Wx-B1* allelic variants identified in cultivated emmer wheat and comparison with those of durum wheat. *Molecular Breeding*, *28*, 403-411.

Guzmán, C., Caballero, L., Alvarez, J. B. and Yamamori, M. (2011b). Amylose content and starch properties in emmer and durum wheat lines with different waxy proteins composition. *Journal of the Science of Food and Agriculture*, *91*, 1625-1629

Guzmán, C., Caballero, L., Martín, M. A. and Alvarez, J. B. (2012b). Molecular characterization and diversity of the *Pina* and *Pinb* genes in cultivated and wild diploid. *Molecular Breeding*, (in press, DOI: 10.1007/s11032-011-9599-1).

Guzmán, C., Caballero, L., Moral, A. and Alvarez, J. B. (2010). Genetic variation for waxy proteins and amylose content in Spanish spelt wheat. *Genetic Resources and Crop Evolution*, *57*, 721-725.

Guzmán, C., Caballero, L., Yamamori, M. and Alvarez, J. B. (2012a). Molecular characterization of a new *waxy* allele with partial expression in spelt wheat. *Planta*, (in press, DOI: 10.1007/s00425-011-1577-7).

Hammer, K., Arrowsmith, N. and Gladis, T. (2003). Agrobiodiversity with emphasis on plant genetic resources. *Naturwissenschaften*, *90*, 241-250.

Hammer, K., Heller, J. and Engels, J. (2001). Monographs on underutilized and neglected crops. *Genetic Resources and Crop Evolution*, *48*, 3-5.

Harberd, N. P., Bartels, D. and Thompson, R. D. (1986). DNA restriction fragment variation in the gene family encoding high-molecular-weight (HMW) glutenin subunits of wheat. *Biochemical Genetics*, *24*, 579-596.

Harlan, J. R. (1981). The early history of wheat: earliest traces to the sack of Rome. In: L. T. Evans and W. J. Peacock, (Eds.), *Wheat science: today and tomorrow*. Cambridge, U.K.: Cambridge University Press. 1-29.

Jackson, E. A., Holt, L. M. and Payne, P. I. (1983). Characteization of high-molecular-weight gliadin and low-molecular-weight glutenin subunits of wheat endosperm by two-dimensional electrophoresis and the chromosomal localization of their controlling genes. *Theoretical and Applied Genetics*, *66*, 29-37.

James, M. G., Denyer, K. and Myers, A. M. (2003). Starch synthesis in the cereal endosperm. *Current Opinion in Plant Biology*, *6*, 215-222.

Kerber, E. R. and Rowland, G. G. (1974). Origin of the free-threshing character in hexaploid wheat. *Canadian Journal of Genetics and Cytology*, *16*, 145-154.

Kuroda, A., Oda, S., Miyazawa, S. and Seko, H. A. (1989). Methods of measuring amylose content and its variation in Japanese wheat cultivars and Kanto breeding lines. *Japanese Journal of Breeding*, *39*, 142-143.

Lafiandra, D., Benedettelli, S., Margiotta, B., Spagnoletti-Zeuli, P. L. and Porceddu, E. (1990). Seed storage-protein and wheat genetic resources. In: J.P. Srivastava, and A.B. Damania, (Eds.), *Wheat genetic resources: meeting diverse need*. Aleppo, Syria: ICARDA. 73-87.

Li, W., Huang, L. and Gill, B. S. (2008). Recurrent deletions of puroindoline genes at the grain *Hardness* locus in four independent lineages of polyploid wheat. *Plant Physiology*, *146*, 200-212.

Li, Z., Chu, X., Mouille, G., Yan, L., Kosar-Hashemi, B., Hey, S., Napier, J., Shewry, P., Clarke, B., Appels, R., Morell, M.K. and Rahman, S. (1999). The localization and expression of the class II starch synthases of wheat. *Plant Physiology*, *120*, 1147-1156.

Li, Z., Sun, F., Xu, S., Chu, X., Mukai, Y., Yamamoto, M., Ali, S., Rampling, L., Kosar-Hashemi, B., Rahman, S. and Morell, M. K. (2003). The structural organisation of the gene encoding class II starch synthase of wheat and barley and the evolution of the genes encoding starch synthases in plants. *Functional and Integrative Genomics*, *3*, 76-85.

Lillemo, M. and Morris, C. F. (2000). A leucine to proline mutation in puroindoline b is frequently present in hard wheats from Northern Europe. *Theoretical and Applied Genetics*, *100*, 1100-1107.

Liu, C. Y. (1995). Identification of a new low Mr glutenin subunit locus on chromosome 1B of durum wheat. *Journal of Cereal Science*, *21*, 209-213.

Martín, M. A., Martín, L. M., and Alvarez, J. B. (2008). Polymorphisms at the *Gli-A^{u}1* and *Gli-A^{u}2* loci in wild diploid wheat (*Triticum urartu*). *Euphytica*, *163*, 303-307.

Massa, A. N., Morris, C. F. and Gill, B. S. (2004). Sequence diversity of puroindoline-a, puroindoline-b, and the grain softness protein genes in *Aegilops tauschii* Coss. *Crop Science*, *44*, 1808-1816.

McFadden, E. S. and Sears, E. R. (1946). The origin of *Triticum spelta* and its free-theshing hexaploid relatives. *Journal of Heredity*, *37*, 81-87.

Metakovsky, E. V., Novoselskaya, A. Y., Kopus, M. M., Sobko, T. A. and Sozinov, A. A. (1984). Blocks of gliadin components in winter wheat detected by one-dimensional polyacrylamide gel electrophoresis. *Theoretical and Applied Genetics*, *67*, 559-568.

Miflin, B. J., Field, J. M. and Shewry, P. R. (1983). Cereal storage proteins and their effect on technological properties. In: J. Daussant, J. Mossé, and J. Vaughan, (Eds.), *Seed Proteins*. London, UK: Academic Press. Inc. 255-319.

Morita, N., Maeda, T., Miyazaki, M., Yamamori, M., Miura, H. and Ohtsuka, I. (2002). Dough and baking properties of high-amylose and waxy wheat flours. *Cereal Chemistry*, *79*, 491-495.

Morris, C. F. (2002). Puroindolines: the molecular genetic basis of wheat grain hardness. *Plant Molecular Biology*, *48*, 633-647.

Morris, C. F. and Rose, S. P. (1996). Wheat. In: R. J. Henry, and P. S. Kettlewell, (Eds), *Cereal grain quality*. New York, USA: Chapman and Hall. 3-54.

Nieto-Taladriz, M. T., Rodríguez-Quijano, M. and Carrillo, J. M. (2000). Polymorphism of waxy proteins in Spanish durum wheats. *Plant Breeding*, *119*, 277-279.

Padulosi, S., Hodgkin, T., Williams, J. T. and Haq, N. (2002). Underutilized crops: trends, challenges and opportunities in the 21st Century. In: J. M. M., Engels, V., Ramanatha-Rao, A. H. D. Brown, and M. T. Jackson, (Eds.), *Managing Plant Genetic Diversity*. Wallingford, UK: CABI Publishing. 323-338.

Payne, P. I. (1987). Genetics of wheat storage proteins and the effects of allelic variation on bread-making quality. *Annual Review of Plant Physiology*, *38*, 141-153.

Payne, P. I., Holt, L. M., Jackson, E. A. and Law, C. N. (1984). Wheat storage proteins: their genetics and their potential for manipulation by plant breeding. *Philosophical Transactions of the Royal Society of London B*, *304*, 359-371.

Peña-Chocarro, L. (1996). In situ conservation of hulled wheat species: the case of Spain. In: S. Padulosi, K. Hammer, and J. Heller, (Eds.), *Hulled wheats*. Rome, Italy: International Plant Genetic Resources Institute. 129-146.

Pflüger, L. A., Martín, L. M. and Alvarez, J. B. (2001). Variation in the HMW and LMW glutenin subunits from Spanish accessions of emmer wheat (*Triticum turgidum ssp. dicoccum* Schrank). *Theoretical and Applied Genetics*, *102*, 767-772.

Pomeranz, Y. and Williams, P. C. (1990). Wheat hardness: its genetic, structural and biochemical background, measurement, and significance. *Advances in Cereal Science and Technology*, *10*, 471-546.

Qualset, C. O., McGuire, P. E. and Wargurton, M. L. (1995). Agrobiodiversity: key to agricultural productivity. *California Agriculture*, *49*, 45-49.

Rahman, S., Jolly, C. J., Skerritt, J. H. and Wallosheck, A. (1994). Cloning of a wheat 15-kDa grain softness protein (GSP). GSP is a mixture of puroindoline-like polypeptides. *European Journal of Biochemistry*, *223*, 917-925.

Rodríguez-Quijano, M., Lucas, R. and Carrillo, J. M. (2003). Waxy proteins and amylose content in tetraploid wheats *Triticum dicoccum* Schulb, *Triticum durum* L. and *Triticum polonicum* L. *Euphytica*, *134*, 97-101.

Rodríguez-Quijano, M., Nieto-Taladriz, M. T. and Carrillo, J. M. (1998). Polymorphism of waxy proteins in Iberian hexaploid wheats. *Plant Breeding*, *117*, 341-344.

Rodríguez-Quijano, M., Vázquez, J. F. and Carrillo, J. M. (1990). Variation of high molecular weight glutenin subunits in Spanish landraces of *Triticum aestivum* ssp. *vulgare* and ssp. *spelta*. *Journal of Genetics and Breeding*, *44*, 121-126.

Rodriguez-Quijano, M., Vázquez, J. F. and Carrillo, J. M. (2004). Waxy proteins and amylose content in diploid *Triticeae* species with genomes A, S and D. *Plant Breeding*, *123*, 294-296.

Shewry, P. R., Parmar, S. and Field, J. M. (1988). Two dimensional electrophoresis of cereal prolamins: applications to biochemical and genetic analyses. *Electrophoresis*, *9*, 727-737.

Shimbata, T., Nakamura, T, Vrinten, P., Saito, M., Yonemaru, J., Seto, Y. and Yasuda, H. (2005). Mutations in wheat starch synthase II genes and PCR-based selection of a SGP-1 null line. *Theoretical and Applied Genetics*, *111*, 1072-1079.

Singh, N. K. and Shepherd, K. W. (1988). Linkage mapping of genes controlling endosperm storage proteins in wheat. I. Genes on the short arms of group 1 chromosomes. *Theoretical and Applied Genetics*, *75*, 628-641.

Strehlow, W., Hertzka, G. and Weuffen, W. (1994). Aspetti nutrizionali. In: P. Perrino, D. Semeraro, and G. Laghetti, (Eds.), *Le caratteristiche dietetiche del farro. Un cereale della salute*. Bari, Italy: CNR. 52-66.

Tellez-Molina, R. and Alonso-Peña, M. (1952). *Los trigos de la Ceres Hispanica de Lagasca y Clemente*. Madrid, Spain: Instituto Nacional de Investigaciones Agronómicas.

Urbano, M., Margiotta, B., Colaprico, G. and Lafiandra, D. (2002). Waxy proteins in diploid, tetraploid and hexaploid wheats. *Plant Breeding*, *121*, 465-469.

Wilson, E. O., ed. (1988). *Biodiversity*. National Academy Press. Washington, DC.

World Commission on Environment and Development. (1987). *Our common future*. Oxford UK: Oxford University Press.

Wrigley, C., Békés, F. and Bushuk, W. (2006). Gliadin and glutenin. *The unique balance of wheat quality*. St Paul, MN, USA: AACC International.

Wu, X., Zhao, R., Wang, D., Bean, S. R., Seib, P. A., Tuinstra, M. R., Campbell, M. and O'Brien, A. (2006). Effects of amylose, corn protein and corn fiber contents on production of ethanol from starch-rich media. *Cereal Chemistry*, *83*, 569-575.

Yamamori, M. (2009). Amylose content and starch properties generated by five variant *Wx* alleles for granule-bound starch synthase in common wheat (*Triticum aestivum* L.). *Euphytica*, *165*, 607-614.

Yamamori, M., Fujita, S., Hayakawa, K., Matsuki, J. and Yasui, T. (2000). Genetic elimination of a starch granule protein, SGP-1, of wheat generates an altered starch with apparent high amylose. *Theoretical and Applied Genetics*, *101*, 21-29.

Yamamori, M., Kato, M., Yui, M. and Kawasaki, M. (2006). Resistant starch and starch pasting properties of a starch synthase IIa-deficient wheat with apparent high amylose. *Australian Journal of Agricultural Research*, *57*, 531-535.

Yamamori, M., Nakamura, T., Endo, R. and Nagamine, T. (1994). Waxy protein deficiency and chromosomal location of coding genes in common wheat. *Theoretical and Applied Genetics*, *89*, 179-184.

Yamamori, M., Nakamura, T. and Nagamine, T. (1995). Polymorphism of two waxy proteins in the emmer group of tetraploid wheat, *Triticum dicoccoides*, *T. dicoccum*, and *T. durum*. *Plant Breeding*, *114*, 215-218.

Yamamori, M. and Yamamoto, K. (2011). Effects of two novel *Wx-A1* alleles of common wheat (*Triticum aestivum* L.) on amylose and starch properties. *Journal of Cereal Science*, *54*, 229-235.

Zeng, M., Morris, C. F., Batey, I. I. and Wrigley, C. W. (1997). Sources of variation for starch gelatinization, pasting, and gelation properties in wheat. *Cereal Chemistry*, *74*, 63-71.

Zohary, D. and Hopf, M. (1988). *Domestication of plants in the Old World*. Oxford, UK: Oxford Science Publications.

In: Genetic Diversity: New Research
Editors: J. A. C. Amaya and M. M. F. Jimenez
ISBN: 978-1-62081-598-4
© 2012 Nova Science Publishers, Inc.

Chapter 5

THE GENETIC HISTORY OF MOROCCO

Kaoutar Bentayebi[1,2], Antònia Picornell[2] and Saaïd Amzazi[1]
[1]University of Mohammed V. Agdal, Faculty of sciences,
Forensic Genetic unit, Morocco
[2]University of Balearec Islands, Laboratory of Genetic, Spain

ABSTRACT

Morocco is a country located in northwest Africa. It has a long coastline on the Atlantic Ocean that reaches past the Strait of Gibraltar into the Mediterranean Sea. This layered country has a large history that left a mark on its current makeup. It's spread initially from Carthaginians and Romans to the coming of Islam that introduced the dominant religion and language, which are still both religious and political issues. This chapter reflects that layering, through population genetic studies. A wide range of molecular markers have been used to study the Moroccan population as well as its three ethnical groups (Arabic, Berbers and Sahrawis) and to compare it to other populations, in order to well understand and reconstitute the history of this challenging area.

The results lead to an important conclusion about the origins of the Moroccan population. Thus the male gene pool, studied by means of the Y chromosomal STR and SNP, seems to have a Middle East provenance while the feminine lineage studied through the mitochondrial DNA present a European and North African kinship. Those results were supported by some studies on the X chromosome wish spends two thirds of its life time in females. The differentiated contribution of males and females to the genetic landscape of the Moroccan population was observed with a higher migration rate in males than in females.

The autosomal markers shows a Middle East and North African predominance in the genetic pool of the Moroccan population, wish reflect the importance of the Gibraltar Strait as a genetic Barrier. In the other hand, the results show that there is a small difference between Arabs, Berbers and Sahrawis. The Arabic population is closer to Middle East and North-African populations than Berber, the autochthonous population, which is closest to Europeans.

THE GEOGRAPHICAL AND CULTURAL CONTEXT OF MOROCCO

Kingdom of Morocco is strategically placed at the geographical intersection between Europe, Asia, and the rest of Africa. Morocco is situated in northwestern Africa with coasts on the Mediterranean, stretching northwards from Tangier to the Algerian borders. The Atlantic Ocean is running for over 1500 km from the Straits of Gibraltar to the Mauritanian boarder. Just 15 km separate the northern coast of Morocco from the Gibraltar, implying that the kingdom is the most occidental and westerly of the North African and Arabic states. It is one the key African countries of the Maghreb (Morocco, Algeria, Tunisia, Libya and Mauritania). Morocco's area is 722,550 square km makes it slightly bigger than two times the Germany's area. The population of just over 32 million compares with just eight million at independence in 1956 [1].

The country's geography is diverse and divided into five regions by a stretch of rugged mountainous landscapes. The Rif Mountains parallel the Mediterranean coast from Ceuta to Melilla, with large areas of flanking plateau, valleys, and rich coastal plains. The coastal plains of western Morocco stretch from the northern city of Tangier southward to Essaouira, where the High Atlas Mountains interrupt the plains that reach the sea at that point. The mountains are divided into three areas, the Middle Atlas, the High Atlas, and the Anti-Atlas ranges. To the south of the Rif and the Sebou (or Sebu) River basin is the Middle Atlas, separated from the eastern fringes of the High Atlas farther south by the valley of the Abid River. The High Atlas extend some 450 miles in length and forty miles in width and divide the country into two distinct climatic zones—one that receives the westerly winds from the Atlantic and one that is influenced by the proximity of the Sahara. The Anti-Atlas to the south, the homeland of the Amazigh Berbers, is connected with the High Atlas by means of the volcanic massif of Siroua.

Farther to the south of the Anti-Atlas are oases *(smara),* seasonal rivers, and little island towns. This complex network of mountains, plateaus, valleys, rivers, seas, and oceans brings about climatic diversities that are difficult to summarize.

In broad terms, the north is considered warm and temperate and the south semiarid. The Mediterranean climate in the western and northern Morocco is strongly modified by local factors such as the mountain ranges and proximity to the seas.

The Rif and the Atlas chains act as a barrier between western and eastern parts of the country. The westerly winds that come off the Atlantic in the winter bring snow and rain to the western flanks of the mountains, leaving little to the eastern steppes and plains.

The temperature does not fall below 46 degrees except in the Middle Atlas ranges, where the temperature may reach zero degrees and the weather may be harsh. Winter weather conditions may vary from severely cold and rainy to inadequate rainfall, which is usually accompanied by poor agricultural production.

Rainfall in spring is usually slight and irregular, and the mean temperature is about 68 degrees. Around the Mediterranean coast, the climate is generally mild and sunny. The land around the Atlantic coast has enough rainfall to sustain land fertility and good agricultural yield. Heavy rainfalls occur in some parts of the Rif Mountains, the Middle Atlas ranges, and some areas of Tangier closer to the Strait of Gibraltar, but most parts of the Sahara in the south experiences an average rainfall of less than 4 inches per year.

Summer is generally hot and dry, except along the Mediterranean coast, where it is humid, and in the Rif and Atlas ranges, where the temperatures are cool, although the sun may be fierce. Summer rain occurs only in the mountains and is accompanied by violent storms in June and July—similar to weather often experienced in Appalachia of the United States between June and August. Although it is somewhat cool during winter, the Moroccan desert is unbearably hot in summer. This is largely because of the anticyclone from the North Atlantic close to Portugal, the winds from the northeast *(levante),* and the occasional desert winds, such as the sirocco and chergui. The driest months are between April and October. The climate, of course, influences the overall culture of the inhabitants—their agricultural production, settlement patterns, and so on [2].

Historical Background

The First Tracks of the Moroccan Population

Men left many tracks during the prehistorically period, arguing for a very old population, obviously supplied by a good climate. In the Acheuléen (Paleolithic inferior), some tracks of at least 700 000 years old evidenced the presence of an earlier human activity (*Neanderthalian*). The main source of life and livelihood of these Neanderthal people was picking and hunting. In Casablanca-Salé region, many primitive hand stone as the pebble, bifacial, cleaver… were discovered. The Moustérien (Middle Paleolithic) between 40 and 120 mil years ago BC (*Homo sapiens*) was characterized by the conception and development of some tools. During this period, some scrappers were found especially in Jbel Irhoud in Safi region [3] where we find all the lithic industry. The Aterien period (the provenance of this name is Bir el-Ater in Algeria) [4], developed between 40.000 and 20.000 (*Homo sapiens sapiens*) years before the present, is a civilization originating from the North of Africa. This period is characterized by the harnessing of some resistant cutting edge tools like the flint. A broad climatic variation happened in this era, since the fauna's extinction and the flora's drying, turning the land into the desert cutting Africa in two. Only in the Paleolithic superior, when the Homo sapiens managed the iberomaurusian industry that a true tracks of populations were found in Taforalt. The tools found are from 30 to 20 mil years old BC and the funeral rituals were identified: died persons had the body in a horizontal position with painted bones.

These populations stayed until 9000 BC, and then they became extinct by the arrival of the first ancestors of the actual Berber population: the Capsien (name derived from the old city of Capsa or Gafsa today) came from the east (as unearthed by the linguistically studies, classifying the Egyptian and Berbers in the same cluster). Some Neolithic sites, showing settlement and beginning of agriculture, were discovered near to Skhirat (Necropole of Rouazi-Skhirat) [5] and to Tetouan (cave of Kaf Taht el Ghar and of Ghar Kahal) [6].

Empires Succession

The Phoenician, traders, set up their first business in the Moroccan coasts since the XIth century BC and put a port-counter like Tingi (Tanger) or Lixus (Larache). The region emphasized, since the foundation of Carthage (in Tunisia, East of Maghreb). The punic influence was noticed about mil years in Morocco, in its relation with the local Berber tribes leaders: indeed since the VIth century, the Carthaginians gold diggers (gold from Atlas), murex (shell found in Mogador for example, that produce a dye of the same colour), makes business with the Moroccan residents.

From the IVth century BC, the first politic organization of the country was the Kingdom of Mauritania in north of Morocco, resulting from the federation of different Berber tribes that benefits from the punic influence.

When Romans arrived about the IInd century BC, after the Carthage's destruction they formed an alliance with the kingdom of Mauritania, which allows fighting against the numidian leader Jugurtha. Mauritania became a friend kingdom, "customer state" that if depend tightly of Roma and take part of all the intern quarrel of the empire, stays independent. In 40 BC, the kingdom of Maures lost its king. Caligula his assassin copes with war of Aedemon: the revolution was put down in four years and in 46 BC, the emperor Claude annexes the kingdom which becomes the Mauritania tingitane (Tanger from Tingi). The roman domination is limited to the northern plains (until the region of Volubilis near to Meknes) and the empire does not look for controlling tightly the region: it seems that the autonomous and peaceful Berber tribes overlapped in the roman possessions. For all that Roma must fight constantly against the highlanders Berbers.

As the rest of North Africa, Mauritania tangitane will be Christianized. In the IIIrd century the empire remote. It is the case of North Africa, particularly Morocco: The Mauritania tangitane was reduced only to Tingi city in the northern region. The southern cities are left, including Volubilis [7].

The Arabic Conquest

The Arab conquest led to the extension of the Arab-Islamic Empire from Persia to Morocco in the early eighth century. The Arab conquest put an end to the Christian Byzantine power in North Africa, and as a result converted most of the Jewish and Christian natives to Islam.

The spread of Islam in the region was not accompanied by Arabisation at the beginning. The Berbers, who constituted the major population, continued to speak their language. In fact, the Arabs who first arrived in the region were generally sedentary urban people; according to Laroui, they were about 20.000 people. The second wave of Arab conquerors were tribes that had been expelled from Egypt in 1050 CE, namely the Banu Hilal, the Banu Maaqil and the Banu Suleim [8]. Because these tribes were nomads like Berbers, they were accepted by the Berber community, and this engendered the assimilation of Berbers to Arab culture.

Islam was established in Morocco in the eighth century, following the vast Islamic conquest which spread as far as Spain in the ninth and tenth centuries. Subsequently, Moroccans adopted Islam as their religion. Later, Arabic became the main language used in the coastal areas, while Berber was limited to the mountainous and rural regions.

The Arab-Muslim conquerors adopted a language policy that enabled them to spread Arabic and Islamic cultural values. The remarkable relation between Arabic and Islam, as mentioned in the Qur'an itself, made this spread and dominance of Arabic unavoidable. In fact, to understand the Qur'an, one has to be literate in Arabic. Qur'anic schools were opened to train prospective religious and political leaders. These facts led gradually to the domination of Arabic over Berber.

In fact, with the advent of Islam, many non-Arab countries adopted this religion. In contrast with the pre-Islamic period, which was full of tribal and civil wars, the Islamic period was characterized by the integration and assimilation of many nations into Islam. This golden age of Arab-Islamic culture lasted until the invasion of Baghdad in 1258 by the Ottoman Empire, and the isolation of the Islamic world, characterized by religious conservatism and dogmatism. In that period, Arab-Islamic culture suffered a deep stagnation which was followed by a real regression, represented by a closed society with no innovation in its cultural production. Even technical and scientific production regressed between the twelfth and the thirteenth centuries.

We are dealing with Arab-Islamic culture, and not Arab culture, because the most important remnant of the pre-Islamic culture is the Arabic language; with the advent of Islam, the important thing is the Qur'an, not Arabic. The Arab-Islamic culture is relatively vigorous to the extent that it has been open to the influences of Greek, Persian, Hindu, Judaic and Christian cultures.

Subsequently, due to religious dogmatism and conservatism, this culture suffered a serious decline. Its re-birth took place after the campaign of Bonaparte in Egypt in 1798 and the openness of the Arab-Islamic culture to the modern world [9]. The Nahda period (renaissance) started in Syria and Lebanon before it gained Egypt, as a result of the occupation of the Ottoman empire (of the Middle East and North Africa, except for Morocco) and of the influence of Bonaparte. The Nahda movement, which is based on mythical and religious fundamentals, is equivalent to the Western renaissance, which is a literary, artistic and scientific movement which marked the revival and renewal of European civilization in the 16th century. During the Nahda period, intellectuals like Mohamed Abdou in Egypt established schools for the army, encouraged translation, created newspapers to the extent that Egypt became like a part of Europe because of the re-birth of arts and science and because of the presence of Europeans in Egypt.

The Nahda period, which lasted from 1845 to 1905, sought to modernize Islam and society, and started with the generalization of education. Thus, thinkers and religious reformers like Mohamed Abou, Jamal Eddine El Afghani and Mohamed Iqbal rejected backward views and traditions and encouraged the modernization of the Arab-Islamic society through al-/tihad (a reinterpretation of Islam taking into account modernity and social change).

The second half of the nineteenth century and the first half of the twentieth century were marked by colonization and the struggle for independence and by a social revolution against decline. After independence, a new Nahda was launched as the ruling elite endeavored to upgrade arts, sciences, and education and to develop the economy and modernize society.

Starting from 1574, Tunisia and Algeria, unlike Morocco, were occupied by the Ottoman Empire and stayed so until the French colonization era. Note that the predominance of Islam has been a fact since 647AD, hence the adoption of the Muslim lifestyle by the majority of the population.

Despite many European invasions, Morocco has succeeded throughout history in assimilating other cultures without losing its own authenticity.

Demolinguistic Insight

The country's population stands at 29.1 million in 2000. Morocco has the most complex linguistic situation compared to the three other North African countries: the classical Arabic as the administrative language and the dialectal or Moroccan Arabic as the colloquial language, the Berber language called Amazighe (the Tarfit in Rif region, the Tamazight in the Middle east, the Tachelhit in Souss) talked for about 40% of Moroccan, the Hassania dialect spoken in Sahara; the French as the second official language and sometimes the Spanish in the northern and southern regions of the kingdom and last but not least English as third language of an increasing important through years. However, the only official language in Morocco is the Arabic [10]. The repartition of the Moroccan population is remarkably unequalled: 90% of residents settle in the north part of Morocco.

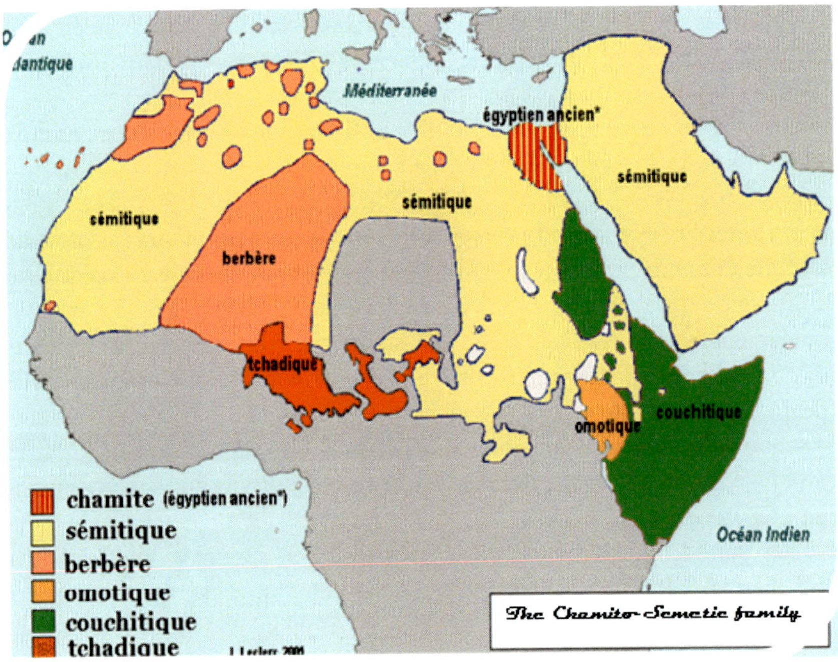

Figure 1. The linguistic diversity of the African continent and western Asia [12].

The country's capital Rabat (1.7 million resident in 2007) is classified after Casablanca and its suburbs (3.2 million of resident), but before Fes (719 000 resident), Marrakech (644 000 residents), Meknes (484 000 residents), Tetouan (484 000 resident), Agadir (420 resident) and Tangier (410 000 resident). The casi totality of the Moroccan population is Sunit muslims (98.7%) [11].

Genetic Polymorphism

The nature is characterized by the wide range of organism's diversity from the desert or glacial landscapes to the deeper points of oceans or higher points of mountains. The life settles in all those places through a continuous process of adaptation over time, during mils or millions of years. This continuous process of variation and adaptation, come sometimes along with an increasing biologic complexity, called the *biologic evolution*. This process is related to the Big Bang explosion, 15 million years ago. The beginning of life on earth was 4.500 million years ago and remained microscopic. In the energetic chemistry of early Earth, a molecule gained the ability to make copies of itself (*replication*) and to be adapted to the environment. Life was developed from the microscopic organisms like viruses and bacteria to the macroscopic organisms as animals and plants. All this diversity, based on the evolutive and replicative ability of nucleic acids, is responsible of life's succession, involving the organism's adaptation to environment.

Within human cells, autosomal markers are distributed throughout all our chromosomes and represent the accumulated inheritance from all our ancestors, unlike Y chromosome (male ancestry) or mitochondrial (female ancestry) markers (Figure 2).

Any species of organism in nature lives in a form of population. A population of organisms is characterized by some sort of cooperative or inhibitory interaction between members of the population. Thus, the rate of growth of a population depends on the population size or density in addition to the physical environment in which the population is placed. When population density is below a certain level, the members of the population often interact cooperatively, while in a high density they interact inhibitorily. In organisms with separate sexes, mating between males and females is essential for the survival of a population.

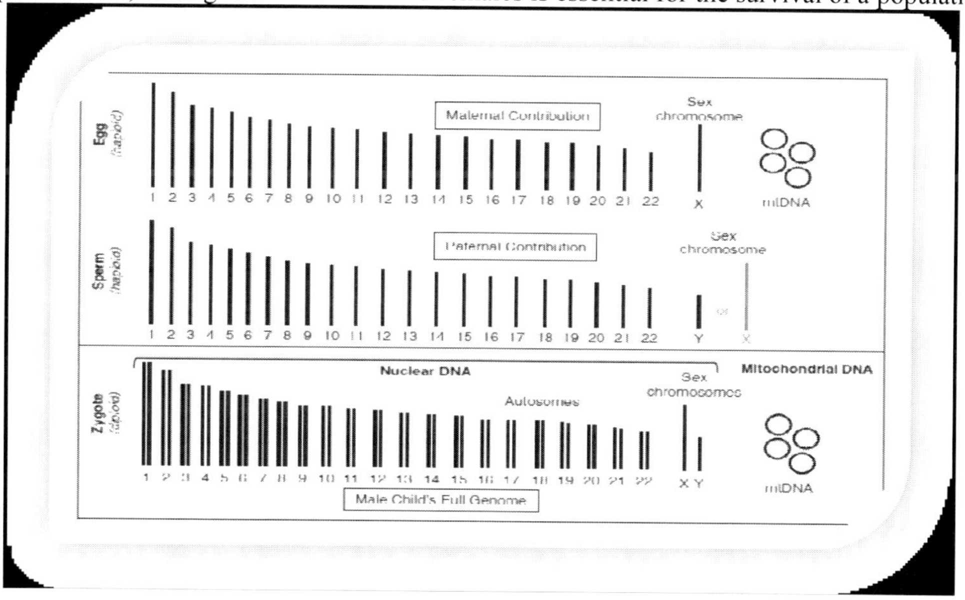

Figure 2. Human genome and inheritance. The haploid complement of chromosomes from a female's egg combines with the haploid chromosomal complement of a male's sperm to create a fully diploid zygote, which eventually develops into a child whose nongamete cells each contain the same genome. Half of the 22 autosomes come from each parent, while mtDNA is inherited only from the mother. The father's contribution of either an X or a Y chromosome determines the child's sex [13].

Interactions between individuals are not confined within a single species but also occur between different species. The survival of a species generally depends on the existence of many other species which serve as food, mediator of mating, shelter from physical and biological hazards, etc.

A population of organisms has properties or characteristics that transcend the characteristics of an individual. The growth of a population is certainly different from that of an individual. The differences between ethnic groups of man can be described only by distributions of certain identifiable genes. All these measurements are characteristics of populations rather than of individuals.

Population genetics is aimed to study the genetic structure of populations and the laws by which the genetic structure changes. By genetic structure we mean the types and frequencies of genes or genotypes present in the population. Natural populations are often composed of many subpopulations or of individuals which are distributed more or less uniformly in an area. In this case the genetic structure of populations must be described by taking into account the geographical distribution of gene or genotype frequencies. The genetic structure of a population is determined by a large number of loci. At the present time, however, only a proportion of the genes present in higher organisms have been identified. Therefore, our knowledge of the genetic structure of a population is far from complete. Nevertheless, it is important and meaningful to know the frequencies of genes or genotypes with respect to a certain biologically important locus or a group of loci [14].

Population Genetics is considered as a dialog between predictions based on principles of Mendelian inheritance and results obtained from empirical measurement of genotype and allele frequencies that form the basis of hypothesis tests. Idealized predictions stemming from general principles form the basis of hypotheses that can be tested. At the same time, empirical patterns observed within and among populations require explanation through the comparison of various processes that might have caused a pattern. Population genetics is based on two concepts: Genetic diversity and population. Chiefly, it tries to find the genetic variability or observable polymorphism in different genetic loci (alleles), in a given population and then, to assess this genetic polymorphism, calculating the frequencies (proportions) of all the alleles present in this population. The results obtained, when compared in different scales of time in the same population, give information about the genetic evolution of this population, through a variation (increase, decrease or constancy) of allelic frequencies. The genetic comparison of different populations allow to assess the observant differences (or similarities) and then to get the key for the genetic history of each population, a group of populations or even a species. This genetic history aimed to characterize the genetic origins of populations but also to redraw their interaction over time (genetic admixture, gene flow). Furthermore to the population behavior (cultural choice, religious, demographic, sanitary, etc), the allelic frequencies variations are associated to other process "evolutionary forces": mutation, migration, natural selection and genetic drift. The main objectives of the population genetics are:

1. To assess the genetic variation of natural populations and to describe the organization basic grounding for this variation.
2. To explain the origins, the evolution and stability of the genetic variation, by the evolutionary forces effect.

A Comparison of DNA Typing Methods

Population Genetics and Biologic Anthropology cannot be distinguished by its own techniques and methods. It is for some such reason that, as most writers on the subject freely admit, these fields-unlike other disciplines, might appear as a "soft" and less exact science than it purports to be. However, its originality lies without doubt in its capacity to gather several approaches used in different fields of research in order to answer the different issues. Indeed, technologies used for performing population genetic analysis differ in their ability to differentiate two individuals and in the speed with which results can be obtained. The speed of the DNA analysis has dramatically improved over the years. DNA testing that previously took 6 or 8 weeks can now be performed in a few hours.

DNA analysis has used a variety of techniques including single-locus probe and multi-locus probe restriction fragment length polymorphism (RFLP) methods and more recently polymerase chain reaction (PCR)-based assays. Numerous advances have been made in the last quarter of a century in terms of sample processing speed and sensitivity. Instead of requiring large bloodstains with well-preserved DNA, tiny amounts of sample — as little as a few cells in some forensic cases — can yield a useful DNA profile.

The gamut of DNA typing technologies used during the past several decades for human identity testing is compared in Figure 3. The various DNA markers have been divided into four quadrants based on their power of discrimination, that is, their ability to discern the difference between individuals, and the speed at which they can be analyzed. New and improved methods have developed over the years such that tests with a high degree of discrimination can now be performed in a few hours.

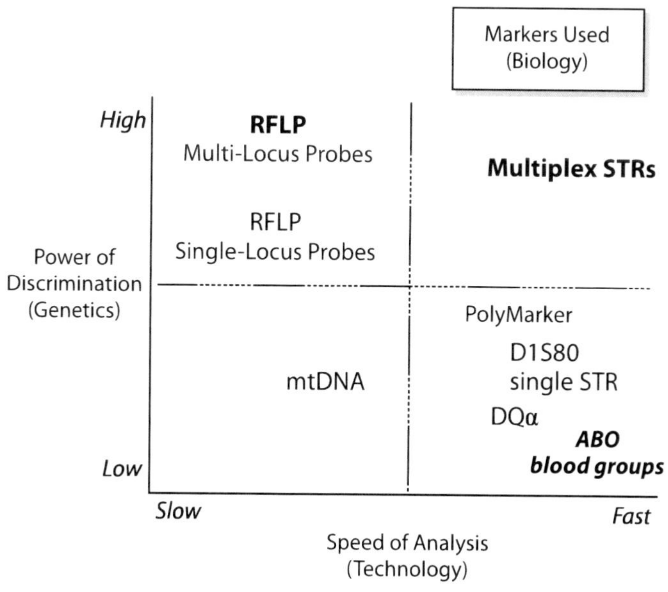

Figure 3. Comparison of DNA typing technologies. DNA markers are plotted in relationship to four quadrants defined by the power of discrimination for the genetic system used and the speed at which the analysis for that marker may be performed. Note that this diagram does not reflect the usefulness of these markers in terms of forensic cases [13].

An ABO blood group determination, which was the first genetic tool used for distinguishing between individuals, can be performed in a few minutes but is not very informative. There are only four possible groups that are typed — A, B, AB, and O — and typically more than 80% of the population is either type O or type A.

Thus, while the ABO blood groups are useful for excluding an individual from being the source of a crime scene sample, the test is not very useful when an inclusion has been made, especially if the sample is type O or type A.

On the other extreme, multi-locus RFLP probes are highly variable between individuals but require a great deal of labor time, and expertise to produce and interpret a DNA profile. Analysis of multi-locus probes (MLP) cannot be easily automated, a fact that makes them undesirable as the demand for processing large numbers of DNA samples has increased. Deciphering sample mixtures, which are common in forensic cases, is also a challenge with MLP - RFLP methods, which is the primary reason that laboratories turned to single locus RFLP probes used in serial fashion.

The best solution in the search for a high power of discrimination and a rapid analysis speed has been achieved with short tandem repeat (STR) DNA markers, shown in the upper right quadrant of Figure 3 and represented in Figure 4. Because STRs by definition are short, three or more can be analyzed at a time. Multiple STRs can be examined in the same DNA test, or 'multiplexed'. Multiplex STRs are valuable because they can produce highly discriminating results and can successfully measure sample mixtures and biological materials containing degraded DNA molecules. This method can significantly reduce the amount of DNA required for analysis, thereby conserving more of the irreplaceable DNA collected from forensic evidence for use by scientists from opposing counselor for additional specialized DNA testing. In addition, the detection of multiplex STRs is automated, which is an important benefit as demand for DNA testing increases.

Mitochondrial DNA (mtDNA), which is shown in the quadrant with the lowest power of discrimination and longest sample processing time, can be very helpful in forensic cases involving severely degraded DNA samples or when associating maternally related individuals. There are instances when nuclear DNA is either so degraded or present in such low amounts in samples (e.g., hair shafts or putrefied and ancestral bones or teeth) that it is either untestable or undetectable and mtDNA is the only viable alternative forensic DNA technology that can produce interpretable data for forensic comparisons.

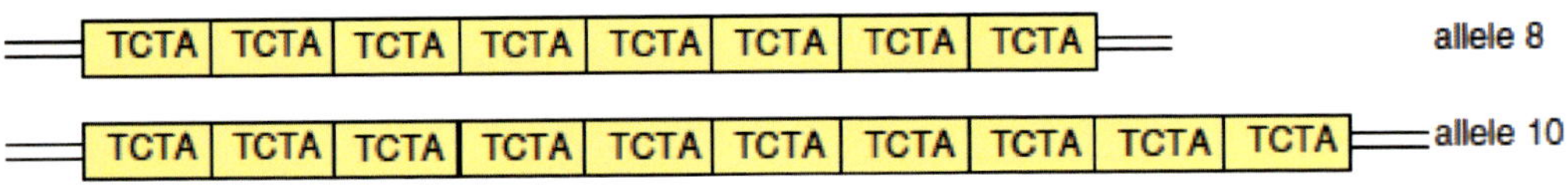

Figure 4. This example shows the structure of two alleles from the locus D8S1179. The DNA either side of the core repeats is called flanking DNA. The alleles are named according to the number of repeats that they contain – hence alleles 8 and 10 [13].

In other cases, such as the identification of skeletal remains, mtDNA is often the preferred method for comparing DNA from the bone evidence to the known reference mtDNA from potential family members to determine whether the mtDNA matches. In many situations, multiple technologies may be used [13].

Genetic Characterization of the Moroccan Population

The strategic position of Morocco and the broad history of the country makes this region very interesting and suitable for population genetic investigations. Many studies using some relevant and validated markers were performed in order to well characterize this multiliguistic and multicultural population.

Autosomal Markers

In view of their high level of variability, autosomal short tandem repeats (STRs) are very useful as markers in both disciplines of forensic and population genetics studies. These systems had proved their efficiency in terms of determination of relationships among related individuals and populations.

Furthermore to the 13 STRs of the American data base CODIS (*COmbined DNA Index System*): TPOX, D3S1358, FGA, D5S818, CSF1PO, D7S820, D8S1179, TH01, VWA, D13S317, D16S539, D18S51 and D21S11, two well selected systems D2S1338 and D19S433 were added in order to increase the power of discrimination. Thus, 15 relevant STRs included in the Identifiler kit (Applied Biosystems) were used to provide the allelic distributions in an Arabic Moroccan population from *Rabat-Sale-Zemmour-Zaer* and estimate their forensic parameters [15].

The results prove the discriminative potential of these markers (table1), with an important value of the D18S51 marker; which matches with Shepard and Herrara (2005) study used on an Iranian population.

Indeed, the allelic frequencies' distribution in the totality of markers can provide information about the population's equilibrium and the genetic structure of the population. In a demographic context, these information unearth the socio-behavioral and historic dynamic of the population. From the 15 markers, 13 verify a genetic equilibrium, unlike the two markers D16S539 and FGA. A genetic drift or the inbreeding (about 20%) in this region may explain the genetic drift of these markers. The Principal Component Analysis allowed the graphic representation of population relationships. It was used to compare the Arabic Moroccan population with four sub-Saharan, six North-African and three Northern Mediterranean populations (Figure 5). Both axes of the graph show 57.53% of the total variance.

The points-scattering shows that the populations are divided in two distinct groups. The first one encloses the sub-Saharan populations, which give an evidence of their genetic homogeneity in relation to their geographical position. While the North African populations' gathers with the North Mediterranean. This fact can be explained by their close geographic position as well as their shared socio-culture and history. In this same group, two clusters are distinguished (Figure 5). The first one (upper) enclose the North African population, excepted the Moroccan Arabic speaking group of Rabat-Sale-Zemmour-Zaer, which lies close to the North Mediterranean group in the second cluster (down). Furthermore to the geographical and historical similarities between Moroccan Arabic and North Mediterranean, this distribution can be explained by the broad migration of the Muslims of Andalusia to Morocco, especially to Rabat-Sale-Zemmour-Zaer region in 1610 (Ibn Khaldoun).

Table 1. Biostatistical parameters of an Arabic ethnical group from Morocco using 15 autosomal STRs.
PD: Power of discrimination; PE: Power of exclusion;
PIC: polymorphism information contenant; TPI: typical paternity indice, H_0: Observed Heterozygosity; H_e: Expected Heterozygosity [11]

Locus STR	PD	PIC	PE	TPI	Ho	He
D5S818	0,888	0,690	0,430	1,67	**0,701**	0,728
FGA	0,962	0,850	0,544	2,17	0,770	0,863
D8S1179	0,944	0,810	0,653	2,91	0,828	0,834
D21S11	0,956	0,840	0,643	2,83	0,824	0,856
D7S820	0,927	0,770	0,561	2,27	0,779	0,797
CSF1PO	0,869	0,680	0,485	1,89	0,735	0,730
TH01	0,936	0,800	0,561	2,27	0,779	0,825
D13S317	0,905	0,720	0,477	1,85	0,730	0,760
D16S539	0,917	0,776	0,544	2,17	0,770	0,795
D2S1338	0,956	0,850	**0,740**	**3,92**	**0,873**	0,862
D19S433	0,933	0,790	0,616	2,62	0,809	0,817
vWA	0,950	0,820	0,606	2,55	0,804	0,846
TPOX	0,875	0,690	0,400	1,57	**0,681**	0,731
D18S51	**0,968**	**0,870**	0,643	2,83	0,824	0,880
D3S1358	0,903	0,730	0,553	2,22	0,774	0,770

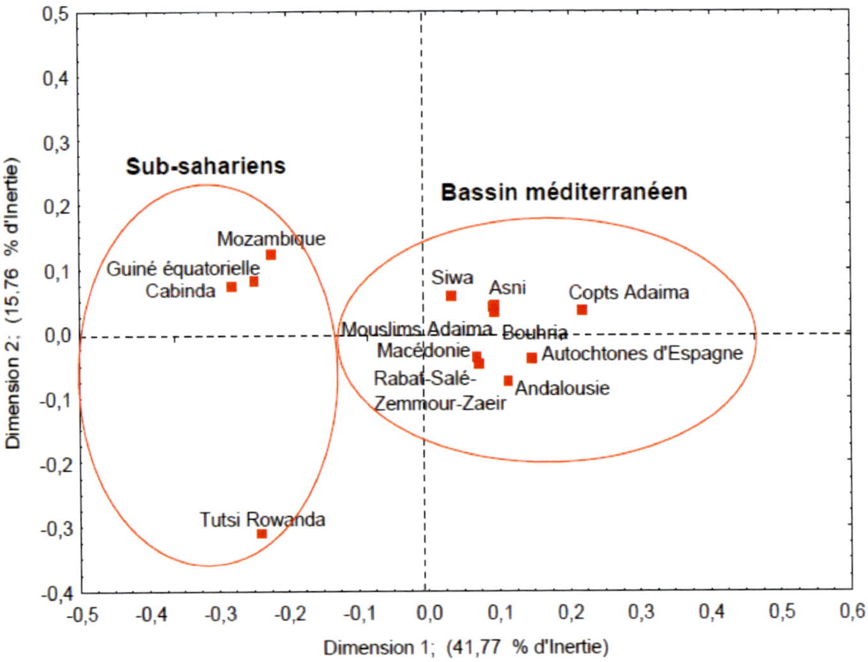

Figure 5. Principal Component Analysis of 15 STRs in a sample of an Arabic-Speaking group of Morocco on an Afro-Mediterranean scale [11].

Additionally, Figure 6 shows a PCA constructed with the second data set based on the allelic frequencies of 18 populations from different area carried out in order to a wider comparaison with the previous samples and to investigate the effect of other populations in the clustering and thereby their genetic affinity and relationship. The two first axes account for 54,88% of the total variance. It can be inferred that the population's structure described here is slightly different compared to the first one. The difference in the genetic distances depends in the socio-cultural, economic and geopolitics context as well as the technologic progress. The Middle Eastern and North African populations show an intermediate position, compared to other populations. This distribution is due to the close geographic position to the shared religion and to the Arabic migration effect from Middle East to North Africa with a view to spread Islam. These results complete Coudray's study [16].

A wider study on the same 15 STRs set was yield in a sample from the three ethnical groups (Arabic, Berber and Sahrawi) of the Moroccan population. The results confirm the previous study, indeed the Moroccan population seems to be closer to the Middle Eastern group. All the statistical values reported in this work showed no significant difference between the three ethnical groups from those reported previously [16].

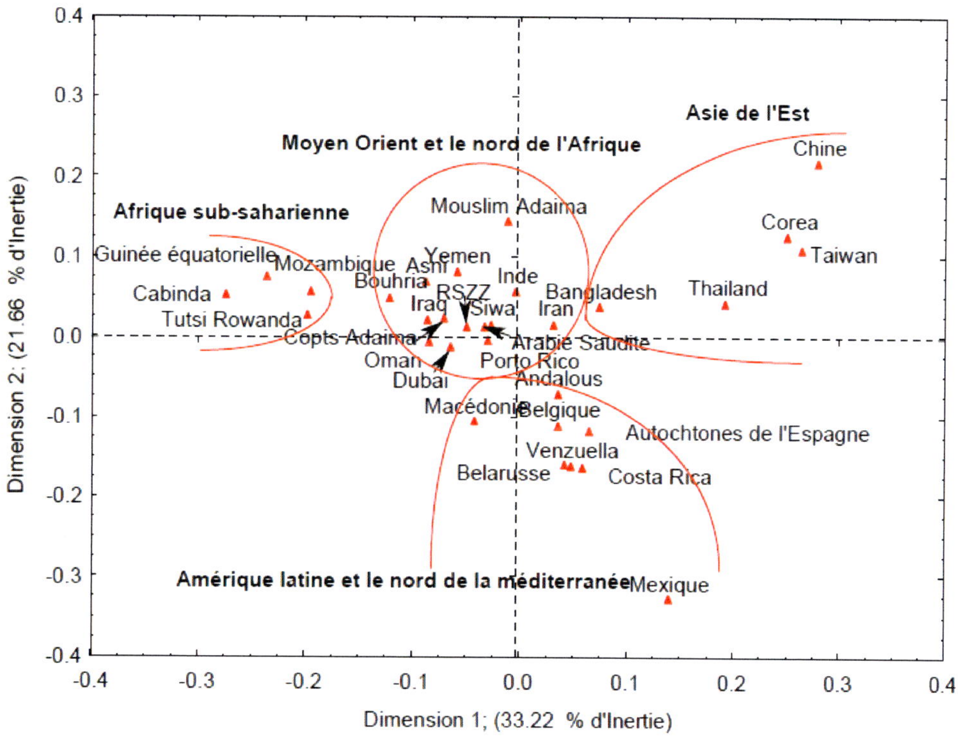

Figure 6. Principal Component Analysis of 15 STRs in a sample of an Arabic-Speaking group of Morocco and a broad word wide population [11].

In the objective to study a level of admixture between Moroccan subpopulations, a set of six polymorphic Alu insertions have been selected; and carried out in group of Arabs, group of Berbers and an admixed heterogeneous Moroccan population resident in Casablanca (the Moroccan industrial capital where all ethnical groups are admixed during migration events)

[17]. Alu repeats are a category of short interspersed nuclear elements widely distributed in the genome of all primates. With a size of less than 500 bp, Alu elements are the most abundant mobile elements in the human genome (B1 100 000 copies). Twenty-five percent of the 'young' Alu human-specific elements have been incorporated in the human genome so recently that they are dimorphic for the presence or absence of the insertion.

Subsequently, populations can be polymorphic for the presence or absence of an Alu element in different individuals at a particular chromosomal location. Alu elements have been shown to be useful in human evolution studies because they offer two important advantages compared with other polymorphic markers: (i) they are identical by descent – that is, individuals that share Alu-insertion polymorphisms have inherited them from a common ancestor and (ii) the ancestral state of each Alu insertion polymorphism known to be the absence of the element, so that they can be used to 'polarize' the evolutionary process.

The relationships between the three Moroccan groups : (1) 53 unrelated individuals randomly sampled, resident in Casablanca where the urban growth is maintained by immigration from all parts of Morocco (the urban residents represent a sample of the populations of the whole area from which the immigrants have originated); (2) 50 unrelated individuals from the Arab Moroccan population (Doukala, Chaouia, Houz, etc.); (3) 40 unrelated Berbers originated from different regions of Morocco (Taroudante, Tafraout, Souss, Amzmiz, Khmissat) representing a pooled Moroccan Berbers population; and other populations are shown by the topology of the neighborjoining tree (Figure 8) [18]. The structure of the tree showed that the populations clustered together with the other African populations. It is also interesting to note that the Arabic population is closer to the African populations and the branch linking them together showed a strong bootstrap (74.52%). This affinity suggested a certain degree of gene flow from Africans.

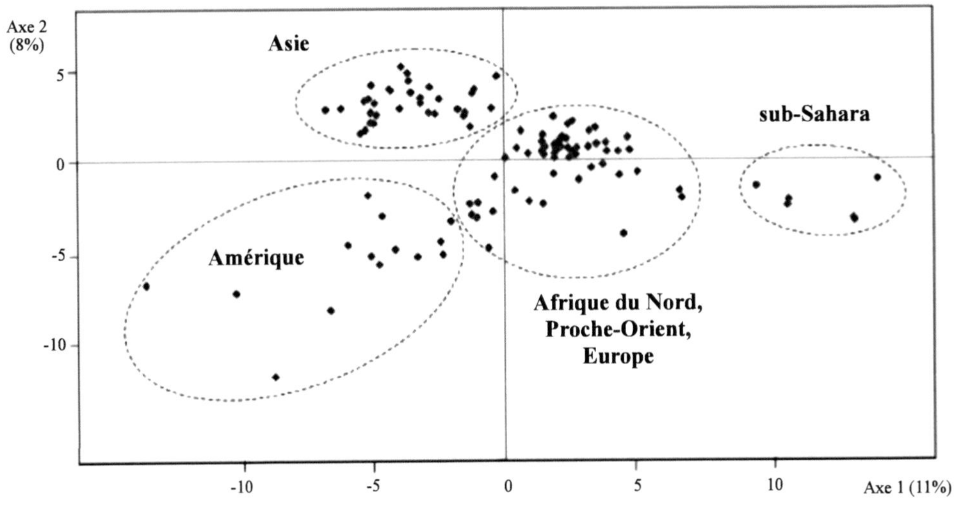

Figure 7. Principal Component Analysis of the STRs CODIS in a worldwide population [16].

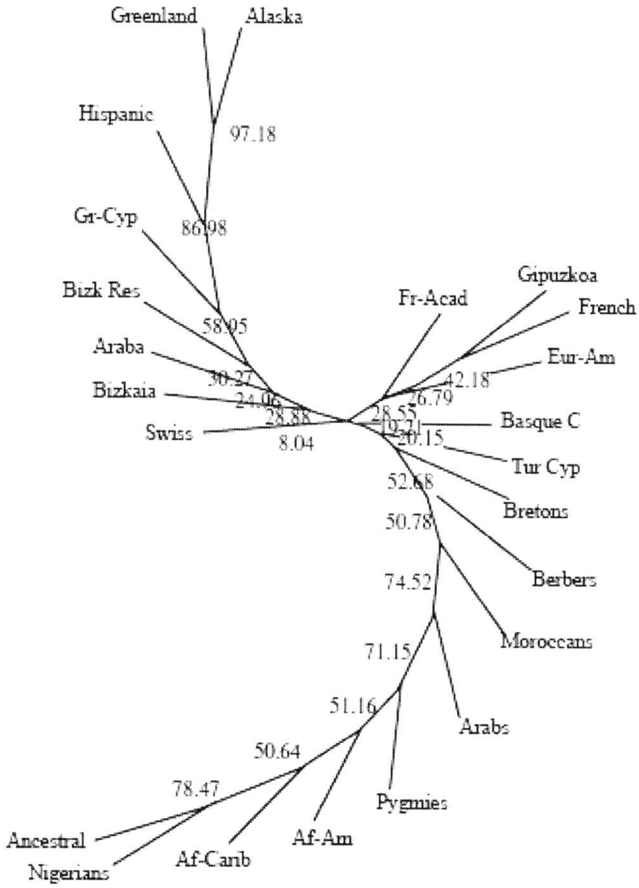

Figure 8. Neighbor-joining tree of population relationship derived from the Nei matrix distance of six polymorphic Alu repeats (APO, PV92, ACE, F13B, D1, and TPA25) using TreeView program. Bootstrap values (in %) are indicated [18].

Otherwise, the Berber population is closer to Europeans than the Arab population (52.6% of bootstrap) and present shorter genetic distance to Europeans than to Africans, which is certainly true because there is a Caucasoid component in its genetic background. The general Moroccan population is placed between Arabs and Berbers (50.7% of bootstrap support). European populations are clustered together but no robust branches existed between them, suggesting a high homogeneity within European populations. The gene flow analysis shows that the populations here studied are plotted above the theoretical line, which signifies that they have received gene flow. Moroccan population and Berbers have experienced less gene flow than the Arab population that seems to have received more gene flow from the African populations.

This is in accordance with the historical data which show that Arabic populations have been admixed during fusion of the immigrants from Arabia with Ethiopians from African origin forming the Aksum empire and their expansion have initiated during development of important trades to West and North Africa. Furthermore, the gene flow in the Casablanca resident population that represent the general Moroccan population seems to be normal because the historical data show that Morocco has been submitted to heavy exchanges during important immigrations events and caravan trades from Africa but also from the Caucasoid populations.

Y chromosome markers

Although autosomal STR loci are the primary marker systems used, Y chromosome STRs (Y-STRs) also have proven useful in human identification studies. Until recently, the Y chromosome seemed to fulfill the role of juvenile delinquent among human chromosomes [19] — rich in junk, poor in useful attributes, reluctant to socialize with its neighbors and with an inescapable tendency to degenerate. The availability of the near-complete chromosome sequence, plus many new polymorphisms, a highly resolved phylogeny and insights into its mutation processes, now provide new avenues for investigating human evolution. The Y chromosomal STRs are also very useful in forensic sciences. They allow amplification of male-specific DNA regions in a male/female mixture without female DNA competition for reagents during the PCR. In addition, these loci reside in the non-recombinant portion of the Y chromosome and barring mutation, are inherited unaltered through the male lineage. These features of Y-STR-based haplotype evidence facilitate demonstration of the presence of male DNA in a mixture of male and female DNA and for ascribing its male lineage origin.

The International Y-STR User Group recommended use of nine Y-STR loci for identity testing. These are: DYS19, DYS385, DYS389I, DYS389II, DYS390, DYS391, DYS392, and DYS393, defined as the European minimal haplotype (EMH). In the United States, the scientific working group on DNA analysis methods (SWGDAM) also recommended use of these nine loci, as well as the loci DYS438 and DYS439. The YFiler System kit (Applied Biosystems, WI, USA) was developed to facilitate the simultaneous analysis of these 11 loci in a single multiplex system. The multiplex includes an additional locus DYS437, DYS448, DYS456, DYS458, DYS635 and Y GATA H4, for a total of 17 Y-STR loci. A sample of 267 unrelated Moroccan males from different ethnic groups (Arabs, Berbers and Sahrawi), was typed for 17 Y-STR loci [20]. Discrimination capacity (96.3%) and haplotype diversity (99.91%) were calculated. A total of 257 haplotypes were identified, of which 237 were unique and 10 were found in two individuals each. DYS385 showed the highest diversity (0.887) followed by DYS458 (0.820) as a single locus marker. The gene diversity values of 17 loci showed the highest diversity at DYS385 (0.887) followed by DYS458 (0.820) as a single locus marker. The haplotype diversity (HD), discrimination capacity (DC) and unique haplotype (UH) were, respectively, 0.9991, 0.963 and 247. Y-STR polymorphisms in Moroccan population, using Y-filer, could be then a powerful discrimination tool for routine forensic applications. Two microvariant alleles (18.2 and 19.2) were observed at the locus DYS458 with a lower frequency (5.99%) than that observed in the Upper (Southern) Egyptian population.

The haplotype data were compared with the populations submitted to the YHRD database, namely Algeria, Tunisia, Upper Egypt, Iran, Greek, Italy, Central Portugal, Namibia and China. As shown in the multidimensional scaling (MDS) (Figure 9) and confirmed by genetic distances, the studied sample seems to be nearest to Tunisian, Algerian and upper Egyptian than other Mediterranean populations like Portuguese, Italian and Greek. A high genetic distance was observed when comparing our population with Namibian. We can conclude from this study that the Moroccan male lineage have rather a Middle Eastern provenance, since during the religious Arabic conquest to North Africa, the leaders and actors were men.

Other type of polymorphisms; biallelic markers; such as single nucleotide polymorphisms (SNPs) and insertion/deletions (indels), represent an important class of markers on the Y-chromosome. Y-chromosome SNPs (Y-SNPs) are mostly used in molecular anthropology for evolutionary research. Moreover, typing a sample's Y chromosome haplogroup allows male ancestry inference. The analysis of 159 male samples from Morocco (including Arabic, Berber and Sahrawi) was carried out [21], a total of 10 haplogroups were observed.

The most common Y chromosome haplogroups in the combined Moroccan population resulted E1b1b1, E1b1b1b and J1 (56%, 49% and 10% respectively). The E-M215 derivative E1b1b1 is observed in significant frequencies in Moroccan samples. It's defined by the E-M35 SNP which appears to have originated in East Africa then migrating to the Near East and then on to North Africa and Europe.

Previous studies have shown that E1b1b1b (E-M81), formerly E3b1b or E3b2, is the predominant haplogroup in northwestern Africa [22]. Bosh et al. [23] obtained a higher frequency (64%) of E-M81 (E1b1b1b) in Northwest African populations than the present study. This disparity could be explained by the difference of geographic area and circumstances of sample collection.

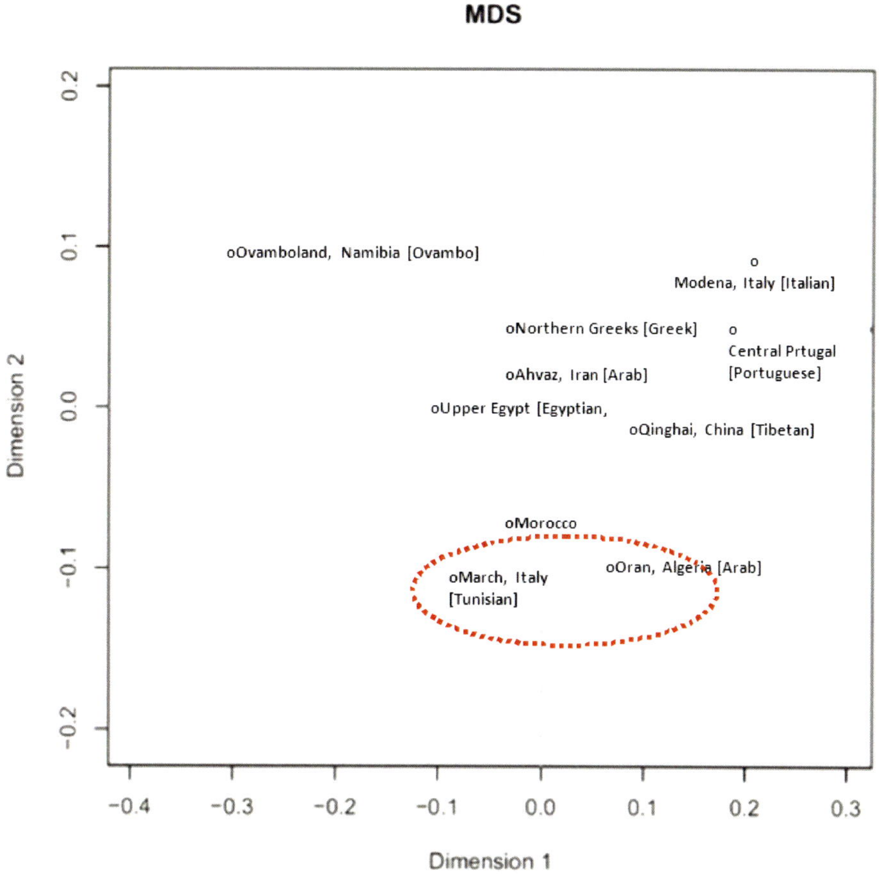

Figure 9. Multidimensional Scale (MDS) representation of 17 Y- STRs in a sample of Moroccan population (Arabic, Berber and Sahrawi) and other Middle Eastern and Mediterranean populations [20].

However, according to the present results and those of Bosh et al., it seems that the E1b1b1b haplogroup is more characteristic of Sahrawi than Berbers. It is also seen, although at low frequency, in the Iberian Peninsula (4%) and Sicily (3%) due to recent gene flow from Northwest Africa through Gibraltar strait. Nevertheless, this haplogroup (E1b1b1b) is not found in sub-Saharan Africa and its frequency sharply declines through the continent towards the east. The J1 haplogroup, defined by the single nucleotide polymorphism (SNP) M267, is most frequent in the Arabian Peninsula especially in Yemen (76%).This could be attributed to the early medieval period during which the Semitic expansion spread J1 out of Arabia into North Africa.

Mitochondrial DNA Markers

Mitochondrial DNA (mtDNA); the maternal lineage indicator; is a powerful tool in reconstructing population history, because a fine-grained phylogeography has been defined for it. The study of Aboukhalid et al [24] on the mitochondrial DNA control region variation of 509 unrelated individuals living in Morocco shows a mosaic composition of West Eurasian and African lineages with frequencies of 55.64% (Figure 10) and 44.35% respectively. In the Moroccan population's dataset, the most frequently observed haplotype was 16519C 263G 309.1C 315.1C 573.1C (Hg R0).

This haplogroup was seen altogether in 32 samples (6.28% of samples). Haplogroup R0, formerly known as haplogroup pre-HV, derives from macro-haplogroup R and the ancestor of haplogroups HV and its subhaplogroups.

Another haplotype with a less frequency 16153A 16298C 72C 195C 263G 309.1C 315.1C (Hg HV0) was found in 11 samples (2.16% of samples), from which 7 belonged to the Moroccan mixed population, 3 belonged to the Moroccan Berber and 1 belonged to the Moroccan Sahrawi.

In other investigation of Coudray et al. [16], a PCA based on the mitochondrial haplogroups frequencies of 39 populations: 18 North African (Morocco, Algeria, Tunisia, Egypt and Mauritania), 5 sub-Saharan (Cameroun, Angola, Kenya, Mozambique, Ethiopia), 6 Middle Eastern (Yemen, Iraq, Israel, Jordan, Turkey, Syria) and 10 South Europeans (Gracie, Italy, Spain, Portugal) gave an insight about the maternal lineage variation of these wide range of populations. Both axes on the graph show 41% of the total variance (Figure 11).

Three clusters are defined; the "group 1" located in the center of the graph represent the Europeans, North African and Middle Eastern populations. The second "group 2" plotted in the axe 1 and 2 represent the upper Egypt, Berbers of Sened (Tunisia) and Ethiopia, while the Yemenian population is located between the two previous groups. The last group "3", lies in the positive variation of the axe 1 and negative of the axe 2, gathering the four sub-Saharan populations (Cameroun, Angola, Mozambique and Kenya).

Analyzing the graph, we can see clearly that the group 1 and 3 are the most different, while the group 2 shows an intermediate position. The European population represented by the group 1 lies in the extremity of this group (negative values if the axe 1) while the North African are located in the other extremity (center of the graph). Thus, if we consider the geographical localization of populations, we can notice as for the precedent genetic systems, that European population and sub-Saharan are the most different genetically. Also, that there is a difference between the Ethiopian population and other sub-Saharan and that the North African populations are in an intermediate position: the North African populations are the nearest from European and Middle Eastern populations, while the upper Egyptian has an affinity to the sub Saharan. All the populations have the same variance considering the axe 3 (11% of the total observed variance).

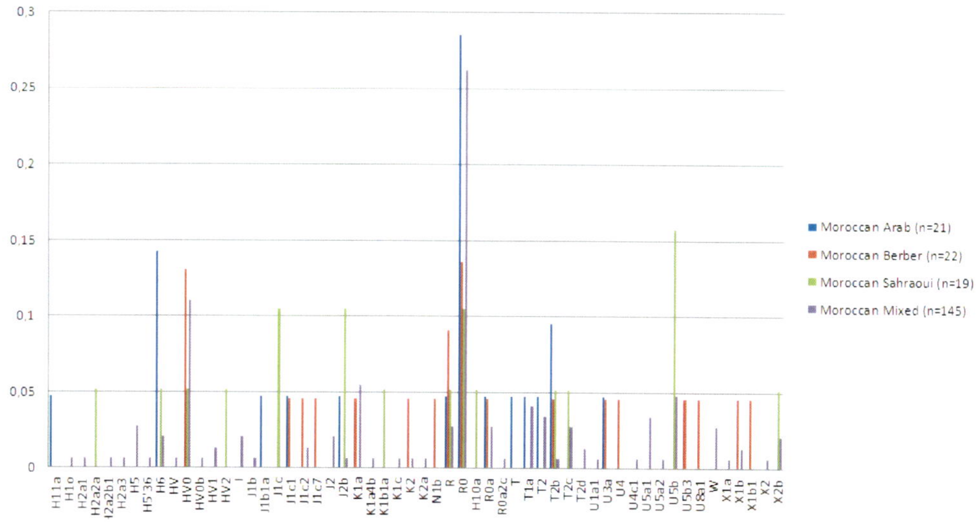

Figure 10. West Eurasian haplogroup distributions in the Moroccan population's samples [24].

X Chromosome Markers

After getting an insight on the genetic variation and affinity of the Moroccan population and other neighbor groups via several polymorphisms through the autosomal chromosomes, Y chromosome and Mitochondrial DNA, we will complete the Moroccan genetic map by the X chromosome genetic analysis. The X chromosome has the unique feature of being present only in a single copy in male subjects, which leads to a series of special characteristics that justify its increasing interest in studies of human population genetics. Mutations occur less frequently in this chromosome because every existing X chromosome spends two-thirds of its lifetime in female subjects where the nucleotide mutation rate is much lower than in male subjects. Additionally, the effective population size of the X chromosome is three quarters that of the autosomes. These facts contribute to the lower diversity of this chromosome (estimated to be about half of that on the autosomes). However, the smaller population size of the X chromosome also causes changes owing to genetic drift being faster than in other chromosomal regions and, therefore, population structure is expected to be more emphasized. Consequently, populations should differ more in their X chromosomes than in their autosomes. Linkage disequilibrium is also greater on the X chromosome, because only two-thirds of this chromosome manages to recombine in each generation. The size of regions with a single genetic history is expected to be larger than in autosomes, once more making it ideal for human population genetic studies. Finally, determining haplotypes on the X chromosome is a simple procedure and therefore it has proved to be a good choice for haplotype-based phylogenetic studies. X chromosome variation has been applied successfully in large-scale geographical studies that cope with the origin of non-African populations pointing in the direction of the 'Out of Africa' model. However, the use of X chromosome for fine-scale geographical studies is negligible when compared with the high number of human population studies focused on Y chromosome and mtDNA variation.

Bentayebi et al. and Picornell et al. studies [25, 26] aimed to build an X-STR database of the Moroccan population for anthropological and forensic purposes, as well as to provide reference data for comparison with other studies based on the analysis of the 4 X-STR trios included in the new kit Investigator Argus X-STR 12 kit on three ethnical groups of the Moroccan population.

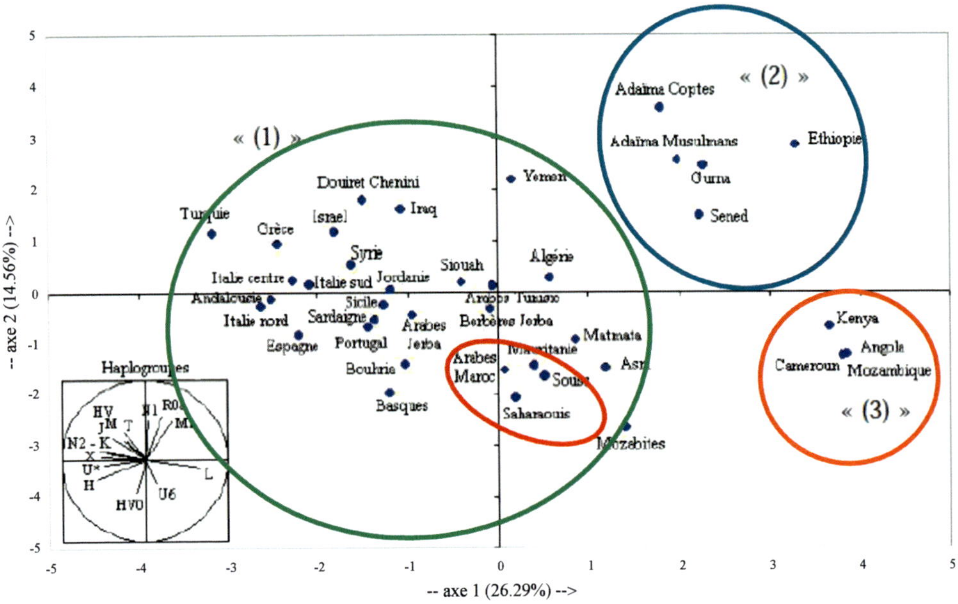

Figure 11. The PCA was elaborated from the mitochondrial haplogroups frequencies of 39 human populations from Africa (North Africa and sub-Sahara), Middle East and South Europe [16].

No significant differences were found between the three ethnic groups. DXS10146 and DXS10135 were the most polymorphic X-STRs, with 23 alleles, and DXS8378 was the least polymorphic one, with 5 alleles. Overall values obtained for the power of discrimination were high (>0.999999) in both females and males. Combined mean exclusion chance for trios and duos were 0.999999999 and 0.9999998, respectively. No deviations from Hardy–Weinberg equilibrium were observed after Bonferroni correction in the Moroccan population. The present study shows that there is a high genetic diversity in the Moroccan population for these 12 X-STRs.

Considering the p-values from locus by locus comparisons between Morocco and other published population data, the Moroccan population seems to be nearest to Algerian and some European populations than to sub-Saharan Africans. Nevertheless, the results showed some differences even with Algerian, in accordance with the X chromosomal SNPs study, where a high overall genetic homogeneity was observed in Mediterranean populations except in Morocco.

These results are consistent with the influence of the Romans and the history of Moroccan colonization, since the largest migratory contingent in this population was European, composed mainly of French, Spanish and Portuguese between the fourteenth and nineteenth centuries. However, the Moroccan population showed significant differences from the Asian populations in almost all loci implying genetic differences according to the geographic location. To well assess the X chromosome contribution in the evolutionary studies, X chromosomal Alu insertions were analyzed in the investigations of Bentayebi et al. and Georgious et al. [27, 28].

Alu elements have been shown to be useful in human evolution studies because they offer two important advantages compared with other polymorphic markers: (i) they are identical by descent – that is, individuals that share Alu-insertion polymorphisms have inherited them from a common ancestor and (ii) the ancestral state of each Alu insertion polymorphismis known to be the absence of the element, so that they can be used to 'polarize' the evolutionary process. For this issue 525 individuals from six different populations, each one originating from a different country were analyzed for 12 Alu insertions by Georgious et al. The hierarchical AMOVA analysis in the five Mediterranean groups (Basques included) showed that only a small and non-significant part of the genetic variance could be attributed to the variation between North–South groups (F_{CT} = 1.24%), indicating no particular genetic differentiation between the two sides of the Mediterranean Sea. However, the markers examined are consistent with a more important diversity within North Africa (F_{ST} = 3.45%, $P < 0.001$) than in South Europe (0.38%, $P < 0.001$). Comas et al [29] and Gonzalez-Perez et al [30] had previously studied population relationships in the western-Mediterranean basin using polymorphic autosomal Alu elements. These two surveys indicated a North versus South differentiation (F_{CT} values: 1.80 and 1.96%, respectively) slightly higher than in these results, but a population variation within groups clearly lower (F_{SC} values of 2.30 and 0.47%, respectively) than that evidenced from the X chromosome Alu markers (F_{SC} of 3.06%.).

Moreover, the 13 Alu markers of this study reveal a population variation within North Africa (F_{ST} = 3.45%) six fold higher than that obtained from a similar number of autosomal Alu elements (F_{ST} = 0.57%) [29]. The X chromosome markers reveal a higher population differentiation in comparison with the same kind of genetic markers in autosomes. This possibly reflects the effect of the reduced population size of the X chromosome on population variation, which, as we mentioned in the introduction, makes populations differ more in their X chromosomes than in autosomal markers.

The PCA used in the study of Bentayebi et al. [31] on 9 Alu insertions, revealed generally accepted relationships among the Mediterranean populations. It is remarkable that the Arab-speaking sample from Morocco shows a particular genetic position as compared with other North African groups. In fact, the Moroccan genetic distances to European samples are smaller than those to North African groups. This close position of Morocco to the Europeans also appears in the population distribution in the PCA graph (Figure 12). This could be explained by the history of the Moroccan population as explained previously.

Conclusion

Morocco is a layered country. Each phase of its history has left a mark on its current makeup, and this chapter reflects that layering through the genetic investigation described below. On to a geographical base, where the disposition of the mountains and rivers, the seas and deserts, influenced where people chose to live, successive cultural and political structures were erected. Morocco was peopled by Caucasoid populations that extended from Iberia to Morocco 20000–70000 BC a late Paleolithic culture called Ibero–Marusian. The contribution of Arabs and Berbers to the genetic background of Moroccan population is certainly evident during the first and second millennium.

Berbers represent the first Caucasian population that has peopled Morocco. They are believed to have their local ancestors among Capsian Mesolithic's and their "Neolithic" descendants, possibly with genetic contribution from the important Neolithic migration from Near East [32]. The Berber language has survived since then showing slight differences between tribes. Furthermore, the Berber identity has been developed in south Morocco with a Negroid component. The Arabs have their origin from the Arabian Peninsula, they have conquested North Africa (Egypt, Tunisia, Algeria, and Morocco) between 7th and 11th century during the Islamic expansion and have exposed Berbers to heavy pressures that forced them to be installed in small niches especially mountainous regions.

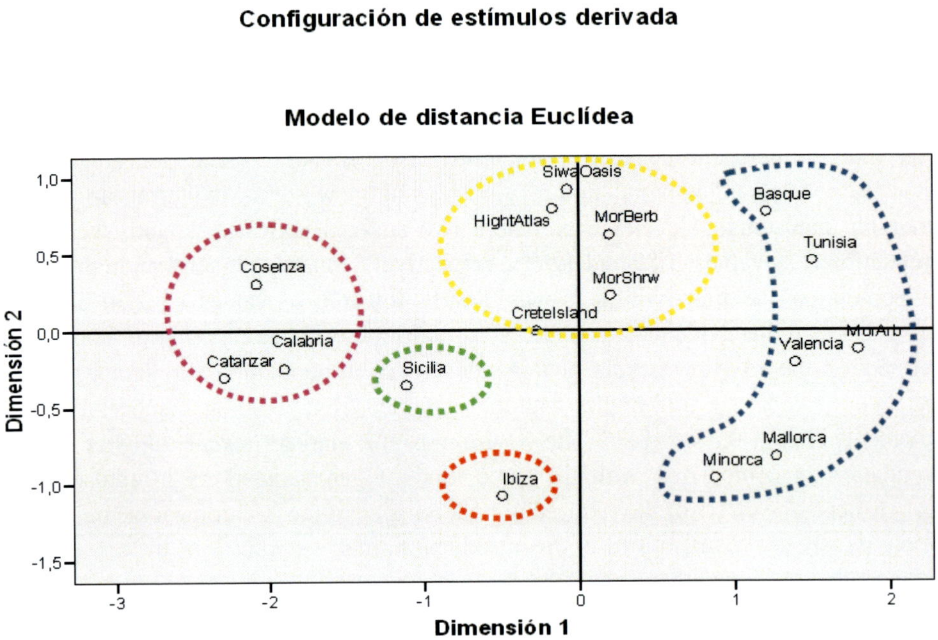

Figure 12. The Principal Component Analysis of a Western Mediterranean group, elaborated from 9 X chromosomal insertions [31].

The Carthagians and Romans brought the first cities, and first linked Morocco into the wider economic markets of the Mediterranean. The coming of Islam continued that process, and introduced the dominant religion, Islam, and the Arabic language, which are still both religious and political issues. The settlement of Arabic men in Morocco yielded to a breeding with the autochthonous population which explain the Middle Eastern ancestry of Moroccan men argued by the Y chromosomal STR and SNP analyses. The wide mixture of the Arabic and autochthonous population through many generations and the high genetic relationship of Berbers with European can explain the genetic profile of Moroccan population when the autosomal markers were analyzed and the PCA was established. This potential European origin of autochtonous population was tested by the maternal lineage indicator "Mitochondrial DNA" and confirmed by the analysis of X chromosome markers. Since this chromosome spend two third of its lifetime in women and because women indicate the ancestry as they find difficulties to immigrate compared with men. The results was that the mtDNA and X chromosomal STR and ALU genetic pool is closer to western Mediterranean than to Middle Eastern populations which confirm the previous hypothesis.

All the studies related bellow give an insight of the genetic history and modern ancestry of the Moroccan population.

References

[1] Moha Ennaji. Multilingualism cultural identity and education in Morocco. *Springer,* 2005.

[2] Jamie Woodward. The physical geography of the Mediterranean. *Oxford University Press*. 2009.

[3] Hublin JJ, Tillier AM, and Tixier J. L'humérus d'enfant mousterién (Homo 4) du Jebel Irhoud (Maroc) dans son contexte archéologique. 1987. Bulletins et Mémoires de la Société d'Anthropologie de Paris, Series 25 4:115–142.

[4] Debenath A, Raynal JP, Texier JP, Roche J and Ferembach D. Stratigraphie, habitat, typologie et devenir de l'Atérien marocain: dones récentes. 1987 L'Anthropologie 90, 233-246.

[5] Jean-Paul Lacombe. Anthropology of the moroccan Neolithic. The necropolis of Skhirat: chrono-geographic approach of the parietal dystrophy. 2004. *Antropo*, 7, 155-162.

[6] Debenath M., Le peuplement préhistorique du Maroc: données récentes et problèmes. 2000. *L'Anthropologi*e. 104, 131-146.

[7] Penell C.R, Morocco: From Empire to Independence. *One world Oxford*. 2003.

[8] Julien C-A. Histoire de l'Afrique du Nord. 1986. Vol 2. Paris: *Payot.*

[9] Fitouri C. Biculturalisme, bilinguisme et education. 1983. Paris/Neuchatel: *Delachaux & Niestlé.*

[10] Njoku RC. Culture and customs of Morocco. *Green Wood Press*. 2006. London.

[11] El Ossmani H. La génetique de Rabat entre l'Andalousie et le Moyen Orient. Editions universitaires. 2010.

[12] Source: http://www.tlfq.ulaval.ca/axl/monde/famarabe.htm

[13] Butler MJ. *Fundamentals of forensic DNA typing*. Elsevier: Science & technology books. 2010.

[14] Hamilton MB. Population Genetics. Willy-Blackwell. 2009.

[15] El Ossmani H, Bouchrif B, Talbi J, El Amri H, Chafik A. Gene diversity of 15 STR among the Arab speaking population of Rabat-Salé-Zemmour-Zaer. *Antropo*. 2007. 15, 55-62.

[16] Coudray C. Histoire génétique et evolution des populations berbérophones nord-africaines. Université Toulouse III-Paul Sabatier. Ecole doctorale Biologie, santé, *biotechnologie*s. 2006.

[17] Bouabdellah M, Ounzar F, Aboukhalid R, Elmzibri M, Squalli D, Amzazi S. STR data for 15 AmpFISTR Identifiler loci in the Moroccan population, *Forensic Sci. Int. Gent. Suppl*. 2008. 1, 306-308.

[18] Chbel F, Pancorbo M M, Martinez-Bouzas C, Azeddoug H, Alvarez-Alvarez M, Rodriguez-Tojo M J, Nadifi S. Polymorphism of six Alu insertions in Morocco:

Comparative study between Arabs, berbers and Casablanca Residents. *Human Genetics*. 2003. 39 (10), 1398-1405.

[19] Jobling MA, Tyler-Smith C. The human Y chromosome: an evolutionary marker comes of age. Nature Reviews Genetics 2003. 4, 598-612 | doi:10.1038/nrg1124.

[20] Aboukhalid R, Bouabdellah M, Abbassi M, Bentayebi K, Elmzibri M, Squalli D, Amzazi D, Amzazi S. Haplotype frequencies for 17 Y-STR loci (AmpFISTR Y-filer) in a Moroccan population sample. *Forensic Sci. Int. Gent*. 2009. 4, e73-e74.

[21] Aboukhalid R, Bouabdellah M, Lai R, Bradford N, EL Ossmani H, Talbi J, Bentayebi K, Squalli D, Podini D and Amzazi S. Y chromosomal SNP analysis using the minisequencing strategy in a Moroccan population samples. *Journal of forensic research*. 2010. 1 (2), 1000113.

[22] Cruciani F, La Fratta R, Santolamazza P, Sellitto D, Pascone R, et al. Phylogeographic Analysis of Haplogroup E3b (E-M215) Y Chromosomes Reveals Multiple Migratory Events Within and Out Of Africa. *Am J Hum Genet*. 2004. 74, 1014-1022.

[23] Bosch E, Calafell F, Comas D, Oefner PJ, Underhill PA, et al. High- Resolution Analysis of Human Y-Chromosome Variation Shows a Sharp Discontinuity and Limited Gene Flow between Northwestern Africa and the Iberian Peninsula. *Am J Hum*. 2001. *Genet*. 68, 1019-1029.

[24] Aboukhalid R, Andreaggi K.S, Bouabdellah M, Bentayebi K, El Mzibri M, Squalli D, Irwin J A, and Amzazi S. Mitochondrial DNA control region variation from samples of the Moroccan population (submitted).

[25] Bentayebi K, Picornell A, Bouabdeallah M, Castro J A, Aboukhalid R, Squalli D, Misericordia M M, Amzazi S: Genetic Diversity of 12 X-chromosomal short tandem repeats in the Moroccan population, *Forensic Sci. Int. Gent*. 2012. 6, e48-e49.

[26] Picornell A, Bentayebi K, Bouabdeallah M, Castro JA, Bouabdellah M, Squalli D, Misericordia M, Amzazi S. X-chromosomal haplotype frequencies of four linkage groups in a North african population, *Forensic Sci. Int. Gent*. Supp Series. 2011, e19-e20.

[27] Bentayebi K, Ramon M, Castro J, Barbaro A, Aboukhalid R, Amzazi S and Picornell A. Inferring ethnicity from the X-chromosome ALU insertions: data from Western Mediterranean human groups, *Forensic Sci. Int. Gent. Supp Series* 2011, e27-e28.

[28] Athanasiadis G, Esteban E M, Via M et al. The X chromosome ALU insertions as a tool for human population genetics: data from European and African human groups, *Eur J Hum Genet*. 2007. 15, 578–583.

[29] Comas D, Calafell F, Benchemsi N et al. Alu insertion polymorphisms in NW Africa and the Iberian Peninsula: evidence for a strong genetic boundary through the Gibraltar Straits. *Hum Genet* 2000. 107, 312–319.

[30] Gonzalez-Perez E, Via M, Esteban E et al. Alu insertions in the Iberian Peninsula and north west Africa – genetic boundaries or melting pot? *Coll Antropol*. 2003. 27, 491–500.

[31] Bentayebi K, Ramon M.M, Castro J.A, Barbaro A, Aboukhalid R, Amzazi S and Picornell A. Population genetic inference from X chromosome *Alu* insertions in Western-Mediterranean region (submitted).

[32] Cavalli, Menozzi L, Piazza A. The history and geography of human genes. *Princeton Univ. Press*. 1994.

In: Genetic Diversity: New Research
Editors: J. A. C. Amaya and M. M. F. Jimenez
ISBN: 978-1-62081-598-4
© 2012 Nova Science Publishers, Inc.

Chapter 6

DIFFERENT APPROACHES TO ASSESS STRUCTURE AND GENETIC DIVERSITY IN FOREST SPECIES

M. A. Martín*[*] *and L. M. Martín
Departamento de Genética, Escuela Técnica Superior de Ingeniería Agronómica y de Montes, Edificio Gregor Mendel, Universidad de Córdoba Campus de Excelencia Internacional Agroalimentario, Córdoba, Spain

ABSTRACT

Forest systems provide timber, food, non-food goods and externalities, being one of them the contribution to the genetic diversity safeguard. Conservation of forest genetic resources is essential for sustaining the environmental and productive values of forests. The knowledge of such diversity is particularly relevant in the case of woody species, which have been named "ecosystem engineers". The objective our study is to show the way to establish the safeguard of genetic diversity through landscape genetics discipline. We present how the results obtained in some forest species using different approaches of this discipline has provided a better understanding on the species genetic resources.

Keywords: Forest genetic resources, sustainability, *Castanea sativa*, *Araucaria araucana*

INTRODUCTION

Forest systems are key ecosystems that provide timber, food, as well as a large number of other goods and services (prevention of soil erosion, flood control, restoration of soil fertility, hunting, eco-tourism, environmental and aesthetic values). The goods and services, which have non-marketed value, are named "externalities" (Pigou 1920; Merlo & Rojas-Briales 2000). During the last years, there has been a crescent interest in these types of values, whose

[*] Departamento de Genética, Escuela Técnica Superior de Ingeniería Agronómica y de Montes, Edificio Gregor Mendel, Universidad de Córdoba Campus de Excelencia Internacional Agroalimentario (ceiA3), ES-14071 Córdoba, Spain, * Corresponding author: Phone: +34-957-212575; E-mail: ge2macum@uco.es.

potential is to provide additional benefits while conserving forests. In this respect, the biodiversity maintained by forests can be considered an example of externality.

Biodiversity is defined as the variation of all living organisms at different levels: ecosystems, species and genetic diversity (Wilson 1988). In this context, genetic diversity is defined as the basis of all biodiversity and is widely considered as the main requirement for the long-term survival of species on an evolutionary time scale (Namkoong 1997; Booy et al. 2000). It has been demonstrated that the persistence of populations is positively linked to their genetic variability (Frankham & Ralls 1998; Saccheri et al. 1998), thus, the knowledge about levels of genetic diversity and population genetic structure are key elements that determine species ability to respond to selection, either natural and artificial procedures. All this is particularly relevant in the case of forest trees, which are considered "ecosystem engineers" (Wright & Jones 2006; Shackak et al. 2008). This concept is defined as "organisms that directly or indirectly modulate the availability of resources to other species by causing state changes in biotic and abiotic materials" (Jones et al. 1994). In terms of their effects, woody plants can be considered primary landscape modulators and have significant environmental impacts (Perry 1998).

Threats to Genetic Resources

Landscape structures and the evolution, maintenance, and loss of biodiversity are tightly linked. Landscape structures have been increasingly modified by the climatic changes and the accelerating dynamics of global land use, leading to an unprecedented loss and fragmentation of habitats, which in turn is one of the major drivers of the current global loss in biodiversity.

Forest tree species have faced long-term changes in environmental conditions during the last glaciations by migration and/or adaptation to the most suitable habitats. Nevertheless, their genetic diversity is seriously threatened. Inappropriate forest management can turn into the fragmentation of habitats and their effects are translated in the decrease of habitat availability, changes in spatial configuration and quality of habitats and the erosion of species genetic diversity (Henle et al. 2004).

Since life appeared on Earth, the number of species has increased, usually by a continuous diversification process. This store of genetic diversity provided an enormous adaptation capacity, giving stability and balance within the biosphere.

About ten thousand years ago, agriculture and livestock began to be practiced in a few places in Asia, Middle East and America. The reasons that led to these practices are still being debated, but there is a consensus on the fact that they represent a new way of relationship between man and nature. The expansive nature of agricultural society produced a deep alteration of systems, which undoubtedly had a negative effect on the wild environment. It also produced a mutual adaptation between humans and the plants they grew and managed, and between these plants and their environment, leading to a change in domestic species, that thereby also were enriched in genetic diversity.

The expansion of the agriculture led to higher rates of population growth. From *Homo sapiens* appearance to the origin of agriculture (about 100,000 years), population was scarcely constituted by 10 million people, what means that the mean growth was 100 people per year. Conversely, since the origin of agriculture to the industrial revolution, in the late nineteenth century, population was about 1,000 million people, that is, the mean growth of population was 100,000 people per year (one thousand times more than in the hunter-gatherer era). Since the industrial revolution to nowadays (about 150 years), the population grew to seven billion, which represented an annual population growth of forty million, resulting in four hundred times higher than in the agricultural era and four hundred thousand times more than in the hunter-gatherer era.

This explosive growth has been because crops have occupied the most productive half of the land, thereby halving the forest. In addition, there has been a drastic change in fauna and flora. When analyzing the list of the three hundred most important crops in the world, they are represented only 55 of the 300 families of phanerogams. Furthermore, a large portion of them belong to a very small number of families, thus for example, 40 are legumes and 29 grasses (Harlan 1992). Furthermore, the nature of this flora suffered a profound change as a result of the change in the selection that acted on it. Natural selection was surpassed by the artificial selection, by farmers and technicians, and by automatic selection, which occurs naturally as a result of continuation of man's intervention. Agriculture has also led to changes in non-cultivated plants that have adapted to environments altered by man, such as weeds or species adapted to livestock systems. Only a small number of species have adapted well to these new ecological niches, increasing explosively their demographics and, in addition, changing genetically to better meet these new features. One example of this are species associated to the *dehesa* system, as *Quercus ilex* L. and *Quercus suber* L., which are of great value for agriculture, livestock and forestry. These systems were included in the Nature 2000 Network for the European Union for landscape and environmental importance.

These gradual changes have allowed, until relatively recently, the situation to be perceived as a new equilibrium, but in relation to genetic diversity, it should be noted that there has been a significant loss in the number of wild and forest species, with the extinction of many of them, and the reduction of the distribution areas of many others. All this has been translated in a decline of their genetic resources, and therefore, of their ability to adapt to new changes and situations.

In recent times, dynamics even more dangerous for the sustainability of the planet have appeared as a consequence of the technological development that has accompanied the demographic explosion previously described. The Industrial Revolution introduced a particularly relevant aspect. It was a qualitative leap in the availability of energy and in the development of other technologies that have brought man to achieve the power over the rest of the ecosystem, something unimaginable in earlier times. This has permitted high levels of economic growth, the welfare state in advanced countries being the most spectacular achievement. However it has also led to a crisis in the global ecosystem. The critical point is that this technological development has led to a narrowing of the genetic basis of the most productive agricultural and forestry species. This fact must be understood in a double sense: the new agriculture and selviculture are based on fewer species and a lower genetic diversity of each of these species.

The sustainability is defined as the use of natural resources without risking their exploitation by future generations (World Commission on Environment and Development 1987). Both changes in the flora and the narrowing of the genetic basis of the species most used by man are a risk to the sustainability of the planet, therefore plants management can only be considered as sustainable if it includes a suitable system of plant genetic resources conservation.

Thus, forest systems have been drastically reduced, due to the increasing demand for goods and services required. All this necessitates that the production of such goods and services must be reconciled with the protection of genetic resources of the involved species.

The conservation of forest genetic resources can and should integrate in *ex situ* and *in situ* systems. *Ex situ* conservation is carried out in specialized institutions called "gene banks" that store a high genetic diversity in form of live plants, seeds or other types of explants (Brown & Hardner 2000). Conversely, *in situ* conservation takes place in forest systems based on high degree of genetic diversity and are reproduced naturally (Young & Boyle 2000).

Ex situ conservation is essential to maintain and increase the efficiency of intensive systems, through which the principal tree species supply the products that society demands. Nevertheless, it does not answer a fundamental question: to allow the selection to continue and forest species to continue evolving. The maintenance of forest, with greater or lesser levels of management, covers this objective and at the same time, provides other goods and externalities.

The problem that arises in forests is to assess their contribution to conservation, that is, to know the amount of genetic diversity they store, their structure and dynamics. Many evolutionary processes are influenced by environmental variation over space and time, including population genetic divergence, speciation and evolutionary change in morphology, physiology and behavior. Studies on genetic structure in natural populations, including within and among population genetic diversity and genetic differentiation, have been a major topic in evolutionary ecology and genetics. In this respect, the international literature has a new and integrating approach to the study and management of natural resources: landscape genetics. This approach integrates the tools provided by molecular genetics and ecology with the new statistical tools (eg. geostatistics, maximum likelihood and Bayesian approaches). The aim is to provide information about the interaction between landscape features and evolutionary processes, such as gene flow, genetic drift and natural selection. Furthermore, it enables the spatial mapping of e.g. allele frequencies from one or more species (or populations) and subsequently the correlation of such patterns with the current landscape (Manel et al. 2003; Latta 2006). The new field of landscape genetics of adaptive genetic variation establishes the relationship between adaptive genomic regions and environmental factors which is prominent in plant studies (Holderegger et al. 2010).

In this way, accurate information on the status of any species genetic resources in a given area could be had, and thus its contribution to the conservation of its diversity could be assessed. This approach seems particularly suitable to respond to the society's demand on forest and agricultural sustainable management.

Although landscape genetics approach can give an adequate view of the situation of the genetic resources in a particular species, the next step to design a strategy for its conservation and sustainable use, implies to undertake other studies focused on the productive organization, both its technical and socioeconomic factors. Technical aspects to consider are the productive potential, the varieties quality, the management practises and the possibility of classifying the product. In the socioeconomic aspect, the organization of the sector and the human communities involved in its management should be analysed. A key aspect in a project of this nature is the analysis of the traditional knowledge associated to the management of this resource. In this respect, the traditional knowledge associated with the genetic resources is an integral part of the genetic resource itself (Esquinas-Alcázar 2005).

MOLECULAR MARKERS TO ASSESS GENETIC DIVERSITY

There are several types of markers for the evaluation of genetic diversity. The morpho-physiological characters have been the first to be used because they are easy to measure and allow a first and fast separation of the material. One universal characteristic shown by species and individuals is the phenotypic differentiation related to different traits. Nevertheless, the measurement and characterisation of genetic diversity through morphologic variables shows as main disadvantage that phenotype is the result of the joint action of genotype and environment. This implies that in some cases, it is not known to which degree the observed trait is due to genetics or environmental factors. This means that many of the traits are strongly influenced by environmental conditions and/or that populations evaluated are homogeneous for such traits (Nuez et al. 2000).

Modern studies have developed another type of markers based on molecules that are part of the hereditary material or on the direct result of the gene action. In general, they display simple gene regulation and high variation (Müller-Stark & Schubert 2000).

A molecular marker is a fragment (sequence) of DNA that can be physically located in the genome of an organism. These fragments can be found near a gene that codes a characteristic of interest or in non-coding regions that show structural attributes. Furthermore, they are specific of each individual, group of individuals and species. This fact makes molecular markers a tool. Polymorphisms at DNA level can be detected through molecular techniques.

For molecular markers to be useful in genetic diversity, studies must have several requirements. Firstly, they must be polymorphic, so diversity can be detected among individuals and populations compared. Secondly, they must be co-dominant, so all alleles can be detected, thus avoiding masking for dominance. They also should be widely distributed throughout the genome of the organisms, so variability analyses are representative of the entire genome of the species. Finally, they have to be repetitive, efficient and not too expensive.

Molecular markers are mainly based on two types of molecules: proteins and nucleic acids. The first, based in proteins, includes isozymes and seed storage proteins. Isozymes are defined as the multiple forms of an enzyme with the same catalytic activity but different kinetic, structural and migration coefficient properties (Market & Moller 1959). For years they were thought to be an interesting tool for the studies of variability in plants, providing a wealth of information in forest species (Hamrick & Godt 1990; Premoli et al. 2000). Although they exhibit Mendelian inheritance, can be applied without extensive technical development and are cost effective, one of their main problems is their relative few polymorphism (Gepts 1990).

Seed storage proteins are present in all plant species seeds and have proved to be useful in evaluation of genetic variability in many species (Gepts 1990). These proteins are present in seeds of all plant species, because they are sources of amino acids for synthesis processes that occur during germination. Moreover, their variations have no repercussion on any physiological function, thus allowing these random changes to be maintained. The seed storage proteins are classified in four fractions according to their solubility: albumins, globulins, prolamins and glutelins (Osborne 1924). Their main advantages as markers are a high polymorphism level, simple genetic control, environmental independency and their economic, easy and fast analysis (Gepts 1990).

The role of these proteins as genetic marker in forest species has been scarcely explored despite their demonstrated effectiveness in evaluating genetic diversity both in broad-leaved and conifers. In this respect, the protein fraction used as a marker is different depending of the species: albumins are used in *Abies pinsapo* Boiss. (Martín et al. 2010a), globulins in *Pinus pinea* L. (Alvarez et al. 2004) and *Castanea sativa* Mill., (Alvarez et al. 2003) and glutelins in *Quercus ilex* L. (Martín et al. 2009a) and *Nothofagus* spp. (Martín et al. 2010b). Furthermore, in conifers this technique is particularly favourable given that seed storage proteins constitute an important component of the megagametophyte, a haploid tissue derived from the female gametophyte that surrounds the embryo in a mature seed. Thus, analysis of the segregation pattern in parental trees may enable the determination of the genetic control and linkage relationships without making crosses (Neale & Adams 1981; El-Kassaby et al. 1987).

Microsatellite markers (SSRs) have become the most used markers for studying forest genetics, because they are highly polymorphic, codominant and widespread across the genome (Glaubitz & Moran 2000). From the evolutionary point of view, they are interesting because are present in all species genomes, although distributed in low frequency in coding regions (Tautz & Renz 1984; Powell et al. 1996). These markers are neutral and have proven very useful in studies of genetic diversity, however, are not suitable for estimating adaptive genetic diversity.

Recently, the increased availability of DNA sequences has permitted the development of EST-based SSR markers. ESTs (expressed sequence tags) are expressed in different physiologic conditions of plants. It has been stated that the generation of SSRs from EST is relatively easy and inexpensive because they are sub-products of sequence data from genes or ESTs that are publicly available. Their main advantages compared with genomic SSRs are that they are easy to obtain and are present in expressed regions of the genome, showing the potential of having known functions (Varshney et al. 2005). They are also more than three times as transferable across species as compared with anonymous SSRs (Peakall et al. 1998). Studies based on gene expression hold great potential for shedding light on complex ecological phenomena such as phenotypic plasticity. Such studies can be used to identify candidate genes and to provide a genome wide means of studying the genetic basis of the mechanisms by which organisms respond to environmental cues (Grivet et al. 2008).

Case of Study

Although it is inevitable that land use changes will influence the structure of forests, such changes should be planned to help ensure conservation as a major component in land use planning and management strategies. Currently, different degrees of management can be found in forest trees, from natural forests with no intervention (for example *Araucaria araucana*, *Nothofagus* sp., *Abies pinsapo*, *Taxus bacatta*) to highly altered systems (*Olea europaea*, *Quercus* sp., *Juglans reggia*, *Castanea sativa*). Because of this, a strategy to preserve these gene pools has to be adapted to each particular species.

In this section, we argue how different conservation strategies must be adopted to the maintenance of genetic resources according to differences in the management. For this aim, we describe the genetic resources situation of two different species according to their nature and management.

Chestnut

Among the thirteen species belonging to the genus *Castanea*, sweet chestnut (*Castanea sativa* Mill.) is considered the only native species in Europe. It is therefore one of the multipurpose tree species of most economic importance for the Mediterranean region. This species has characteristics of interest, being grown extensively for its fruit and timber and is important not only for rural diversification, local economies and cultures but also for its contribution to the landscape and environment. Furthermore, this species is considered a good model of integration between natural and man driven distribution of biodiversity under changing environmental.

Due to the above described multipurpose characteristics, man has influenced over many centuries the European chestnut populations through propagation and transplanting of plant material, intensive cultivation of grafted (clonal) plants, silvicultural practices and fragmentation of populations due to changes in land use. Thus, this large human handling may has lead to a reduction of population size and to the depletion of extant genetic stocks, leading to a population structure far from an expected situation purely natural. Consequently, chestnut genetic structure is complex due to its different management. It can be distinguished: a) chestnut populations that come from seeds (high forests) and present mainly environmental value. In this case, each tree has a different genotype; b) chestnut stands for timber production that regenerate by stump (coppice). In ancient formations, each of these stumps leads to a different set of feet, arranged in circular form. In any case, the resulting formation has a single genotype; and c) grafted chestnut dedicated to fruit production (Orchards). The clonal varieties are grafted onto seedling rootstocks coming from seeds. In this case, the genetic structure of rootstocks is different to the grafted varieties. Thus, in the grafted part there is a clone mixture and in the rootstocks, each tree has its own genotype (Figure 1). The reproduction systems followed in these traditional plantations makes that new rootstocks are the result of the chestnut varieties germination (Martín et al. 2009b), in which, given the species self-incompatibility (McKay 1942), pollen is from exogenous origin: other varieties or trees that comes from sexual reproduction (Figure 2). This implies that the dynamics of the generational change is very different to the expected in situations a) and b). At this respect, the evaluation of the level of gene flow among stands and maintenance of substantial population size allow the adoption of strategies for the conservation and preservation of the species.

Given these considerations, and that an important part of chestnut stands are in protected places, the improvement in the knowledge of the distribution of the species genetic variability, and its integration in landscape will be a support for management systems that ensure its sustainability. In this respect, conservation and utilization strategies of genetic resources require fundamental knowledge about levels of genetic diversity, population genetic structure and effective population size, since these are the key elements that determine a species ability to respond to selection, either natural or artificial procedures.

Figure 1. Example of rootstocks where chestnut traditional varieties are grafted.

In Europe it has been carried multidisciplinary projects aimed to integrate different research fields to devise strategies for the conservation of chestnut genetic resources (Villani et al. 2006). In this projects different genetic markers specific for the species were developed and used to predict their behaviour to global change effects (Buck et al. 2003; Marinoni et al. 2003; Mattioni et al. 2008; Martín et al. 2010c). This latter aspect is of great importance for the future distribution of the species along with its selviculture and vulnerability against perturbations. In this respect, the assessment of the genetic structure of chestnut across Europe using intersimple sequence repeat (ISSRs) markers has showed that long-term management techniques have influenced the genetic makeup of the species. Furthermore, the results have revealed the existence of five distinct gene pools across Greece, the northwestern coast of the Iberian Peninsula and a large gene pool covering the rest of the Mediterranean basin (Mattioni et al. 2008).

During the last years the consequences of global warming on the shifts in plants phenology, namely in the anticipation of budburst and blooming dates, has been widely discussed. In relation to adaptive diversity, chestnut is considered a good bioindicator of the influence of global warming on the anticipation of the flowering time because blooms at a time of low meteorological instability (Penel et al. 2002). In this respect, a study of European chestnut populations from contrasting climatic conditions in the Mediterranean basin detected significant differences in the level of differentiation among populations using functional microsatellite markers. This analysis revealed different clustering pattern between northern and southern populations for EST-SSRs (Martín et al. 2010c).

Figure 2. Detail of chestnut pollination by bees.

In Spain, natural populations were evaluated to improve the knowledge about both the ecological and genetic stock knowledge using the landscape genetics approach. For this, sixteen chestnut populations, covering the distribution range of the species in Spain, were analysed using microsatellite markers. Results revealed a high level of genetic diversity, although the distribution of such variability was not homogeneous. The analyses of differentiation between populations detected three different groups in northwestern, northeastern and southeastern Spain, which followed a geographical pattern. Nevertheless, the genetic distance between populations within each of these groups was low, indicating that populations belonging to each of these groups, probably share a common gene pool due to their geographic proximity. Conversely, a remaining set of populations situated in the central part of southern Spain were the most divergent, showing the highest genetic differentiation between them and with respect to the rest of the populations. All these considerations have permitted the identification of areas particularly rich in diversity that must have been taken in consideration to plan the conservation of the species genetic resources. Furthermore, it has allowed formulating a hypothesis regarding the history of chestnut in the Iberian Peninsula and suggesting that its current genetic structure is the result of a natural colonization history and human driven domestication (Martin et al 2012a).

In a traditional plantation, given that varieties are grafted, the aerial part is expected to be composed of a mixture of clones and this fact is actually found when using molecular markers (Pereira-Lorenzo & Fernández-López 1997; Pereira-Lorenzo et al. 2001; Gondard et al. 2007). However, there is considerable confusion when some of these studies use the concept of "cultivar population" to define different genotypes associated to a varietal denomination

(Pereira-Lorenzo & Fernández-López 1997). In our opinion, the new diversity found in such material would be insufficient to affirm that there is a variety composed by different clones. Conversely, this statement should be accompanied by additional studies that demonstrate that the denomination that encompasses such "cultivar population" shows, in fact, distinct integrity traits. In this respect, these studies do not provide an interpretation in this sense, given the genetic and environmental influences that affect the morphological traits used and the limitations of isozymes as molecular markers.

In Andalusia (southern Spain), the study of traditional varieties comprised the following steps: 1) the compilation of varietal denominations from the existing literature and the information provided by local farmers; 2) the cataloguing of trees that belongs to each varietal denomination; 3) the tracking system used by farmers to decide which variety to use for grafting; 4) the identification of varieties using morphologic qualitative traits; and 4) the identification of varieties by means of microsatellite markers (Martín et al. 2007; Martín et al. 2009b). This study has permitted to identify 36 clonal varieties and identify cases of homonyms (accessions that share the same name but have a different genetic profile) and synonyms (individuals with the same SSR profile but a different traditional name). Furthermore, differences between trees belonging to the same varietal name were detected, as "*Portuguesa*" and "*Bravia Pilonga*" (Martín et al. 2009b), being the nature of the genetic material in each case is different. "*Portuguesa*" is a real polyclonal variety that shows similar morphological traits and different microsatellite profile. Thus, a similarity analysis indicated that trees from this variety were grouped together and with a close genetic relationship. On the contrary, each tree from "*Bravia Pilonga*" denomination had a different morphological and genetic pattern. In our opinion, this reveals that this denomination does not constitute a variety, but this term is applied to trees from seeds with nuts that peeled well. Therefore, we understand that only the combined use of qualitative morphological traits, a highly discriminate marker and the local information provided by farmers can shed light on the complexity of the varietal situation.

In any case, and according to other authors (Bounous 2002; Gobbin et al. 2007; Martín et al. 2010d), it is clear that chestnut traditional systems contain a high genetic diversity that must be preserved.

In addition, the characterisation of chestnut varieties by morphologic and chemical traits showed high genetic diversity and significant differences among varieties from different areas in Andalusia (Martín et al. 2008) (Figure 3). Important genetic differences in fatty acids composition were detected, being the characters with greater heritability and breeding potential (Ramos 2011). Furthermore, it has been stated the existence of a group of varieties represented by few individuals, that constitutes an important part of the species genetic diversity in Andalusia, but are in danger of disappearing (Martín et al. 2007; Martín et al. 2009b).

Cotyledon storage proteins have shown to be a high polymorphic marker in the species, highlighting an important genetic diversity in the Andalusian populations (Álvarez et al. 2003). Likewise, a first approximation to the inheritance of these proteins was assessed in the offspring of a controlled cross between two local chestnut varieties (*Corriente* and *Pilonga*) from southern Spain. The analysis carried out in 15 polymorphic bands corresponding to the albumin fraction displayed one case of allelism and four of linkage, demonstrating their usefulness to measure genetic diversity in the species (Martín 2012b).

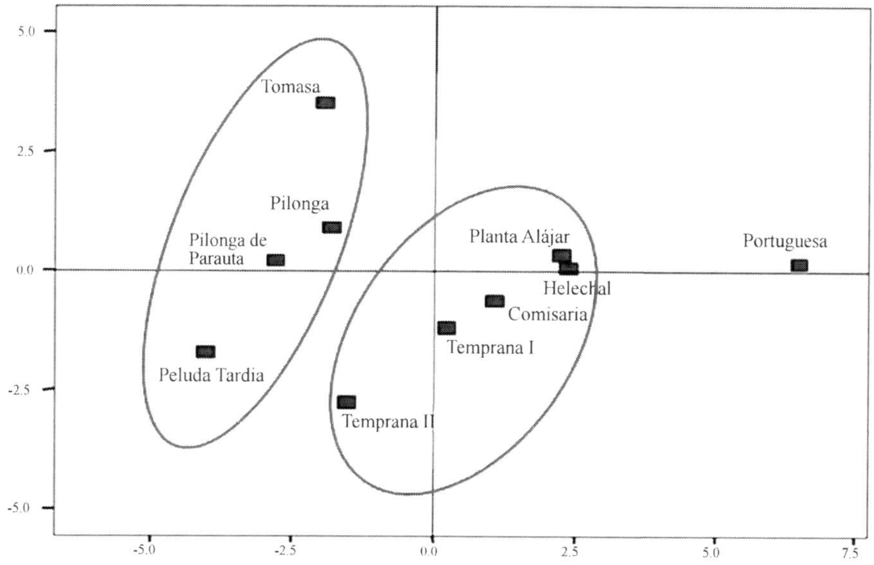

Figure 3. Multivariate analysis showing differences between varieties according to chemical components. Group I, varieties from Malaga; Group II, varieties from Huelva.

In contrast to the high variability described, our studies have stated the existence of threats that are endangering chestnut genetic resources. Sexual nature stands are in progressive recession due to its abandonment and poor phytosanitary state. The importance of diseases as *Phytophthora* spp. and *Mycosphaerella masculiformis* were stated and the pest presence was evaluated, highlighting incidence C*idia splendana* (Fernández-Chaves & Vargas-Osuna 2006). The level of damage caused by *C. splendana* differed by region, ranging from 5.12 - 20.32% in Huelva and 0.28 - 2.16% in Malaga. Damages caused by *C. elephas* were much more localized, focusing on a single plot in Huelva, being 15.05%. In the rest of the cases, its incidence was testimonial (less than 1%) or null. In addition, our results have detected fragmentations problems in some of the populations that displayed higher degree of allelic richness, given as an example the area of Sierra Norte (Sevilla). Moreover, the varietal substitution process, the low crop profitability and the farmer ages, have led chestnut to a critical situation. Thus, in Huelva there is a reduction of production due to possible fails in pollination, because as well as the self-incompatibility of the species there is a proliferation of astaminates varieties and a reduction of pollinator populations.

To sum up, it has been stated that chestnut production is based on a high number of varieties that constitute a real "on farm" conservation system, and growers hold considerable knowledge; however, many of these varieties are at risk of disappearing in a short time (Martín et al. 2009b; Alvarez et al. 2010a).

This situation could be corrected, as these production systems have many possibilities to increase their economic viability through improvement of product characterisation, their incorporation to organic farming and the possibility of aid for the conservation of the biodiversity involved in their maintenance. Moreover, the quantification of genetic diversity contained in natural/naturalised populations, may suppose the put in value of this resource due to its contribution to the genetic resources of the species.

Araucaria Araucana

Araucaria araucana (Mol) K. Koch, or monkey-puzzle tree, is considered a representative symbol of Chilean forest biodiversity and is part of the impressive landscape in native forest ecosystems because of its umbrella-shaped adult crown (Figure 4). It is one of the oldest conifers of South America and stands out as the dominant species given its particular characteristics of endemicity and longevity (Hoffmann et al. 2001). Besides being part of the impressive landscape in native forest ecosystems it is important as food source and other cultural values for indigenous Pehuenche community (Herrmann 2006).

Figure 4. Examples of *A. araucana* forest in A) Malacahuello N.P. in the Andes Cordillera; B) Nahuelbuta N.P. in the Coastal Cordillera (umbrella-shaped crown).

The species is dioecious with pollen predominantly dispersed by wind and gravity seed (Heusser et al. 1988) (Figure 5). It also regenerates vegetatively by sprouting and root suckering (Schilling & Donoso 1976, Veblen 1982).

A. araucana occurs in either pure or mixed species stands along with *Nothofagus* spp. at an elevation of between 1000 and 1600 m. It is associated with coigüe (*Nothofagus dombeyi* (Mirb. Oerst.) and seldom with oak (*Nothofagus obliqua* (Mirb. Oerst. var.) in the lower altitudes, with coigüe and lenga (*Nothofagus pumilio* (Poepp. et Endel. Krasser) in medium altitudes and only with lenga at upper forest limits (Veblen et al. 1995).

The current distribution of the species is fragmented and restricted to some specific areas in Chile and Argentina. In Chile, it covers over 261,000 ha in the Andean region at the frontier of Chile and Argentina (between 37° 30' and 39° 30' S), and in two disjunct populations in the Nahuelbuta Cordillera (between 37° 30' and 39° 30' S) (Veblen 1982) (Figure 5). In Argentina the species covers a restricted area at latitudes between 37° 45' and 40° 20' S.

Figure 5. Details of female (A) and male (B) trees of *A. araucana*.

The current populations are remnants of a more extensive former distribution that was severely diminished by the intense exploitation of the species in the past, the impact of domestic herbivores, seed harvest and natural perturbations (Veblen 1982; Burns 1993; Aagesen 2004). Furthermore, the restricted distribution, slow growth and poor regeneration capacity of the species, make it particularly susceptible to external pressures and its conservation is a subject of great concern (Montaldo 1974; Mutarelli & Orfila 1992). In this respect, the seeds of *A. araucana* are large, heavy and have not wings, so they fall near the maternal tree and are therefore not widely dispersed (Figure 6).

The species is included in Appendix I of the Convention on International Trade of Endangered Species of Wild Flora and Fauna (CITES) and is listed as a vulnerable species by the International Union for Conservation of Nature (IUCN 1996). Although this species is not under imminent risk of extinction, its structure and natural forest dynamic show deep perturbations (Drake et al. 2005; Donoso 2006).

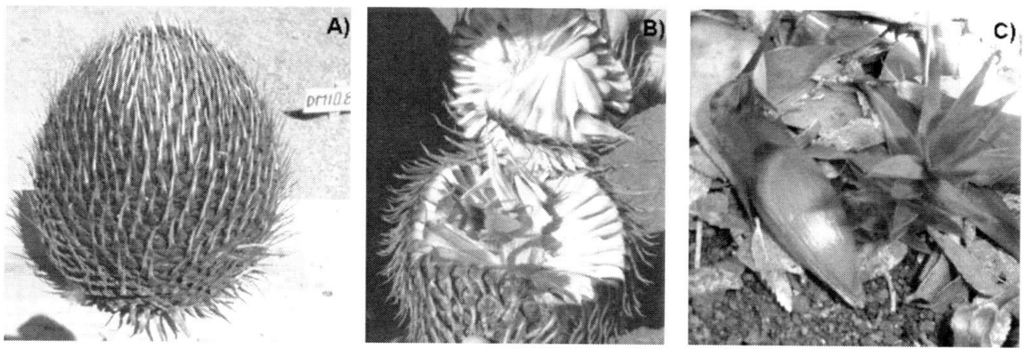

Figure 6. Details of the *A. araucana* fruit: A and B) size of the cone; C) size of the pinion.

An expedition to the areas of *A. araucana* distribution was conducted to determine the current state of the species. For this, eight locations were selected across both the Andean and the Nahuelbuta Cordilleras, and several experimental plots were established according to climatic, ecological and geographical attributes within each location. According to the stand composition, the association *Araucaria-Nothofagus* was verified in six of the eight locations, while in the two remaining there was found pure stands. In the Andes Cordillera, araucaria stands grew in different physiographical conditions (slope and aspect), showing in general low regeneration rates. Furthermore, the low density of the species in some areas made plot establishment difficult (less than 50 trees per hectare). Likewise, it was stated that some area were severely altered since no signs of regeneration were detected, and the remaining individuals showed old grown state (Drake et al. 2009). In Nahuelbuta Cordillera, araucaria showed medium levels of natural regeneration forming even aged stands associated with *N. dombeyi*. Furthermore, due to the high forest density in these stands a higher proportion of trees were found with umbrella form. Likewise, it was detected a bias in the expected proportion of male/female trees (50:50), being the number of female trees very low in some stands from the Andes Cordillera (Drake et al. 2009). These results have revealed high fragmentation in some Araucaria forests, indicating low or nonexistent regeneration and corroborating the hypothesis that these Araucaria stands are in decline phase and will be lost in a short period of time (Drake et al. 2005).

In order to assess whether ongoing patterns of habitat fragmentation threat the species genetic resources and its evolutionary potential, it is necessary to explore the genetic diversity and ecological attributes across the range of the species. Even more, the correlation of genetic data with ecological and geographic variables may show more complete information about the state of the species. Turner (2001) mentioned that by combining all these aspects areas of potential conservation may be identified.

Until now, there is no a complete diagnosis about the situation and distribution of the genetic diversity of the species and the available information is contradictory, in some cases. Studies using different markers as gas liquid chromatography (Raffi & Dodd 1998), isozymes (Gallo et al. 2003; Ruiz et al. 2007) and RAPDs (Bekessy et al. 2002, 2003) have indicated that genetic diversity is high and unaffected by the ecological heterogeneity of the species. Conversely, a recent study based on chloroplast and mitochondrial markers reported the low variation in the species (Marchelli et al. 2010). Likewise, seed storage proteins were use to determine whether these proteins could be used as a new marker of genetic diversity in *A. araucana*. The results revealed that globulin was the best fraction for measuring the variability in the species, due to their high level of variation (20 identified bands, eleven of them polymorphic) (Figure 7). To confirm Mendelian inheritance, segregation of all polymorphic bands was fit to the expected segregation ratio (1:1), and no significant deviation was detected in any case. Further, 31 cases of independence (fit a 1:1:1:1 segregation ratio) and 24 cases of linkage were detected (Alvarez 2010b). Furthermore, a high degree of genetic diversity was detected, with more than 92% of the variation within populations (Drake et al. 2012). This study highlighted that these proteins can be used to measure the genetic diversity in *A. araucana*, giving an additional tool to complement the information provided by other genetic markers.

As above mentioned, microsatellite have proved to be the most suitable markers for studying population genetic diversity. In this respect, our group has developed in recent times sixteen microsatellites in the species, being the first specific developed in the species (Marconi et al. 2011; Martín et al. in press). Furthermore, these primer pairs have shown high polymorphism level and a reliable and easily interpretable pattern. Given that the extent and distribution of genetic diversity are key elements for the species evolutionary potential, these primers may provide the baseline for the establishment of conservation strategies.

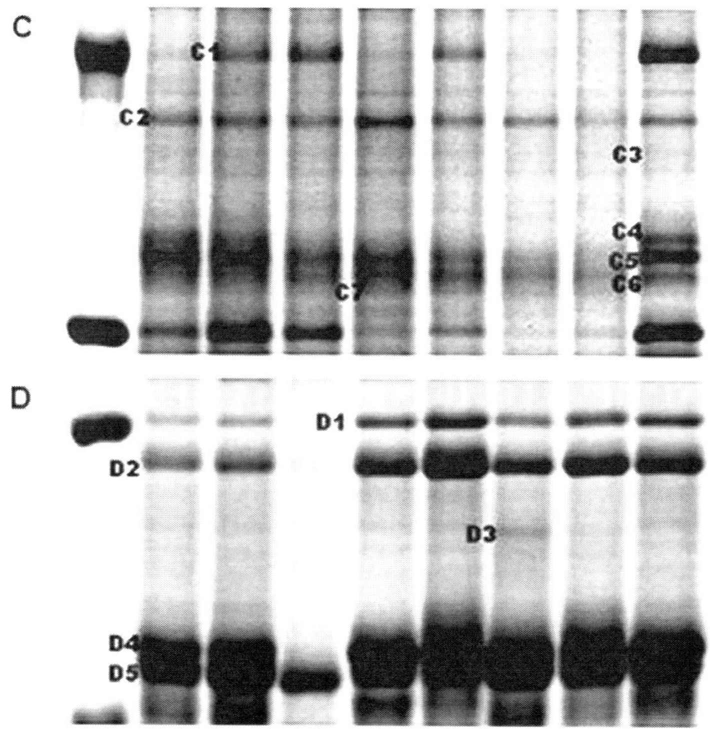

Figure 7. SDS-PAGE gel of the variation found in *A. araucana*. The first line is a molecular marker used as reference. C-D, zones classified according to the molecular marker. C1-D5, different bands found in the study.

To sum up, it is widely accepted that forest species have been deeply affected by human pressure. In some cases, their distribution has been drastically reduced and in others their management has affected the structure of the populations and the distribution of the genetic diversity. For all this and in this time of global crisis, it is urgent the protection of biodiversity as an externality of forests.

The persistence of populations is positively linked to their genetic variability, which is widely considered the main requirement for the long-term survival of species on an evolutionary time scale. For this, the understanding of the genetic diversity presents in a species and the distribution of this variation among populations is important to set up appropriate management strategies. In this respect, molecular genetic techniques, powerful statistical tools and the knowledge provided by managers of the resources could shed light on the evaluation of such genetic diversity, facilitating the prioritization in conservation decisions and monitoring forest genetic resources.

In this chapter, we have described the progress made in two very different species, such as *C. sativa* and *A. araucana*, in terms of biology, ecology and management.

Acknowledgments

This research was supported by grants AGL2009-07931 and AGL2010-15147 from the Spanish Ministry of Science and Innovation and the European Regional Development Fund

(FEDER) from the European Union and by grants A/023099/09 and A/030789/10 of the Spanish Agency of International Cooperation for the Development from the Spanish Ministry for Foreign Affairs and Cooperation. The first author is grateful to Agrifood Campus of International Excellence (ceiA3) from the Spanish Ministry of Education and the Ministry of Science and Innovation for the financial support.

REFERENCES

Aagesen, D.L. (2004). Burning monkey-puzzle: Native fire ecology and forest management in northern Patagonia. *Agriculture Human Values*, 21, 233-242.

Alvarez, J.B.; Muñoz-Diez, C.; Martín-Cuevas, A.; López, S.; Martín, L.M. (2003). Cotyledon storage proteins as markers of the genetic diversity in *Castanea sativa* Miller. *Theoretical and Applied Genetics*, 107, 730-735.

Álvarez, J.B.; Toledo, M.J.; Abellanas, B.; Martín, L.M. (2004). Use of megagametophyte storage proteins as markers of the genetic diversity in stone pine (*Pinus pinea* L.) in Andalusia, Spain. *Genetic Resources and Crop Evolution*, 51, 621-627.

Alvarez, J.B; Martín, M.A; Caballero, L.; Martín, L.M. (2010a). The role of plant genetic resources in the sustainable agriculture. In: Salazar A.; Rios I. (Eds.), Sustainable Agriculture: Technology, Planning and Management. *USA: Nova Science Publishers, Inc. pp. 145-176.*

Alvarez, A.M. (2010b). Desarrollo de un marcador polimórfico basado en las proteínas de reserva del megagametofito para *A. araucana*. Tesis final de Master. Córdoba University.

Bekessy, S.A.; Allnut, T.R.; Premoli, A.C.; Lara, A.; Ennos, R.A.; Burgman, M.A.; Cortes, M.; Newton, A.C. (2002). Genetic variation in the vulnerable and endemic Monkey Puzzle tree, detected using RAPDs. *Heredity*, 88, 243-249.

Bekessy, S.A.; Ennos, R.A.; Burgman, M.A.; Newton, A.C.; Ades, P.R. (2003). Neutral DNA markers fail to detect genetic divergence in an ecologically important trait. *Biological Conservation*, 110, 267-275.

Booy, G.; Hendricks, R.J.J.; Smulders, M.J.M.; van Groenendael, J.M.; Vosman, B. (2000). Genetic diversity and the survival of populations. *Plant Biology*, 2, 379-395.

Bounous, G. (2002). Il castagno. *Coltura, ambiente ed utilizzazioni in Italia e nel mondo.* Edagricole. Bologna.

Brown, A.H.D.; Hardner, C.M. (2000). Sampling the gene pools of forest trees for ex situ conservation. In: Young, A.; Boshier, D.; Boyle, T. (Eds.), *Forest conservation genetics. Principles and practise.* New Zeland: CABI publishing. pp. 185-196.

Buck, E.J.; Russell, K.; Hadonou, M.; James, C.J.; Blakesley, D. (2003). Isolation and characterization of polymorphic microsatellites in European chestnut (*Castanea sativa* Mill.). *Molecular Ecology Notes*, 3, 239-241.

Burns, B.R. (1993). Fire-induced dynamics of *Araucaria araucana- Nothofagus antartica* forest in the southern Andes. *Journal of Biogeography*, 20, 669-685.

Donoso, C. (2006). Las especies arbóreas de los Bosques Templados de Chile y Argentina. Autoecología. Marisa Cúneo Ediciones. Valdivia (Chile), pp. 678.

Drake, F.; Herrera, M.A.; Acuña, E. (2005). Propuesta de manejo sustentable de *Araucaria araucana* (Mol.) C. Koch. *Bosque*, 26, 23-32.

Drake, F.; Martín, M.A.; Herrera, M.A.; Molina, J.R.; Drake-Martin, F.; Martín, L.M. (2009). Networking sampling of *Araucaria araucana* (Mol.) K. Koch in Chile and the bordering zone of Argentina: implications for the genetic resources and the sustainable management. *iForest,* 2, 207-212.

Drake, F.; Martín, M.A.; Alvarez, A.; Molina, J.R.; Alvarez, J.B.; Herrera, M.A.; Martín L.M. (2012c). Development of a new polymorphic genetic marker in *Araucaria araucana* (Mol) K. Koch. *Spanish Journal of Agricultural Research (in press).*

El-Kassaby, Y.A..; Meagher, M.D.; Parkinson, J.; Portlock, F.T. (1987). Allozyme inheritance, heterozygosity and outcrossing rate among *Pinus monticola* near Ladysmith, British Columbia. *Heredity*, 58, 173-181.

Esquinas-Alcázar, J. (2005). Protecting crop genetic diversity for food security: political, ethical and technical challenges. *Nature Reviews Genetics*, 6, 946-953.

Fernández-Chaves, L.; Vargas-Osuna, E. (2006). El control de la carpocapsa en la castaña: la experimentación en la Serranía de Ronda. In: GDR Serranía de Ronda (Eds.)., *3ª Feria Andaluza de la Castaña*. Serranía de Ronda, Málaga. pp. 125-133.

Frankman, R, Ralls, K. (1998) Conservation biology: inbreeding leads to extinction. *Nature,* 392, 441-442.

Gallo, L. (2003). Conservación, manejo y uso sustentable de los recursos genéticos de la *Araucaria araucana* en Argentina, Comunidades Aucapan y Chiuquilihuin, San Carlos de Bariloche. *Report of the Project Conservation, Management and Sustainable Use of Forest Genetic Resources with reference to Brazil and Argentina.* International Plant Genetic Resources Institute, Rome, Italy.

Gepts, P. (1990). Genetic diversity of seed storage proteins in plants. In: Brown, A.H.D.; Clegg, M.T.; Kahler, A.L.; Weir, B.S. (Eds.)., *Plant population genetics, breeding and genetic resources.* Massachusetts, USA: Sinauer Associates Inc Publishers, Suderland. pp. 64-82.

Glaubitz, J.C.; Moran, G.F. (2000). Genetic Tools: The use of Biochemical and Molecular Markers. In: Young, A.; Boshier, D.; Boyle, T. (Eds)., *Forest Conservation Genetics.* Australia: CSIRO Publishing. pp. 39-52.

Gobbin, D.; Hohl, L.; Conza, L.; Jermini, M.; Gessler, C.; Conedera, M. (2007). Microsatellite-based characterization of the *Castanea sativa* cultivar heritage of southern Switzerland. *Genome*, 50, 1089-1103.

Gondard, H.; Regina, I.S.; Salazar, S.; Peix, A.; Romane, F. (2007). Effects of forest management on plant species diversity in *Castanea sativa* stands in Salamanca (Spain) and the Cevenes (France). *Scientific Research and Essays*, 2, 62-70.

Grivet, D.; Sork, V.L.; Westfall, R.D.; Daviss, F.W. (2008). Conserving the evolutionary potential of California valley oak (*Quercus lobata* Née): a multivariate genetic approach to conserving planning. *Molecular Ecology*, 17, 139-156.

Hamrick, J.L.; Godt, M.J.W. (1990). Allozyme diversity in plant species. In: Brown, H.D.; Clegg, M.T.; Khaler A.L.; Weir B.S. (Eds.)., *Plant Population Genetics, Breeding, and Genetic Resources*. Sunderland, MA: Sinauer Press,. pp. 43-63.

Harlan, J.R. (1992). *Crops and Man. American Society of Agronomy/Crop Science Society of America*, Madison, Wisconsin. Second Edition.

Henle, K.; Lindenmayer, D.B.; Margules, C.R.; Saunders, D.A.; Wissel, C. (2004). Species survival in fragmented landscapes: where are we now? *Biodiversity and Conservation*, 13, 1–8.

Herrmann, T. (2006). Indigenous knowledge and management of Araucaria araucana forest in the Chilean Andes: implications for native forest conservation. *Biodiversity and Conservation*, 15, 647-662.

Heusser, CJ..; Rabassa, J.; Brandant, A.; Stuckenrath, R. (1988). Late-Holocene vegetation of the Andean Araucaria region, Province Neuquen, Argentina. *Mountain Research and Development*, 8, 53-63.

Hoffmann, A.; Sierra, M.; Prosser, C.; Valle, M. (2001). *Enciclopedia de los Bosques Chilenos*. Gráfica Andes, Santiago de Chile, Chile.

Holderegger, R.; Buehler, D.; Gugerli, F.; Manel, S. (2010). Landscape genetics of plants. *Trends in Plant Science*, 15, 675-682.

IUCN. (1996). World list of threatened trees. International Union for Conservation of Nature, Gland, Switzerland. *http://www.iucn.org/themes/ssc/redlist/redlist.htm*.

Jones, C.G.; Lawton, J.H.; Shachak, M. (1994). Organisms as ecosystem engineers. *Oikos*, 69, 373-386.

Latta, R.G. (2006). Integrating patterns across multiple genetic markers to infer spatial process. *Landscape Ecology*, 21, 809-820.

Manel, S.; Schwartz, M.K.; Luikart, G., Taberlet, P. (2003). Landscape genetics: combining landscape ecology and population genetics. *Trends in Ecology and Evolution*, 18, 189-197.

McKay, J.W. (1942). Self-sterility in the Chinese chestnut (*Castanea mollisima*). *Proceedings of the American Society of Horticultural Sciences*, 41, 156-160.

Marchelli, P.; Baier, C.; Mengel, C.; Ziegenhagen, B.; Gallo, L.A. (2010). Biogeographic history of the threatened species *Araucaria araucana* (Molina) K. Koch and implications for conservation: a case study with organelle DNA markers. *Conservation Genetics*, 11, 951-963.

Marconi, G.; Martín, M.A.; Cherubini, M.; Raggi, L.; Drake, F.; Villani, F.; Albertini, E.; Mattioni, C. (2011). Microsatellite-AFLP development for *Araucaria araucana* (Mol.) K. Koch, an endangered conifer of Chilean and Argentinean native forests. *Silvae Genetica*, 60, 285-288.

Marinoni, D.; Akkak, A.; Bounous, G.; Edwards, K.J.; Botta, R. (2003). Development and characterization of microsatellite markers in *Castanea sativa* (M.). *Molecular Breeding*, 11,127-136.

Market, C.L.; Moller, F. (1959). Multiple forms of enzymes: tissue, ontogenetic and species specific pattern. *Proceedings of the Natural Academy of Sciences USA,* 45, 753-763.

Martín, M.A.; Moral, A.; Martín, L.M.; Alvarez, J.B. (2007). The genetic resources of European sweet chestnut (*Castanea sativa* Miller) in Andalusia, Spain. *Genetic Resources and Crop Evolution*, 54, 379-387.

Martín, M.A.; Martín, L.M.; Álvarez, J.B. (2008). Nut characterization of the main traditional chestnut varieties from Andalusia. *Acta Horticulturae*, 784, 71-76.

Martín, M.A.; Navarro-Cerrillo, R.; Ortega, P.; Alvarez, J.B. (2009a). The use of cotyledon proteins to assess the genetic diversity in sweet holm oak. *Journal of Forest Science,* 55, 526-531.

Martín, M.A.; Alvarez, J.B.; Mattioni, C.; Cherubini, M.; Villani, F.; Martín, L.M. (2009b). Identification and characterization of traditional chestnut varieties of Southern Spain using morphological and simple sequence repeat (SSRs) markers. *Annals of Applied Biology*, 154, 389-398.

Martín, M.A.; Alvarez, J.B.; Martín, L.M. (2010a). Genetic diversity of Spanish fir (*Abies pinsapo* Boiss.) populations by means of megagametophyte storage proteins. *Annals of Forest Science*, 67, 603 (7p).

Martín, M.A.; Muñoz, S.; Muñoz, F.; Uribe, M.; Molina, J.R.; Herrera, M.A.; Martín, L.M.; Alvarez, J.B. (2010b). Primeros resultados en el desarrollo de un marcador genético basado en las proteínas de reserva en especies del género *Nothofagus*. *Bosque*, 31, 252-257.

Martín, M.A.; Mattioni, C.; Cherubini, M.; Taurchini, D.; Villani, F. (2010c). Neutral and adaptive genetic diversity in European chestnut populations by means of genomic and genic microsatellite markers. *Tree Genetics & Genomes*, 6, 735-744.

Martín, M.A.; Mattioni, C.; Cherubini, M.; Taurchini, D.; Villani, F. (2010d). Genetic characterisation of traditional chestnut varieties in Italy using microsatellite (SSRs) markers. *Annals of Applied Biology*, 157, 27-44.

Martín, M.A.; Mattioni, C.; Molina, J.R.; Álvarez, J.B.; Cherubini, M.; Herrera, M.A.; Villani, F.; Martín, L.M. (2012a). Landscape genetic structure of chestnut (*Castanea sativa* Mill.) in Spain. *Tree Genetics & Genomes,* 8, 127-136.

Martín, M.A.; Alvarez, J.B.; Gutierrez, J.C.; Martín, L.M. (2012b). Inheritance of cotyledon storage proteins in European sweet chestnut (Castanea sativa Miller). *Forest Systems* (in press).

Martín, M.A; (2012c)sjar.

Martín, M.A.; Mattioni, C.; Lusini, I.; Drake, F.; Cherubini, M.; Herrera, M.A.; Villani, F.; Martín, L.M. Microsatellite development for the relictual conifer *Araucaria araucana* (Araucariaceae) using next-generation sequencing. *American Journal of Botany* (in press).

Mattioni, C.; Cherubini, M.; Micheli, E.; Villani, F.; Bucci, G. (2008). Role of domestication in shaping *Castanea sativa* genetic variation in Europe. *Tree Genetics & Genomes*, 4, 563-574.

Merlo, M.; Rojas-Briales E. (2000). Public goods and externalities linked to Mediterranean forests: economic nature and policy. *Land Use Policy*, 17, 197-208.

Montaldo, P.R. (1974). La bio-ecología de *Araucaria araucana* (Mol.) Koch. *Instituto Forestal Latino-Americano de Investigación y Capacitación Boletín*, 46-48, 1-55.

Müller-Stark, G.; Schubert, R. (2000). Genetic markers as tool for bioindication in forest ecosystems. In: Young, A.; Boshier, D.; Boyle, T. (Eds.)., *Forest Conservation Genetics: principles and practise*. CSIRO-CABI. pp. 227-237.

Mutarelli, E.; Orfila, E. (1992). Ensayo de tratamientos experimentales en bosques de *Araucaria araucana* (Mol.) C. Koch. en la zona del Lago Moquehue, provincia de Neuquén, Argentina. *Revista Forestal Argentina*, 14, 109-117.

Namkoong, G.; Ouédrago, A.S. (1997). *Developing decision making support strategies for conservation and sustainable use of forest genetic resources*. Research proposal, Roma, Italia: IPGRI-UBC, IPGRI.

Neale, D.; Adams, W.T. (1981). Inheritance of isozyme variants in seed tissues of balsam fir (*Abies balsamea*). *Canadian Journal of Botany*, 59, 1285-1291.

Nuez, F.; Carrillo, J.M.; De Ron, A.M. (2000). Introducción. In: Nuez, F.; Carrillo, J.M.; (Eds.)., *Los Marcadores Genéticos en la Mejora Vegetal*. Universidad Politécnica de Valencia. pp. 1-23.

Osborne, T.B. (1924). *The vegetable proteins*. Longmans, Green & Co. London.

Peakall, R.; Gilmore, S.; Keys, W.; Morgante, M.; Rafalski, A. (1998). Cross-species amplification of soybean (*Glycine max*) simple sequence repeats (SSRs) within the genus and other legume genera: Implications fro the transferability of SSRs in plants. *Molecular Biology and Evolution*, 15, 1275-1287.

Penel, V.; Dechamp, C.; Calleja, M. (2002). Evolution of the dates of pollination of nine taxons of trees in the Lyon-Bron region (Rhone) from 1985 to 2000, with particular attention to chestnut trees. *Review for Allergy Immunology Clinic,* 42, 551-555.

Pereira-Lorenzo, S.; Fernández-López, J. (1997). Description of 80 cultivars and 36 clonal selection of chestnut (*Castanea sativa* Mill.) from Northwestern Spain. *Fruit varieties Journal*, 51, 13-27.

Pereira-Lorenzo, S.; Ramos-Cabrer, A.M.; Diaz-Hernandez, B.; Ascasibar-Errasti, J.; Sau, F.; Ciordia-Ara, M. (2001). Spanish chestnut cultivars. *Hortscience*, 36, 344-347.

Perry, D.A. (1998). The scientific basis of forestry. *Annual Review of Ecology and Systematics*, 29, 435-486.

Pigou, A.C. (1920). *The Economics of Welfare*. Macmillan, London.

Powell, W.; Machray, G.C.; Provan, J. (1996). Polymorphism revealed by simple sequence repeats. *Trends in Plant Science,* 1, 215-222.

Premoli, A.C.; Kitzberger, T.; Veblen, T.T. (2000). Isozyme variation and recent biogeographical history of the long-lived conifer *Fitzroya cupressoides*. *Journal of Biogeography*, 27, 251–260.

Rafii, Z.A.; Dodd, R.S. (1998). Genetic diversity among coastal and Andean natural populations of *Araucaria araucana* (Molina), K. Koch. *Biochemical Systematics and Ecology*, 26, 441-451.

Ramos, R. (2011). Caracterización de variedades tradicionales de castaño (Castanea sativa Mill.) en Andalucía por caracteres productivos. Tesis de Máster, Universidad de Córdoba.

Ruiz, E.; González, F.; Torres-Diaz, C.; Fuentes, G.; Mardones, M.; Stuessy, T.; Samuel, R.; Becerra, J.; Silva, M. (2007). Genetic and differentiation within and among Chilean populations of Araucaria araucana (Araucariaceae) based on allozyme variabilty. *Taxon*, 56,1221-1228.

Saccheri, I.; Kuussaari, M.; Kankare, M.; Vikman, P.; Fortelius, W.; Hanski, I. (1998). Inbreeding and extintion in a butterfly metapopulation. *Nature*, 392, 491-494.

Schilling, R.; Donoso, C. (1976). Reproducción vegetativa natural de *Araucaria araucana* (Mol.) C. Koch. *Investigaciones Agricultura*, 2, 121-122.

Shachak, M.; Boeken, B.; Groner, E.; Kadmon, R.; Lubin, Y.; Meron, E.; Neeman, G.; Perevolotsky, A.; Shkedy, Y.; Ungar, E.D. (2008). Woody species as landscape modulators and their effect on biodiversity patterns. *Bioscience*, 58, 209–221.

Tautz, D.; Renz, M. (1984). Simple sequences are ubiquitous repetitive components of eukaryotic genomes. *Nucleic Acids Research*, 12, 4127-4138.

Turner, M. (2001). Landscape Ecology in Theory and Practise. Springer-Verlag.

Varshney, R.K.; Graner, A.; Sorrells, M.E. (2005). Genic microsatellite markers in plants: features and applications. *Trends in Biotechnology*, 23, 48-55.

Veblen, T.T. (1982). Regeneration patterns in *Araucaria araucana* forest in Chile. *Journal of Biogeography,* 9: 11-28.

Veblen, T.T.; Burns, B.R.; Kitzberger, T.; Lara, A.; Villalba, R. (1995). The ecology of the conifers of southern South America. In: Enright, N.; Hill, R. (Eds.)., *Ecology of the Southern Conifers*. Melbourne, pp. 120-155.

Wilson, E.O., ed. (1988). *Biodiversity*. National Academy Press. Washington, DC.

World Commission on Environment and Development. (1987). *Our common future*. Oxford UK: Oxford University Press.

Wright, J.P.; Jones, C.G. (2006). The concept of organisms as ecosystem engineers ten years on: progress, limitations, and challenges. *BioScience*, 56, 203–209.

Young, A.; Boyle, T. (2000). Forest fragmentation. In: Young, A.; Boshier, D.; Boyle, T. (Eds.), *Forest conservation genetics. Principles and practise*. New Zeland: CABI publishing. pp. 185-196.

In: Genetic Diversity: New Research
Editors: J. A. C. Amaya and M. M. F. Jimenez
ISBN: 978-1-62081-598-4
© 2012 Nova Science Publishers, Inc.

Chapter 7

NEW GENETIC PROVISIONS FOR THE REINSTATEMENT OF THE GINSENG NATURAL POPULATIONS

Yuriy Zhuravlev, Galina Reunova, Irina Kats and Tamara Muzarok
Institute of Biology and Soil Science, Russian Academy of Sciences, Vladivostok, Russia

ABSTRACT

Ginseng (*Panax ginseng* C.A. Meyer) belongs to the family Araliaceae and is one of the most important medicinal plants in East Asia. There are almost no wild ginseng plants left in China and Korea, and Russia is the only country where *P. ginseng* is still growing in its natural habitats. All natural populations of *P. ginseng* are now mostly exhausted; therefore, reintroduction is essential to restore this plant to its former range. One hundred sixty-seven (167) *P. ginseng* individuals were collected from the ten administrative areas of Primorsky Krai and studied using AFLP method. A high level of intraspecific genetic diversity ($P = 55.68\%$, $H_S = 0.1891$ at population level; $P = 99.65\%$, $H_t = 0.2857$ at species level) and significant population-genetic differentiation ($\theta_B = 0.363$ from the Bayesian full model; $F_{ST} = 0.360$ from the non-hierarchical AMOVA) were detected for fluorescent AFLPs. The populations with the highest and lowest levels of genetic diversity were determined, and these populations should be conserved for the first time. Two divergent population groups were identified with hierarchical AMOVA, by the UPGMA method, by principal coordinates analysis (PCoA), in minimum spanning tree (MST) and using the STRUCTURE program. These two groups should be recognized as management units (MUs) because the low level of gene flow among them indicates contemporary demographic independence. Using an MST analysis and assignment test, the content of *P. ginseng* populations was verified to differentiate the true population plants from the plants that are not inherent to the population under study. This differentiation can identify the "pure" plants that could be chosen to serve as a stock material for reintroduction.

Keywords: AFLP, genetic diversity, genetic structure, *Panax ginseng* C. A. Meyer, STRUCTURE, assignment test

INTRODUCTION

Ginseng (*Panax ginseng* C.A. Meyer) belongs to the family Araliaceae and is one of the most important medicinal plants in East Asia. The name ginseng is derived from a Chinese term referring to the "man-like" shape of the root. The genus name *Panax* means "cure all" in Greek (Qi et al., 2011). China and Korea have utilized ginseng for more than 2000 years because of its stimulating and tonic properties (In et al., 2010). Particularly in Korea, China and Japan, ginseng is considered the most valuable of all medicinal herbs (Choi et al., 2008). The pharmacological efficacy of *P. ginseng* as identified by modern science includes improved brain function, pain-relieving effects, preventive effects against tumors, anti-tumor activity, enhanced immune system function, anti-diabetic effects, enhanced liver function, improved blood pressure, anti-fatigue and anti-stress effects, improvements to climacteric disorder and sexual functions, and anti-oxidative and anti-aging effects (Choi et. al., 2008).

In the first half of the 20th century, wild-growing ginseng plants could be found yet on the large forestlands of China, Korea and Russian Primorye (Grushwitsky, 1961), but currently, there are almost no wild ginseng plants left in China and Korea. Russia is the only country where *P. ginseng* is still growing in its natural habitats. All natural populations of *P. ginseng* are largely exhausted; therefore reintroduction is essential to restore this plant to its former range. Knowledge of genetic diversity within a population and among populations is important for conservation management, especially in identifying genetically unique structural units within a species and determining the populations that need protection. However, it is important to keep in mind that the genetic resource of wild *P. ginseng* populations might have been influenced by the long history of controlled and unncontrolled ginseng cultivation in the field and, more important, forest farms. Therefore, it is necessary to verify the content of 'indigenous' populations to differentiate the true population members from the plants that are not inherent to the population under study.

Investigation of the population-genetic characteristics is one of the principal tasks in a program of species conservation (Hamrick & Godt, 1996). Choosing suitable genetic markers is an important step in studies of the genetic variability and population structure of a species. The allozyme analysis examines only the protein-coding part of DNA and thus cannot characterize the variability of a whole genome. Multi-locus DNA markers, e.g., random amplified polymorphic DNA (RAPD) (Williams et al., 1990), inter-simple sequence repeats (ISSRs) (Zietkiewicz et al., 1994) and amplified fragment length polymorphisms (AFLPs) (Vos et al., 1995) potentially produce higher values of polymorphism than allozyme analysis because non-coding DNA sequences, which mutate at a higher rate than coding sequences, can also be characterized. Our studies have shown that among the RAPD, ISSR and AFLP methods to investigate the genetic variability, genetic structure and genetic relationships of *P. ginseng* populations, AFLP combined with fragment analysis on an automated DNA sequencer is the most promising (Reunova et al., 2010). Here we used AFLP technology to study the genetic diversity of ginseng populations inhabiting different locations of Primorsky Krai, Russia.

Materials and Methods

Sampled Populations

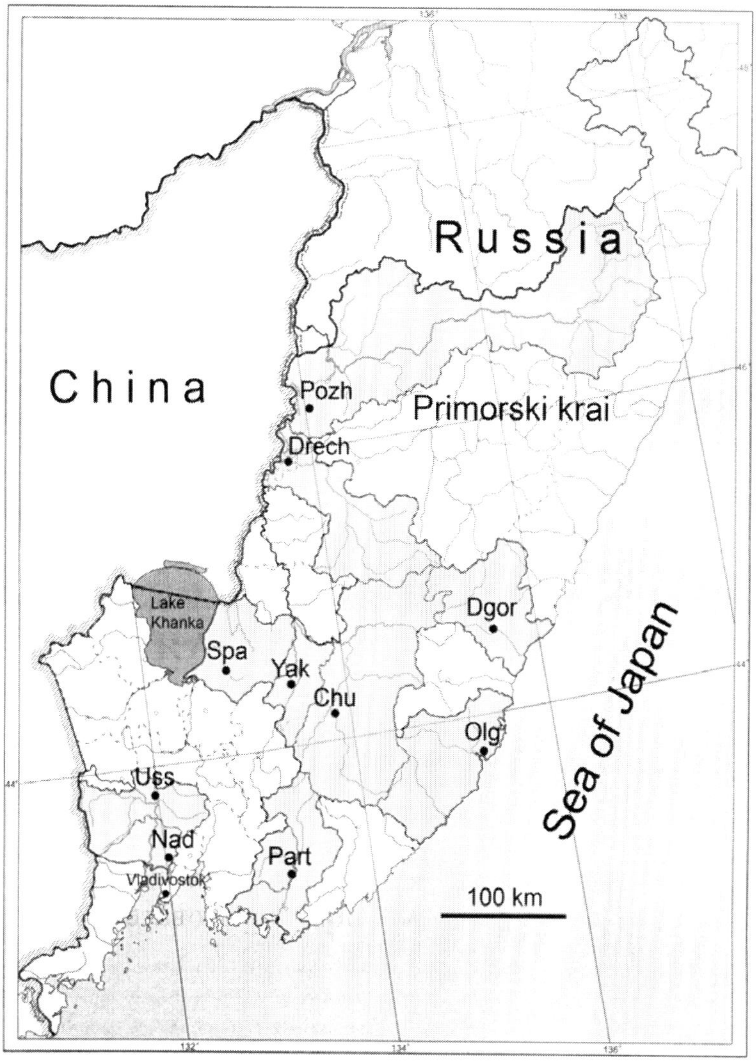

Figure 1. The administrative areas in the territory of Primorskiy krai of Russia where *Panax ginseng* plants were collected. See "Methods" for the names of these areas, the code of populations and the number of collected plants.

One hundred sixty-seven (167) *P. ginseng* individuals were collected from the ten administrative areas of Primorsky Krai (Fig.1) and transferred to a collection nursery. The populations studied were coded according the names of the areas. Twenty (20) *P. ginseng* individuals were collected from the Chuguevsk area (*Chu*), 19 from the Spassk area (*Spa*), 16 from the Ussuriisk area (*Uss*), 13 from the Dalnerechensk area (*Drech*), 16 from the Dalnegorsk area (*Dgor*), 15 from the Olginsk area (*Olg*), 15 from the Pozharsk area (*Pozh*),

24 from the Nadezhdinsk area (*Nad*), 19 from the Partizansk area (*Part*) and 10 from the Yakovlevsk area (*Yak*).

DNA Extraction and AFLP Procedure

Total genomic DNA was extracted from fresh leaf tissue according to Echt et al. (1992). The extracted DNA was purified according to the Murray &Thompson (1980) method.

AFLP genotyping was performed according to Vos et al. (1995) using the *Eco*RI and *Mse*I restriction enzymes. Pre-amplification reactions utilized AFLP primers with two selective nucleotides. After testing nine (9) AFLP selective amplification primers pairs, namely *Eco*(ACG)/*Mse*(CCTC), *Eco*(ACG)/*Mse*(CCTT), *Eco*(ACA)/*Mse*(CCTG), *Eco*(ACA)/*Mse*(CCGG), *Eco*(ACA)/*Mse*(CCAC), *Eco*(ACT)/*Mse*(CCGA), *Eco*(ACT)/Mse(CCTA), *Eco*(ACC)/*Mse*(CCAG), and *Eco*(ACC)/*Mse*(CCGC), the pairs *Eco*(ACA)/*Mse*(CCGG) and *Eco*(ACA)/*Mse*(CCTG) (Table 1) were chosen for the analysis, because they were highly reproducible and easy to score. The reproducibility of each primer pair was checked by carrying out the AFLP procedure twice for four randomly chosen individuals. AFLP adapters and primers were purchased from Syntol (Russia). All the *Eco*RI-NNN selective primers were labeled with fluorescent 6-carboxy fluorescein (6-FAM) at the 5' end. The fluorescence-labelled selective amplification products were separated on a polymer matrix with an internal size standard (GeneScan-500 LIZ, Applied Biosystems) on an automated sequencer (ABI PRISM 3100 Genetic Analyzer, Applied Biosystems). The raw data were collected and aligned with the internal size standard using GENSCAN v.3.7 software (Applied Biosystems). All unambiguous peaks including monomorphic peaks between 50 and 500 base pairs (bp) were analyzed and the scoring results were exported as a presence/absence matrix.

Table 1. AFLP selective primers used in the study of the population genetics of *Panax ginseng*

Eco RI primer	*Mse* I primer	Number of loci
E-ACA (AGACTGCGTACCAATTCACA)	*M-CCGG* (GATGAGTCCTGAGTAACCGG)	149
E-ACA (AGACTGCGTACCAATTCACA)	*M-CCTG* (GATGAGTCCTGAGTAACC TG)	133
Total		282

Data Analysis

The data matrix was generated for each plant of the ten populations in which each locus was treated as a separate character and scored as present (1) or absent (0). The binary matrix was used to calculate genetic diversity indices of the populations and to estimate genetic distances among the populations using standard software.

Genetic Diversity of the Populations

Using POPGENE 32 (POPGENE v. 1.31, Centre for International Forestry Research, University of Alberta and Tim Boyle, Canada), the number and the percentage of polymorphic fragments were determined (Yeh & Boyle, 1997). Gene diversity was calculated with a Bayesian approach using HICKORY v.0.8 (Holsinger & Lewis, 2003), without assuming Hardy-Weinberg equilibrium within populations.

Population Differentiation

To assess genetic differentiation among populations, we conducted analyses of molecular variance (AMOVA; Excoffier et al., 1992) based on pairwise squared Euclidean distances between samples. The same software was used to calculate values for pairwise F_{ST} between populations. Pairwise θ_B values (equivalent of F_{ST}) was calculated with a Bayesian approach using HICKORY (Holsinger & Lewis, 2003). This approach does not assume that genotypes are in Hardy-Weinberg equilibrium and allows direct estimates of F_{ST} from dominant markers. The method also provides some information about the degree of within-population inbreeding f (estimate of F_{IS}), given that there are large numbers of loci and populations. We tested four Bayesian models: (i) a full model, (ii) a model that assumed no inbreeding ($f = 0$), (iii) a model that assumed no population structure ($\theta_B = 0$), and (iv) an f-free model that allowed the incorporation of uncertainty about f into the analysis. The models were evaluated using the measures of deviance information criterion (DIC) to decide which model was preferred (smaller DIC values indicate a better fit) (Spiegelhalter et al., 2002).

The presence of genetic structure at the population and regional levels was tested by an (AMOVA) (Excoffier et al., 1992). We examined three different population structure models with an AMOVA. In the first model, no grouping of populations was defined, and partitioning genetic variation among and within populations was performed regardless of their geographic distribution. In the second and third models, we defined regional groups of populations for further analyses to determine whether genetic differentiation among the regions existed.

From the F_{ST} values obtained using an AMOVA, the indirect estimates of gene flow between populations (N_m) were calculated using on Wright's island model of population structure with the formula $N_m = (1-F_{ST})/4F_{ST}$ (Slatkin & Barton, 1989). To test for isolation by distance, a Mantel test was carried out by regressing $(1-F_{ST})/4F_{ST}$ against the natural logarithm of geographic distance (Mantel, 1967) between all pairs of samples, using ARLEQUIN (Schneider et al., 2000).

AFLP Clustering Analyses and Assignment Tests

Additional analyses were carried out to further investigate the genetic structuring of *P. ginseng*. A principal coordinate analysis (PCoA) based on Euclidean distances between AFLP phenotypes was performed using GENALEX version 6.0 (Peakall & Smouse, 2006) to transform the multidimensional genetic distances between all *P. ginseng* individuals into a 2-dimensional representation that explained as much of the observed variance as possible and to obtain a visualization of the genetic separation of the populations along the axes. However, the explanatory power of the main axes was often only moderate (explaining 40 or 50% of the variation in data). Therefore, we used also other approaches to estimate genetic relationships among the populations. Nei's allele-sharing pairwise genetic distance data (Nei, 1978) were calculated by the program TFPGA (Miller, 1997), and trees were constructed among populations using the UPGMA clustering method. From a matrix of squared Euclidean distances, we reconstructed the minimum spanning tree (MS-tree) between representatives of *P. ginseng* using ARLEQUIN (Schneider et al., 2000).

A Bayesian-based clustering method was also applied to infer how many clusters (K) were most appropriate for interpreting the data without prior information about the number of locations at which the individuals were sampled. The original data matrix of AFLP was imported into the STRUCTURE program, version 2.3.2 (Pritchard et al. 2000), which estimates the log probability of data Pr(X| K) for each given value of K, allowing the estimation of the more likely number of clusters. The program was run under the admixture model, which allows mixed ancestries of individuals, and the option of the correlated allele frequency model, which assumes that allele frequencies in different populations are likely to be similar. These models were considered best by Falush et al. (2003) in cases of subtle population structure. We used 5,000–40,000 burn-in cycles and 5,000– 400,000 Marcov chain Monte Carlo (MCMC) runs for K = 1–10, with 3–8 replicates per K value. K was also selected using the ΔK statistic (Evanno et al., 2005), calculated with the program STRUCTURE HARVESTER v.0.6.8. (Earl, 2009). The ΔK statistic assesses the rate of change in the log probability of the data between successive K values (Evanno et al., 2005), with the maximal ΔK value taken to identify the correct number of clusters. To estimate the immigration rate in the set of analyzed populations, AFLP data were subjected to individual-based assignment tests by applying the maximum likelihood method for dominant data using AFLPOP 1.1 (Duchesne & Bernatchez, 2002). Our goal was to detect genetic outliers (i.e. migrant genotypes) in the sampled populations.

RESULTS

Genetic Diversity within Populations

Using the two primer pairs *Eco*(ACA)/*Mse*(CCTG) and *Eco*(ACA)/*Mse*(CCGG) (Table 1), we detected polymorphic bands among the various samples of *P. ginseng* in this study. Among the 282 scored fragments (Table 1), 281 were polymorphic across the 167 plants of the ten populations (Table 2). Genetic variability was high within populations (Table 2). The intra-population genetic polymorphisms ranged from a prevalence of 38.65% (*Uss*) to 69.15% (*Chu*) with an average of 55.68%. At the species level, the mean prevalence of polymorphisms was 99.65%. The highest genetic diversity values (approximately 70%) were obtained in the *Chu*, *Nad*, *Olg* and *Pozh* populations, whereas the lowest values (approximately 40%) were found in the *Uss* and *Dgor* populations (Table 2).

Based on the polymorphic markers, Bayesian gene diversity ranged from 0.1204 to 0.2330 with an average of 0.1891 at the population level and 0.2857 at the species level. The highest genetic diversity value (0.2330) was found in the *Olg* population, whereas the lowest value (0.1204) was observed in the *Uss* populations (Table 2).

Table 2. Sample size and genetic variability parameters of *Panax ginseng* populations calculated from AFLP data for 282 fragments. Standard deviations are given in parentheses

Population code	Number of plants (order numbers of plants)	Bayesian gene diversity	Polymorphic loci	
			Number	% (*P*)
Spa	19 (17-35)	0.2107(0.004)	163	57.80
Yak	10 (158-167)	0.1838(0.006)	145	51.42
Drech	13 (36-48)	0.2009(0.005)	138	48.94
Pozh	15 (100-114)	0.2323(0.005)	186	65.96
Uss	16 (1-16)	0.1204(0.004)	109	38.65
Nad	24 (115-138)	0.1902(0.004)	190	67.38
Chu	20 (49-68)	0.2245(0.004)	195	69.15
Dgor	16 (69-84)	0.1410(0.004)	114	40.43
Olg	15 (85-99)	0.2330(0.005)	188	66.67
Part	19 (139-157)	0.1547(0.004)	142	50.35
Population average	17	0.1891(0.002)	157	55.68
Species-level value	167	0.2857(0.002)	281	99.65

Table 3. Pairwise genetic distance matrix based on genetic differentiation (F_{ST}) (lower diagonal) and geographic distance (km, upper diagonal) for 10 populations of *Panax ginseng*

	Uss	*Spa*	*Drech*	*Chu*	*Dgor*	*Olg*	*Pozh*	*Nad*	*Part*	*Yak*
Uss	0.00000	124	244	183	293	232	368	35	125	124
Spa	0.41235	0.00000	122	110	198	190	247	156	168	45
Drech	0.27212	0.53153	0.00000	124	117	185	150	278	242	128
Chu	0.30808	0.26936	0.47046	0.00000	114	80	272	205	126	72
Dgor	0.35629	0.52259	0.60506	0.36954	0.00000	122	233	318	235	177
Olg	0.30464	0.25556	0.49335	0.09180	0.36057	0.00000	328	250	137	149
Pozh	0.18200	0.42200	0.21348	0.35356	0.40031	0.35103	0.00000	404	391	267
Nad	0.21894	0.48054	0.54275	0.32029	0.27451	0.31409	0.33000	0.00000	124	150
Part	0.38764	0.25381	0.42708	0.24434	0.49424	0.30650	0.35041	0.46318	0.00000	128
Yak	0.34993	0.16600	0.52375	0.27691	0.42249	0.15721	0.38540	0.39335	0.36194	0.00000

P value = 0.00000

Analysis of Population Differentiation and Gene Flow

All pair wise F_{ST} between populations, obtained with an AMOVA, were significant (P = 0.0000) and varied from 0.09180 (*Olg-Chu*) to 0.60506 (*Dgor-Drech*), indicating that all populations may be considered different from the others (Table 3). The *non*-hierarchical *AMOVAs* revealed that 35.54% of the total variation was attributed to the variability among the populations, whereas 64.46% was represented within the populations (Table 4). In the hierarchical *AMOVA* analysis carried out in the model with two main population clusters identified in the UPGMA analysis (group 1 and group 2) (see below), all variance components were highly significant (Table 4), and 17.33% of the variation was found among the groups. In separate AMOVAs of every population cluster, the proportions of the genetic variation observed within populations were 78.26% and 67.86% in group 1 and group 2, respectively (Table 4). In contrast, in the model based on the groups inferred from the *P.*

ginseng habitats (Zhuravlev et al., 2008) of the Blue Mountains (Russia) (populations *Spa*, *Yak*, *Drech* and *Pozh*); Khasan District (Russia) and Jilin and Heilongjiang Provinces (China) (populations *Uss* and *Nad*) and the southern part of Sikhote-Alin (Russia) (populations *Part*, *Chu*, *Dgor* and *Olg*), the variance component among groups was slightly negative (–5.44%), suggesting an absence of genetic structure at this hierarchical level (Table 5). The total variation among groups was not significant (P = 0.862), confirming that the model was not adequate.

Table 4. Analysis of molecular variance (AMOVA) for 167 individuals of *Panax ginseng* based on 282 AFLP markers. The hierarchical *AMOVA* analysis was carried out based on the groups inferred from the UPGMA analyses: group 1 (populations *Spa, Chu* and *Drech*), group 2 (populations *Pozh, Nad, Part, Yak, Uss, Dgor* and *Olg*). d.f = degrees of freedom, MSD = mean squared deviations. The *P*-value is estimated based on 1023 permutations

AMOVA analysis Sources of variation	d.f	MSD	Variance components	% of total variation	N_m	F_{ST}
Non-hierarchical analysis						
Among populations	9	2413.14	14.56	35.54	0.45	0.36
Within populations	157	4145.65	26.40	64.46		
Hierarchical analysis						
Among groups	1	777.00	7.80	17.33	0.40	0.41
Among populations	8	1636.13	10.82	24.02		
Within populations	157	4145.65	26.41	58.65		
Group 1						
Among populations	2	346.52	8.39	21.74	0.90	0.22
Within populations	49	1479.56	30.20	78.26		
Group 2						
Among populations	6	1289.61	11.69	32.14	0.53	0.32
Within populations	108	2666.10	24.69	67.86		

P value = 0.00000

Table 5. Analysis of molecular variance (AMOVA) for 167 individuals of *Panax ginseng* based on 282 AFLP markers. The hierarchical *AMOVA* analysis was carried out based on the groups inferred from the *P. ginseng* habitats (Zhuravlev et al., 2008): the Blue Mountains (Russia) (populations *Spa*, *Yak*, *Drech* and *Pozh*); Khasan District (Russia), Jilin and Heilongjiang Provinces (China) (populations *Uss* and *Nad*); the southern part of Sikhote-Alin (Russia) (populations *Part*, *Chu*, *Dgor* and *Olg*). The *P*-value is estimated based on 1023 permutations

Source of variation	Sum of squares	Variance components	Percentage variation	*P*-value
Among groups	395.796	-2.19598	-5.44084	0.86217
Among populations within groups	2017.343	16.15152	40.01768	0.00000
Within populations	4145.651	26.40542	65.42316	0.00000

Table 6. Results of the mean genetic differentiation based on the Bayesian approach models

θ_B					
Models	Mean	SD[a]	2.5%	97.5%	DIC[b]
Full	0.3634	0.0093	0.3447	0.3812	7951.4778
$f = 0$	0.2838	0.0084	0.2678	0.3001	8053.9076
$\theta_B = 0$	–	–	–	–	19905.5818
$f =$ free	0.3396	0.0151	0.3064	0.3661	8111.9572

[a] SD is standard deviation.

[b] DIC is deviance information criterion.

In the Bayesian approach to infer population structure, the mean genetic differentiation between subpopulations, based on the Bayesian f full model with the smallest deviance information criterion (DIC = 7951.4778), of Holsinger (θ_B), was 0.3634 (Table 6). This result was similar to the F_{ST} value among populations from the *non*-hierarchical AMOVA analysis (Table 4). In the Bayesian analysis of population structure using the free model, the posterior mean estimate for F_{IS} (f) was 0.5008±0.2889, and in the full model it was 0.9569±0.0424.

Using F_{ST}, the estimates of the number of migrants per generation (N_m) were low: 0.45 between the populations (non-hierarchical analysis) and 0.4 between the clusters. Migrations between populations within two main clusters were: N_m = 0.9 in a small cluster (group 1) and 0.53 in a large cluster (group 2) (Table 4).

AFLP Clustering Analyses and Assignment Tests

The UPGMA dendrogram revealed two distinct main clusters, one containing *Spa*, *Chu* and *Drech* populations (group 1) and the second containing *Pozh*, *Nad*, *Part*, *Yak*, *Uss*, *Dgor* and *Olg* populations (group 2) with *Olg* being distant from other populations in this cluster (Fig. 2). The PCoA (Fig. 3) reinforced the outcome of the UPGMA analysis (Fig. 2). The differentiation along the x-axis (positive vs. negative values) was roughly concordant with the 2 genetic groups identified by UPGMA. The first coordinate (x-axis) accounted for 36.99% of the variance in genetic distance among populations, whereas the second coordinate (y-axis) accounted for 22.80% of the variance. The clustering similar to UPGMA and PCoA, produced a MS-tree that visualized the genetic relationships among *P. ginseng* plants (Fig. 4). Calculated in the AMOVA on the basis of Euclidean distances, the length of each line connecting the representatives inside the populations and between the populations in Fig. 4 reflects the intra- and inter-population genetic distances, respectively (Table 7). According to the values of genetic distances (Table 7), all of the studied ginseng plants on the MS-tree formed two groups (Fig. 4), the first group consisting of the plants of the *Drech* and *Chu* populations with some plants of *Spa* population and the second group the *Part*, *Yak*, *Olg*, *Nad*, *Pozh*, *Uss*, *Dgor* and *Spa* populations. These two groups were genetically separated by 50 units of Euclidean distance (Fig. 4). The *Spa*, *Uss*, *Dgor* and *Part*, *Yak*, *Nad*, *Pozh* populations formed two subgroups genetically separated by 33 Euclidean distance units. The plants of the *Olg* population were separated from the *Part*, *Yak*, *Nad*, *Pozh* subgroup by 35 Euclidean distance units (Fig. 4). The *Olg* population was characterized by the highest value of Euclidean genetic distances between plants (36.5 units), whereas the *Uss* population was characterized by the smallest average distance (17.33 units). The average value of Euclidean genetic distance between the plants of different populations (28.78 units) was higher the average intra-population genetic distances (26.35 units) (Table 7). This finding is consistent with the AMOVA conclusion that revealed the population-genetic structure of wild *P. ginseng*.

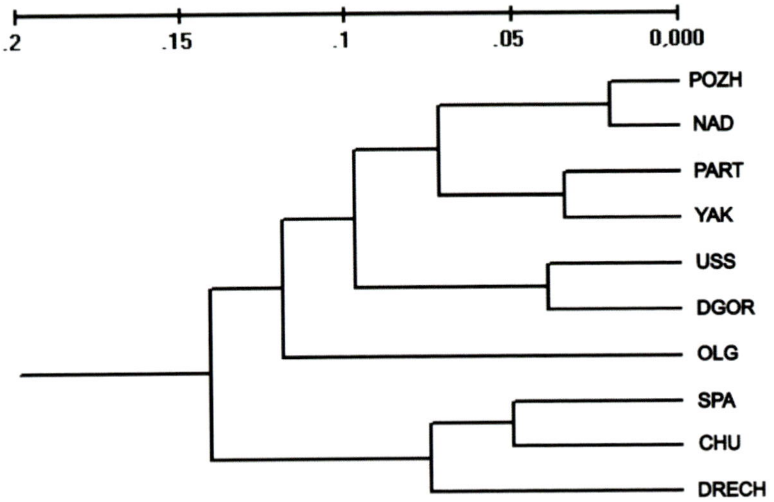

Figure 2. UPGMA dendrogram of genetic relationship among *Panax ginseng* populations based on Nei's genetic distance.

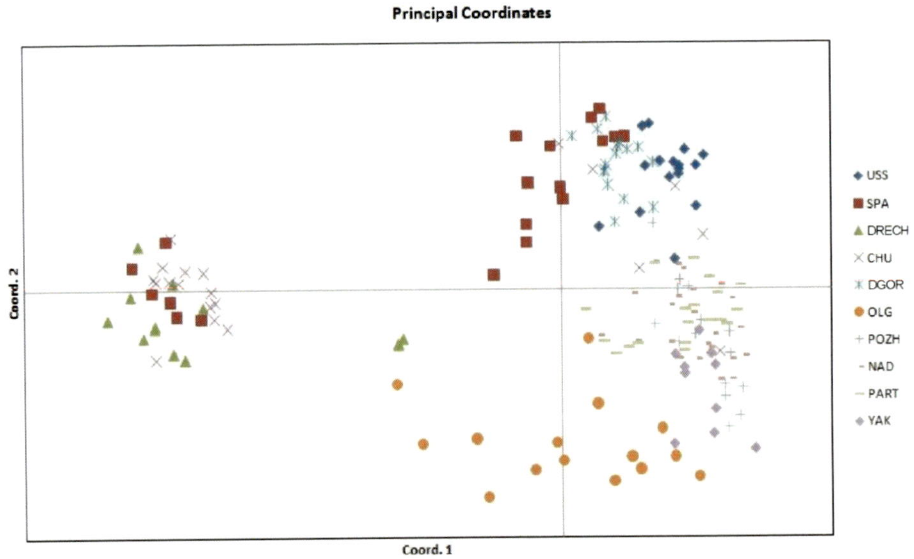

Figure 3. Principal coordinate analysis (PCoA) of 167 *Panax ginseng* individuals from 10 populations, based on a genetic distance matrix of 282 AFLP loci. The first two coordinates explain 59.79% of the variance.

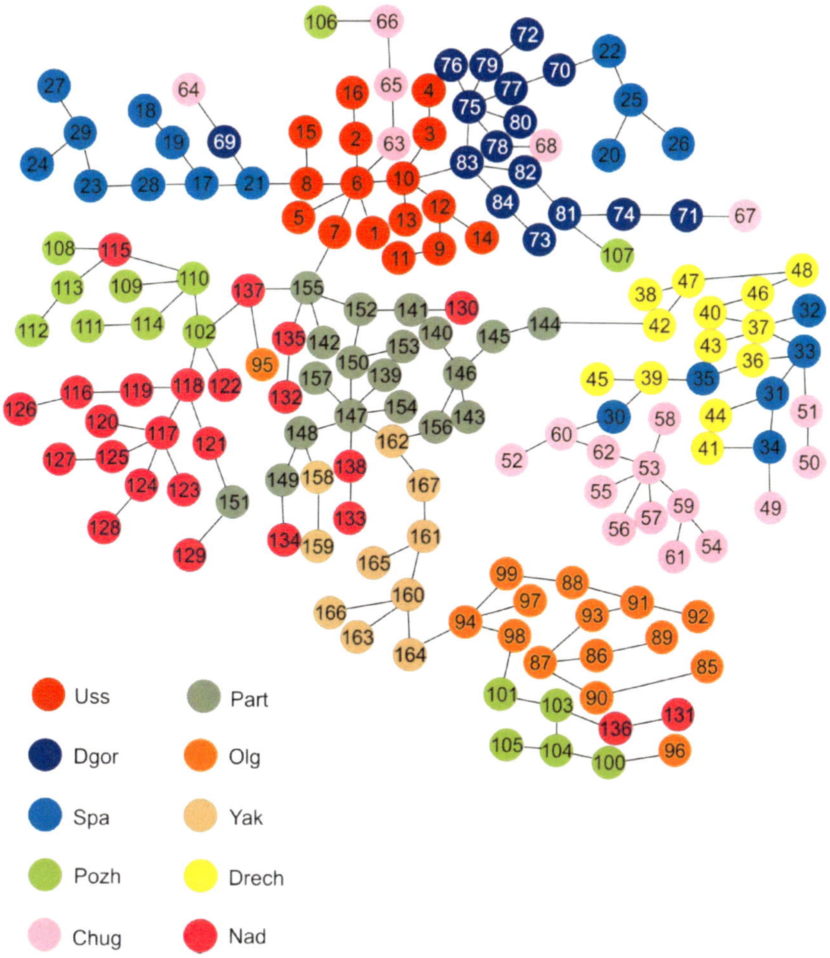

Figure 4. Minimum spanning tree (MS-tree) representing phylogenetic relationships among 167 individuals of *Panax ginseng* populations. The codes are in "Methods". Length of lines corresponds to the Euclidean distances among individuals.

Table 7. The length of lines on MS-tree characterizing the Euclidean genetic distances among plants in populations and among populations of *Panax ginseng*

Among plants in population			Among populations	
Population	Range of length	Average length	Population pair	Length
Uss	8-41	17.33	*Uss – Spa*	24
Spa	9-30	22.57	*Uss – Dgor*	22
Dgor	15-38	23.0	*Uss – Part*	33
Pozh	14-43	28.0	*Part – Drech*	50
Nad	21-36	29.07	*Drech – Chu*	25
Part	12-30	22.06	*Part – Yak*	19
Yak	22-57	35.13	*Yak – Olg*	35
Chu	15-44	25.0	*Pozh –Nad*	24
Drech	12-51	24.88	*Pozh – Part*	27
Olg	14-52	36.5		
Average		26.35		28.78

The ΔK statistic obtained using STRUCTURE HARVESTER analyses indicated the strongest signal of a population subdivision to be K = 2 (Fig. 5*a*), where the majority of the *Chu* and *Drech* populations and some individuals of the *Spa* population were distinct from all other samples (Fig. 5). However, there was a strong secondary peak at K = 8 (Fig. 5*a*), indicative of finer population structuring (Evanno et al., 2005). Therefore, as recommended by Evanno et al. (2005), we re-ran STRUCTURE on each of the clusters indicated at the strongest level of subdivision (K = 2; Fig. 5): (1) *Chug, Drech, Spa* and (2) all the other populations. For the former, the results show a clear single peak at K = 2 (Fig 5*b*), with the *Drech* and *Spa* and the *Chug* samples forming a first and a second cluster, respectively (Fig. 5b). Independent analysis of cluster 2 (*Uss, Dgor, Spa, Pozh, Part, Yak, Nad* and *Olg* populations) detected subdivisions with individuals clustering into two groups (Fig. 5c, *c*). However, a strong secondary peak at K = 4 (Fig. 5*c*) was discovered.

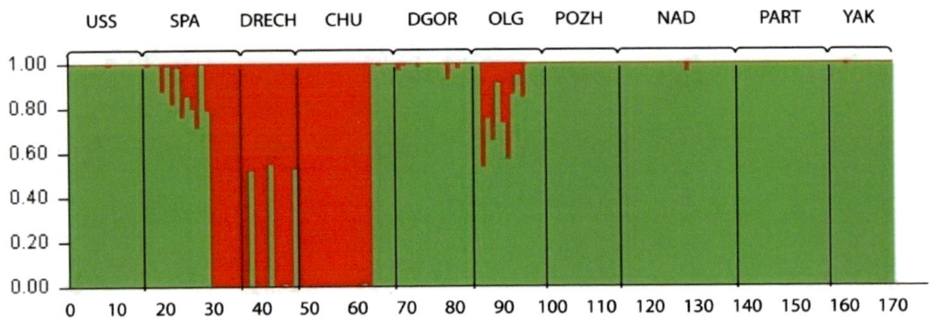

Figure 5. Hierarchical population structure based on STRUCTURE analysis on 282 AFLP loci. The columns in the bar charts represent individuals, with the proportion of an individual's genotype assigned to each cluster (K) denoted by different colours.

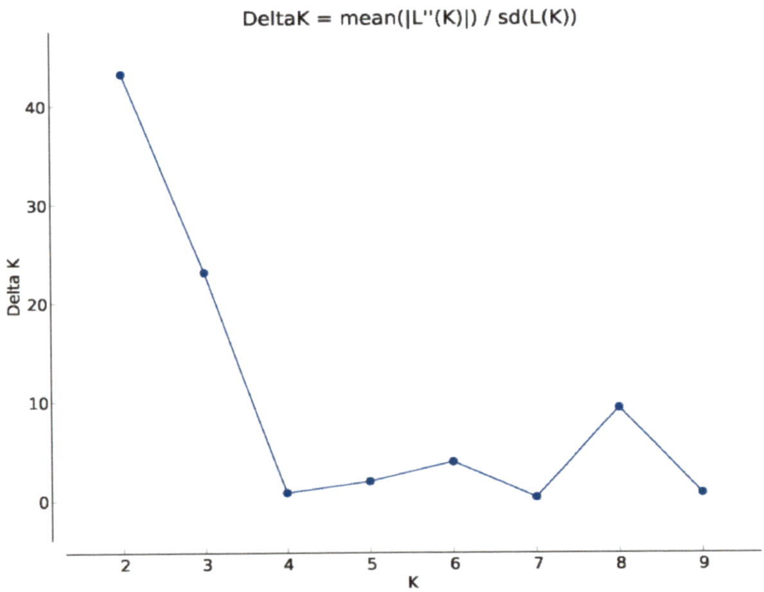

Figure 5 *a*. The most likely number of clusters (K) according to the ΔK statistic (line graphs) at the uppermost level of structuring.

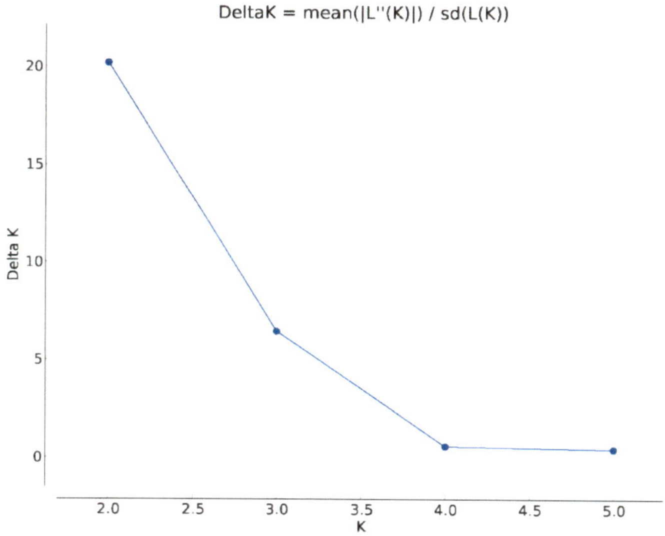

The most likely number of clusters (K) according to the ΔK statistic (line graphs) at the finer scale level of structuring (for red cluster).

Figure 5b. Finer scale structure within cluster 1 (red), K = 2. The lines separate individuals. The figures on horizontal axe are the number of plants. The figures on vertical axe mark the proportion of each cluster.

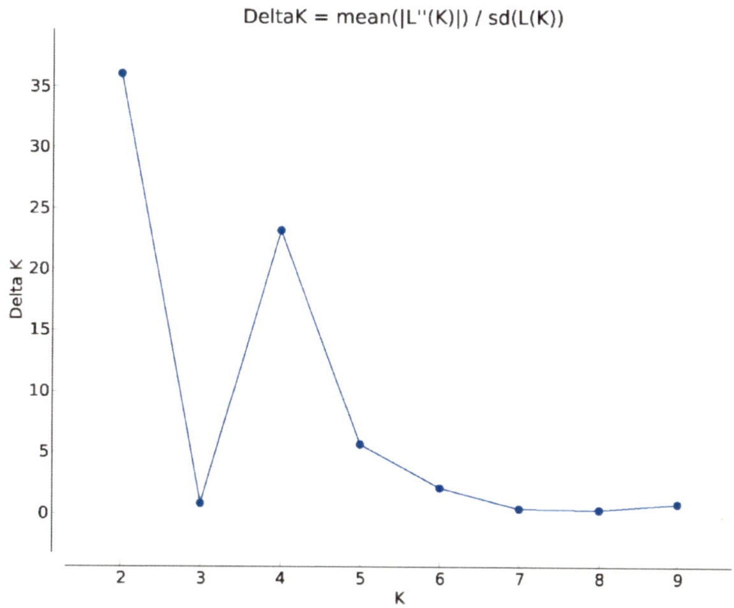

The most likely number of clusters (K) according to the ΔK statistic (line graphs) at the finer scale level of structuring (for green cluster).

Figure 5c. Finer scale structure within cluster 2 (green), K = 2. The lines separate populations. The figures on horizontal axe are the number of plants. The figures on vertical axe mark the proportion of each cluster.

Despite our finding of strong population structure, the presence of potential migrants was indicated by clustering analyses in all studied populations by mis-assignments in PCoA, MST and STRUCTURE. There was considerable overlap among individuals from different populations in PCoA (Fig. 3). The location of a *P. ginseng* in the MS-tree was dependent of the population it belonged to; however, such clustering was not strict, and some populations partially overlapped (Fig. 4). For example, some plants of the *Pozh* population were grouped with those of the *Olg* population. The plants of the *Chu* population were partially mixed with those of the *Uss* and *Dgor* populations, while the plants of the *Nad* population were mixed with those of the *Part* populations. In particular, many plants of the *Spa* population were scattered among the *Drech* and *Dgor* populations.

Additionally, the way the populations tended to cluster seemed to be independent of their geographical origin. Geographically close located populations were not clustered together often (Fig. 2, 4).

Neither the scatterplot (Fig. 6) nor the correlation analysis showed any association between pairwise F_{ST} values and geographical distance among populations. A negative and nonsignificant (Mantel $r = -0.174$; $P = 0.817$) correlation between genetic differentiation (the pairwise F_{ST} distances) and geographic distance (km), was detected (Table 3).

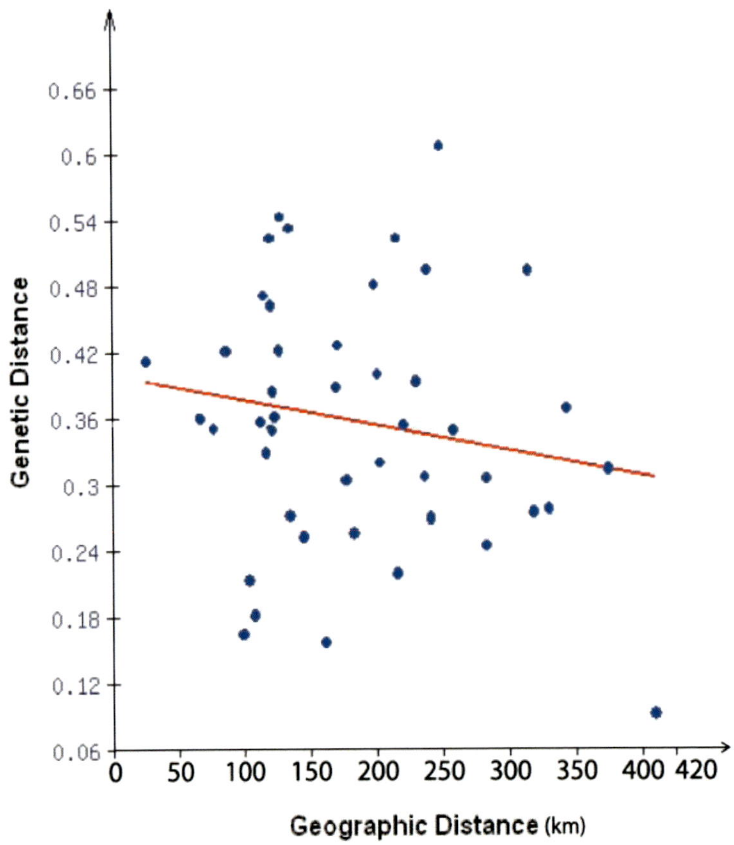

Figure 6. Relationship between genetic distance (F_{ST}) and geographic distance (km) among 10 populations of *Panax ginseng* (r = – 0.174; P = 0.817).

In the assignment analysis, most individuals (129 of the total 167) were assigned to the population from which they were sampled (Table 8), suggesting that each original sample site formed a distinct population. Nevertheless, a considerable proportion of individuals genetically resembled another population much more closely than that from which they were collected, indicating the presence of migrant genotypes in the set of analyzed populations (Table 8). Assignment success was lowest in the *Spa* population (47%) and was very high in the other populations. In the *Drech* and *Dgor* populations, all individuals were assigned correctly. In the *Chu*, *Pozh*, *Nad* and *Yak* populations, 60-70% individuals were assigned to the population from which they were sampled. Only one of the *Olg* individuals was assigned to the *Nad* population, and two individuals of the *Uss* population were assigned to other populations providing 93% and 87% assignment success of the *Olg* and *Uss* populations, respectively (Table 8).

Table 8. Results of the assignment analysis of 167 *Panax ginseng* individuals. The underlined figures represent possible migrants

Given populations (number plants)	Inferred populations										
	Uss	*Spa*	*Chu*	*Drech*	*Dgor*	*Pozh*	*Nad*	*Part*	*Yak*	*Olg*	Assignment success, %
Uss (16)	14				<u>1</u>		<u>1</u>				87
Spa (19)		9	<u>1</u>	<u>5</u>	<u>4</u>						47
Chu (20)	<u>1</u>	<u>1</u>	14		<u>2</u>	<u>2</u>					70
Drech (13)				13							100
Dgor (16)					16						100
Pozh (15)					<u>1</u>	10	<u>3</u>		1		66
Nad (24)						<u>3</u>	17	<u>4</u>			70
Part (19)							<u>3</u>	16			84
Yak (10)								<u>3</u>	6	<u>1</u>	60
Olg (15)							<u>1</u>			14	93

DISCUSSION

Within-Population Genetic Diversity

The problem of conserving biological diversity necessitates the investigation of genetic variability inside and among populations of the species in need of protection. Therefore, sufficiently variable molecular markers should be used to allow investigations of population-genetic structure by partitioning the total gene diversity within and between populations (Crawford, 1997). According to allozyme analysis, our investigations indicated that the variability of *P. ginseng* populations evaluated in a portion of polymorphic loci was about 7% (Zhuravlev et al., 1999; Koren et al., 2003; Zhuravlev et al., 2004). It is expected that molecular DNA markers can potentially provide a higher level of polymorphism detection than allozymes. Nevertheless, our results with RAPD markers correspond to the low genetic variation demonstrated by isozyme gene loci in *P. ginseng* (Zhuravlev et al., 1999; Zhuravlev et al., 2004; Zhuravlev et al., 1996; Artyukova et al., 2000). In contrast, the fluorescence-based automated AFLP method has shown a considerably higher resolving power in *P. ginseng* (Reunova et al., 2010).

The high AFLP genetic variability in *P. ginseng* was revealed both at the species level (99.65% of the bands polymorphic; Bayesian genetic diversity 0.2857; Table 2) and at the population level (55.68% of the bands polymorphic; Bayesian genetic diversity 0.1891; Table 2). Among the ten populations, *Uss* population maintained lower genetic diversity than the other nine (38.65% of the bands polymorphic; Bayesian genetic diversity 0.1204). While genetic diversity is theoretically higher in large populations, the *Uss* population was presumably small in size and therefore has suffered from a greater loss of a genetic diversity than other populations. It is possible that this population has been influenced by inbreeding and genetic drift. Several populations (*Spa*, *Pozh*, *Nad*, *Chu* and *Olg*) were distinguished by having higher levels of variability. For these populations, the average value of polymorphisms was 65.39%. The *Uss* population was characterized by the lowest average genetic distances between individuals (Table 7), which was consistent with the low parameters of variability calculated for this population (Table 2). On the other hand, the *Olg* population demonstrated the highest genetic diversity and genetic distances (Tables 2, 7). Therefore, the *Olg* population will require attention to be conserved *in situ*. The estimates of genetic variation in *P. ginseng* appear to be similar to those in some other *Panax* species. In *Panax notoginseng*, four pairs of fluorescently labeled AFLP primers produced 312 fragments, of which 240 (76.9%) were polymorphic (Zhou et al., 2005). In *Panax stipuleanatus*, the same primers revealed 346 loci, of which 334 (96.5%) were polymorphic (Zhou et al., 2005).

High genetic diversity is believed to be typical for plants with wide geographical ranges and outcrossing (Hamrick & Godt, 1996). At the beginning of the 20th century *P. ginseng* was spread over a vast territories of the Russian Primorye, Korean Peninsula, and East China. The reproductive system of ginseng possesses a highly plastic and stable fertilization process (Zhuravlev et al., 2008) that helps it survive in stress conditions and supports the high genetic diversity inherent in *P. ginseng*. A high level of genetic variation, and a large proportion of which is distributed within populations, was revealed in *P. ginseng* in this study, indicating that the major factors threatening the survival of *P. ginseng* populations are not the genetic variation of this species but human activities (e.g., overexploitation, habitat destruction, urbanization, and pollution).

Population Differentiation, Population Structure and Assignment Tests

All analyses (UPGMA, PCoA, MST, AMOVA, STRUCTURE) showed a clear population differentiation within *P. ginseng* and identified groups of populations. The AFLP data suggest that genetic differentiation (F_{ST} = 0.36; θ_B = 0.36; Tables 4, 6) among populations of *P. ginseng* is substantial. This value of F_{ST} is consistent with estimates of inter-population variability for plant species with mixed propagation types as determined with an AMOVA (F_{ST} = 0.40) (Nybom, 2004). *P. ginseng* is able to produce seeds via autogamy, out-crossing or agamospermy without pollination (Zhuravlev et al., 2008). Our estimates of the degree of within-population inbreeding using the Bayesian approach of Holsinger et al. (2003) (f = 0.5008±0.2889 in free model and f = 0.9569±0.0424 in full model) show that autogamy occurs in wild *P. ginseng* populations. In cultivated *P. ginseng*, inter-population RAPD variability ranged from 1.77% to 42.01% in one study (Ma et al., 2000) and was 31%

in another study (Kim & Choi, 2003). Strong genetic differentiation has also been found in other *Panax* species. The fluorescence-based automated AFLP method has demonstrated that over 40% of the genetic variation of wild *P. stipuleanatus* existed among the populations (Zhou et al., 2005).

The identification of genetically homogeneous groups of individuals is a long-standing issue in population genetics (Evanno et al., 2005). Among the methods that do not assume a predefined population structure, those that identify clusters of individuals are tree-based methods that use genetic distances between individuals and multivariate analyses. However, these graphical methods are only loosely connected to statistical procedures that allow the identification of homogeneous clusters of individuals (Evanno et al., 2005). A Bayesian algorithm implemented in the STRUCTURE program (Pritchard et al., 2000) allows the identification of such groups. STRUCTURE also calculates how likely it is that each individual belongs to each group. This information can be then used to assign individuals to populations.

When we applied STRUCTURE to a variant of the ΔK statistic (Evanno et al., 2005), we discovered the pattern of genetic structure in *P. ginseng* (Fig. 5) that was obtained using UPGMA, PCoA and MST. The clustering patterns presented by UPGMA, PCoA and the MS-tree were similar and identified two main groups of populations with low gene flow (N_m = 0.4) between them; these two groups should be considered two management units (MUs) (Moritz, 1994) for population monitoring and demographic examination.

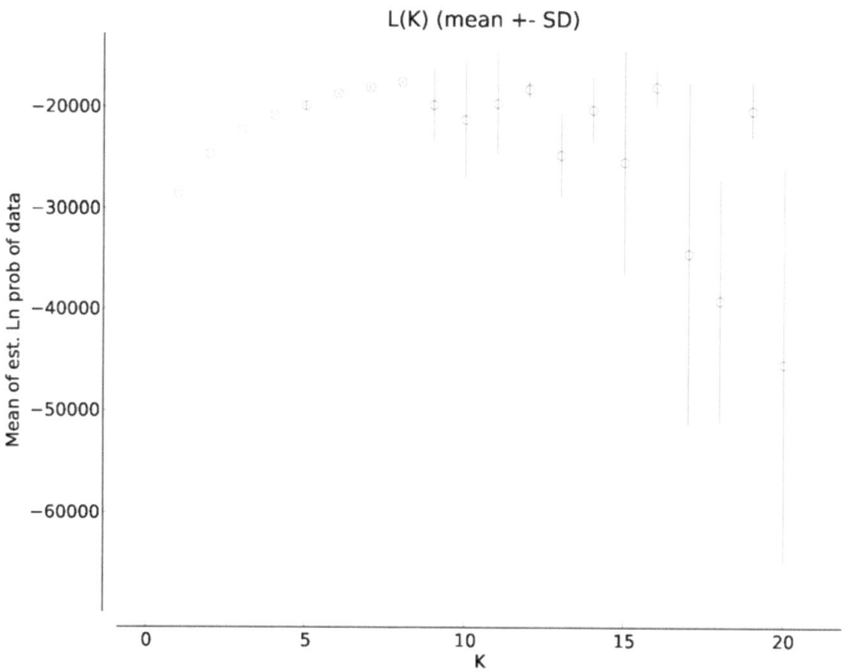

Figure 7. The estimation of number of clusters (K) of *Panax ginseng* with STRUCTURE based on the means of lnPD (the logarithmic value of the probability of the data given K).

However, when we applied the STRUCTURE method that estimates the log probability of data Pr(X|K), the program did not reveal the population structure in *P. ginseng* because the probable number of K (1-20) was not demonstrated (Fig. 7). These data confirm the result of Evanno et al. (2005) that whereas L(K), the estimate for the number of groups given by STRUCTURE, often does not correspond to the real number, ΔK has a mode at the true K for most of the situations investigated.

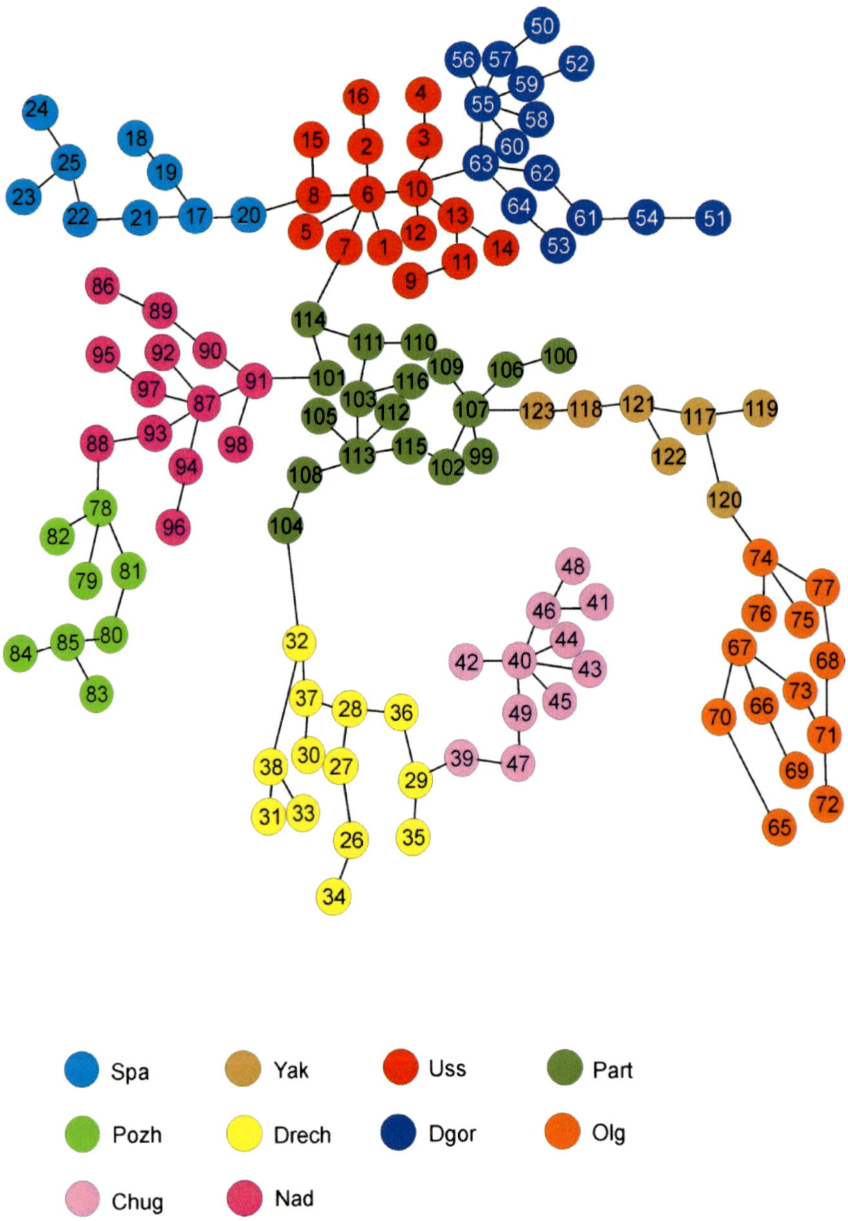

Figure 8. Minimum spanning tree (MS-tree) representing phylogenetic relationships among 123 individuals of *Panax ginseng* populations. The codes are in "Methods". Length of lines *corresponds to* the Euclidean distances among individuals.

The population-genetic structure of *P. ginseng* derived from AFLP data was not consistent with a pattern of isolation by distance (Fig. 6, Table 3), as expected in a system with a patchy distribution of genetic variability (McCauley, 1991). Zoochory may lead to effective long-distance dispersal (Kudoh & Whigham, 1997) which may weaken the regional association between genetic and geographical distances over the entire metapopulation. The expansion of *P. ginseng* occurs exclusively by seed dispersal (Zhuravlev & Kolyada, 1996). The seeds of *P. ginseng* are dispersed presumably by gravity as well as perhaps birds, animals and man. The results of our assignment test support the involvement of a high level of seed-mediated gene flow in the disruption of the historical equilibrium between gene flow and genetic drift. As revealed by the assignment test conducted on the AFLP dataset, some individual plants in every population showed a stronger genetic resemblance to a different population other than their own putative population. The estimated immigration rate in the ginseng populations was 7–53% (Table 8). It is unlikely that effective pollen flow could generate such a high genetic resemblance to another population because the male gamete contains only half the genetic load of an individual. Therefore, the outlier genotypes detected by the assignment test can be considered the products of successful seed dispersal events, i.e., migrants (He et al., 2004). The long-distance seed dispersal by birds, animals and humans could result in the mixing of ginseng plants and also could explain the genetic similarity of geographically distant ginseng populations observed in our study (Figs. 2, 4). At the same time, the fact that the arrangement of *P. ginseng* populations did not always correspond to their geographical areas, may have been due to converging common selection forces in geographically disparate populations (Peters et al., 2009).

Using the minimum spanning tree between the individuals and on the assignment test, the population assignments of ginseng plants were verified. From 167 *P. ginseng* plants the 44 were not placed in the MST in accordance with their population membership (Fig. 4) but rather were mixed with the individuals from some other populations. However, after these plants were removed from consideration, the remaining 123 individuals were arranged in the MST exactly in accordance with the populations they belonged to (Fig. 8). Similar results were obtained with assignment tests using the Aflpop program, which calculated a total of 38 outliers (Table 8). Thus, using the MST analysis and the assignment test the content of *P. ginseng* populations was verified to differentiate the true population plants from the plants that are not inherent to the population under study. This distinction can identify the "pure" plants that could be chosen to serve as a stock material for reintroduction.

ACKNOWLEDGMENTS

This work was supported by grants from the Russian Academy of Sciences (No. 09-I-P23-06; No. 09-I-OBN-02), by the Russian Fund for Fundamental Investigations (No. 11-04-998515-r_vostok_a; No. 11-04-01388-a) and by the Grant Program "Molecular and Cell Biology" of the Russian Academy of Sciences and the "Leading Schools of Thought" grant from the President of the Russian Federation (No. NSH 1635-2008.4).

REFERENCES

Artyukova, E.V., Kozyrenko, M.M., Reunova, G.D., Muzarok, T.I. & Zhuravlev, Yu.N. (2000). RAPD Analysis of Genome Variability of Planted Ginseng, *Panax ginseng*. *Mol Biol* (Moscow), Vol.34, No.2, pp. 297-302.

Choi, K.-t. (2008). Botanical Characteristics, Pharmacological Effects and Medicinal Components of Korean *Panax ginseng* C A Meyer..*Acta Pharmacol. Sin*. Vol.29. No.9. pp. 1109-1118.

Crawford, D.L. (1997). Molecular Markers for The Study of Genetic Variation Within and Between Populations of Rare Plants. *Opera Botanica*, Vol.132, pp. 149-157.

Duchesne, P. & Bernatchez, L. (2002). AFLPOP: A Computer Program for Simulated and Real Population Allocation Based on AFLP Data. *Mol. Ecol. Notes*, Vol.2, pp. 380-383..

Earl, D.A. (2009). Structure Harvester v.0.5. Available from *(http://taylor0.biology.ucla.edu/struct_harvest/).*

Echt, C.S., Erdahl, L.A. & McCoy, T.J. (1992). Genetic Segregation of Random Amplified Polymorphic DNA in Diploid Cultivated Alfalfa. *Genome*, Vol.35, pp. 84-87.

Evanno, G., Regnaut, S. & Goudet, J. (2005). Detecting the Number of Clusters of Individuals Using the Software STRUCTURE: a Simulation Study. *Mol. Ecol*., Vol.14, pp. 2611-2620.

Excoffier, L., Smouse, P.E. & Quattro, J.M. (1992). Analysis of Molecular Variance Inferred From Metric Distances Among DNA Haplotypes: Application to Human Mitochondrial DNA Restriction Data. *Genetics*, Vol.131, pp. 479-491.

Falush, D., Stephens, M. & Pritchard, J.K. (2003). Inference of population structure using multilocus genotype data: linked loci and correlated allele frequencies. *Genetics*, Vol.164, pp. 1567-1587.

Grushwitsky, I.V. (1961). *Ginseng: the aspects of biology*, Nauka Press, Leningrad, Russia.

Hamrick, J.L. & Godt, M.J.W. (1996). Conservation Genetics of Endemic Plant Species. In: *Conservation Genetics, Case Histories fromNnature*, J.S. Avise & J.L. Hamrick, (Eds), 281-304, Chapman and Hall, New York.

He, T., Krauss, S.L., Lamont, B.B., Miller B.P. & Enright, N.J. (2004). Long-Distance Seed Dispersal in a Metapopulation of *Banskia hookeriana* Inferred From a Population allocation Analysis of Amplified Fragment Length Polymorphism Data. *Mol. Ecol*., Vol.13, pp. 1099-1109.

Holsinger, K.E. & Lewis, P.O. (2003). Hickory: a Package for Analysis of Population Genetic Data, ver. 0.8. Storrs, CT: University of Connecticut, Department of Ecology and Evolutionary Biology.

In, J.-G., Kim, M.-K., Lee, O.R., Kim, Y.-J., Lee, B.-S., Kim, S.-Y., Kwon, W.-S. & Yang, D.-C. (2010). Molecular Identification of Korean Mountain Ginseng Using an Amplification Refractory Mutation System (ARMS). *J. Ginseng Res*., Vol.34, No.1, pp. 41-46.

Kim, C. & Choi, H.K. (2003). Genetic Diversity and Relationship in Korean Ginseng (*Panax schinseng*) Based on RAPD Analysis. *Korean J. Genet*., Vol.25, pp. 181-188.

Koren, O.G., Potenko, V.V. & Zhuravlev, Yu.N. (2003). Inheritance and Variation of Allozymes in *Panax ginseng* C.A. Meyer (Araliaceae). *Int J Plant Sci*, Vol.164, pp. 189-195.

Kudoh, H. & Whigham, D.F. (1997). Microgeographic Genetic Structure and Gene Flow in *Hibiscus mocheutos* (Malvaceae) Populations. *Amer. J. Bot.*, Vol.84, pp. 1285-1293.

Ma, X.J., Wang, X.Q., Xu, Z.X., Xiao, P.G. & Hong, D.Y. (2000). RAPD Variation Within and Among Populations of Ginseng Cultivars. *Acta Bot. Sinica*, Vol.42, pp. 587-590.

Mantel, N. (1967). The Detection of Disease Clustering and Generalized Regression Approach. *Cancer Research*, Vol.27, pp. 209-220.

McCauley, D.E. (1991). Genetics Consequences of Local Population Extinction and Recolonization. *Trends in Ecol. and Evol.*, Vol.6, pp. 5-8.

Miller, M.P. (1997). Tools for Population Genetic Analysis (TFPGA) ver.1.3: A Windows Program for the Analysis of Allozyme and Molecular Population Genetic Data. Computer software distributed by author.

Moritz, C. (1994). Defining "Evolutionary Significant Units" for Conservation. *Trends in Ecol. and Evol.*, Vol.9, pp. 373-375.

Murray, M.G. & Thompson, W.F. (1980). Rapid Isolation of High Molecular Weight Plant DNA. *Nucleic Acids Res*, Vol.8, pp. 4321-4325.

Nei, M. (1978). Estimation of Average Heterozygosity and Genetic Distance from a Small Number of Individuals. *Genetics*, Vol.89, pp. 583-590.

Nybom, H. (2004). Comparison of Different Nuclear DNA Markers for Estimating Intraspecific Genetic Diversity in Plants. *Mol. Ecol.*, Vol.13, pp. 1143-1155.

Peakall, R, & Smouse, P.E. (2006). GENALEX 6: Genetic Analysis in Excel. Population Genetic Software for Teaching and Research. *Mol Ecol Notes*, Vol.6, pp. 288-295.

Peters, M.D., Xiang, Q.Y., Thomas, D.T., Stucky, J. & Whiteman, N.K. (2009). Genetic Analyses of the Federally Endangered *Echinacea laevigata* Using Amplified Fragment Length Polymorphisms (AFLP) – Inferences in Population Genetic Structure and Mating System. *Conserv. Genet.*, Vol.10, pp. 1-14.

Pritchard, J.K., Stephens, P. & Donnelly, P. (2000). Inference of Population Structure Using Multilocus Genotype Data. *Genetics*, Vol.155, pp. 945-959.

Qi, L.-W., Wang C.-Z. & Yuan C.-S. (2011). Isolation and Analysis of Ginseng: Advances and Challenges. *Nat. Prod. Rep.*, Vol.28, pp. 467–495.

Reunova, G.D., Kats, I.L., Muzarok, T.I. & Zhuravlev, Yu.N. (2010). Polymorphism of RAPD, ISSR and AFLP Markers of the *Panax ginseng* C.A. Meyer (Araliaceae) Genome. *Russ. J. Genet.*, Vol.46, No.8, pp. 938-947.

Schneider, S, Roessli, D & Excoffier, L. (2000). Arlequin ver. 2.000: A Soft ware for Population Genetic Data Analysis. Genetics and Biometry Laboratory, University of Geneva, Switzerland.

Slatkin, M. & Barton, N.H. (1989). A Comparison of Three Indirect Methods for Estimating Average Levels of Gene Flow. *Evolution*, Vol.43, pp. 1349-1368..

Spiegelhalter, D.J., Best, N.G., Carlin, B.P. & van der Linder, A. (2002). Bayesian Measures of Model Complexity and Fit. *J. of the Royal Statistical Society*, Series B, Vol.64, pp. 583-639.

Vos, P.R., Hogers, M., Bleeker, M., Reijans, M., van der Lee, T., Homes, M., Frijters, A., Pot, J., Peleman, J., Kuiper, M. & Zabean, M. (1995). AFLP: a New Technique for DNA Fingerprinting. *Nucl. Acids Res.*, Vol.23, pp. 4407-4414.

Williams, J.G.K., Kubelik, A.R., Livak, K.J., Rafalski, J.A. & Tingey, S.V. (1990). DNA Polymorphism Amplified by Arbitrary Primers are Useful as Genetic Markers. *Nucl. Acids Res.*, Vol.18, pp. 6531- 6535.

Yeh, F.C., Boyle, T.J.B. (1997). Population Genetic Analysis of Co-dominant and Dominant Markers and Quantitative Traits. *Belg. J. Bot.*, Vol.129, pp. 157.

Zhou, S.L., Xiong, G.M., Li, Z.Y. & Wen, J. (2005). Loss of Genetic Diversity of Domesticated *Panax notoginseng* F H Cheng as Evidenced by ITS Sequence and AFLP Polymorphism: a Comparative Study With *P. stipuleanatus* H. T. et K. M. Feng. *J. Integr. Plant Biology*, Vol.47, pp. 107-115.

Zhuravlev, Yu.N. & Kolyada, A.S. (1996). *Araliaceae: Ginseng and others*, Dalnauka Press, Vladivostok, Russia.

Zhuravlev, Yu.N., Koren, O.G., Kozyrenko, M.M., Reunova, G.D., Artyukova, E.V., Krylach, T.Yu. & Muzarok, T.I. (1999). Use of Molecular Markers to Design the Reintroduction Strategy for *Panax ginseng*. In: *Proceedings of PSA Symposium Biodiversity and allelopathy: From Organism to Ecosystems In the Pacific*, C.H. Chou, G.R. Waller & C. Reinhardt, (Eds.), 183-192, Academia Sinica, Taipei.

Zhuravlev, Yu.N., Koren, O.G., Reunova, G.D., Artyukova, E.V., Kozyrenko, M.M, Muzarok, T.I. & Kats, I.L. (2004). Ginseng Conservation Program in Russian Primorye: Genetic Structure of Wild and Cultivated Populations. *J Ginseng Res*, Vol.28, pp. 60-66.

Zhuravlev, Yu.N, Koren, O.G., Reunova, G.D., Muzarok, T.I., Gorpenchenko, T.Yu., Kats, I.L. & Khrolenko, Yu.A. (2008). *Panax ginseng* Natural Populations: Their Past, Current State and Perspectives. *Acta Pharmacol. Sin.*, Vol.29, pp. 1127-1136.

Zhuravlev, Yu.N., Kozyrenko, M.M., Artyukova, E.V., Reunova, G.D., Muzarok, T.I. & Elyakov, G.B. (1996). Genetic typing of *Panax ginseng* by use RAPD-PCR. *Doklady RAS*, Vol.349, pp. 111-114.

Zietkiewicz, E., Rafalski, A. & Labuda, D. (1994). Genome Fingerprinting by Simple Sequence Repeat (SSR)-Anchored Polymerase Chain Reaction Amplification. *Genomics*, Vol.20, pp. 176-183.

In: Genetic Diversity: New Research
Editors: J. A. C. Amaya and M. M. F. Jimenez

ISBN: 978-1-62081-598-4
© 2012 Nova Science Publishers, Inc.

Chapter 8

GENETIC AND EPIGENETIC VARIATIONS IN DISEASE AND DRUG RESPONSE

Barkur S. Shastry
Department of Biological Sciences, Oakland University, Rochester, MI, US

ABSTRACT

DNA sequence variations (0.1%) between two randomly selected individuals are considered the major factors that determine genetic diversity among individuals, disease susceptibility and response to medication. The most common form of genetic variation is called single nucleotide polymorphism (SNP). Because of their frequency of distribution, stability and simplicity they represent unique genetic markers. They can be used to identify and map complex, common diseases such as high blood pressure and diabetes. SNP markers are also employed to identify drug response, ecology, evolutionary biology, diagnostic test and heterogeneity testing.

Recent progress in genetic technology has also identified a new form of genetic variation known as copy number variation (CNV). They are known to be associated with population diversity and several complex and common diseases including nervous system disorders. Because of their location and nature, many CNVs have functional consequences such as gene dosage alteration, disruption of genes and modulation of activities of other genes. Therefore, these genetic variations will have an influence on phenotypes, susceptibility of an individual to disease, drug response and human genome evolution.

Recently, whole-genome association studies also suggest that epigenetic variation along with nucleotide variation may account for all the heritable phenotypic variation in health and disease. Epigenetics refers to the study of inheritable changes in gene expression without any changes in DNA sequence. These heritable changes include DNA methylation, histone modifications and microRNA expression. Epigenetic control mechanisms may differ among tissues and individuals and they may also change in time during aging. Many genes including enzymes, drug transporters receptors and drug targets are under epigenetic control. It is because of this reason, the importance of epigenetics in clinical medicine has been recently recognized but little is known regarding the role of epigenetics in the pathogenesis of diseases. In the near future, many additional rare polymorphisms, structural variants and tissues specific epigenetic features of the human genome may become significant contributors to disease pathogenesis and medicinal drug response.

INTRODUCTION

The extraordinary progress in genome science has provided unique opportunities to address the questions of genetic basis of diversity among individuals, the most common familial traits, evolutionary processes, adverse drug reactions and complex and common diseases such as diabetes, obesity, hypertension and psychiatric disorders (Emilien et al. 2000). It is generally believed that genomes between two randomly selected individuals contain approximately 0.1% difference or variation. This variation is called polymorphism and it arises because of mutations. Several comprehensive studies on identical and fraternal twins (Martin et al. 1977) and siblings suggest that DNA polymorphism is one of the factors associated with susceptibility to many common diseases, every human trait such as curly hair, individuality and inter-individual difference in drug response. DNA sequence variation is also considered to be responsible for genome evolution. Based on these observations, it has been proposed that by cataloging the DNA polymorphisms in different populations and on different species it may be possible to (a) generate genome based knowledge on the susceptibility of the individual to many common diseases (b) develop safer and more effective individualized medications and (c) to understand evolutionary processes.

SINGLE NUCLEOTIDE POLYMORPHISMS (SNPs)

The simplest form of DNA variation among individuals is the substitution of one single nucleotide for another. This type of change is called single nucleotide polymorphism (SNP) and it occurs at a frequency of 1 in 1000 bp throughout the genome. A recent survey suggests that some variants are common to all population while others are distributed in a more restricted manner. These simple changes can be of transition or transversion type. They can occur either in coding region or outside the coding region. According to one report (Halushka et al. 1999) approximately 50% of SNPs are in the non-coding regions, 25% lead to missense mutations (coding SNPs or non-synonymous) and the remaining 25% are silent mutations. These silent SNPs are called synonymous SNPs (they do not change encoded amino acids). Neutral substitutions in exons are found to occur at a much higher rate (30-60%) compared to that in non-coding regions. Both synonymous and non-synonymous SNPs influence promoter activity and pre-mRNA conformation (or stability). They also alter the translational efficiency, ability of a protein to bind its substrate or inhibitors (Kimchi-Sarfaty et.al 2007) and change the subcellular localization of proteins (nSNPs). Therefore, they may be responsible for disease susceptibility, medicinal drug deposition, and genome evolution. They may also play a direct role with or without other factors in the phenotypic expression of diseases or traits such as tallness, curly hair and individuality (Martin et al. 1997; Krawezak et al. 1992; Lohrer and Tangen 2000; LeVan et al. 2001). However, it is not necessary that all SNPs will change the phenotype.

SNPs have a variety of applications. They can be used to identify genes responsible for disease predisposition by an association study (Riley et al. 2000; Kruglyak L 1999; Shastry 2002; 2003; 2007). SNP markers are also employed to identify drug response, ecology, evolutionary biology, diagnostic test, heterogeneity testing and to identify quantitative trait loci that contribute to polygenic variations (Gary et al. 2000; Schork et al. 2000). Identification of predictive biomarkers may provide a companion diagnostic test and subgrouping of patients that can be used in a clinical setting for treating patients (Sim and Ingelman-Sundberg 2011; Zillak et al. 2011; LaThangue and Kerr 2011). However, one major problem is that many SNPs may not display any variability in the family that is being studied and hence causative role of identified genes must be independently investigated. Additionally, population structure, allelic and non-allelic heterogeneity of phenotype, different levels of linkage disequilibrium (LD) and epistatic interactions of alleles may pose problems (Schork et al. 2000; Weiss and Terwilliger 2000). Despite these limitations there has been some success in identifying disease susceptibility genes (Table 1) and a number of SNPs are now shown to be associated with various disorders (Shastry 2002; 2003).

Table 1. A partial list of diseases associated with SNPs*

Disease	Gene	Disease	Gene
Arrhythmia	KCNQ1	Hyperbilirubinemia	UGT1A1
Atopy and Asthma	Chemokine	Lung cancer	MMP-1
Bipolar affective Disorder	HTR 3A	Obesity Prostate cancer	PAI-1 CDKN1B
Blood pressure	TAFI	Prostate Cancer	COX-2
Calcium stone	E-cadherin	POAG	TIGR
Diabetes type II	GFPT2	Rheumatoid arthritis	PADI-4
Dyslipidemia	Lipase	Urinary bladder cancer	Cyclin D1

KCNQ1 = potassium channel; HTR = serotonin receptor; TAFI = thrombin-activable fibrinolysis inhibitor; GFPT2 = glutamine fructose-6-phosphate amidotransferase; UGT1A1 = UDP-glucoronosyltransferase; MMP-1 = matrix metalloproteinase-1; PAI-1 = plasminogen activator inhibitor-1; COX-2 = cyclooxygenase –2; CDKN1B = a tumor suppressor gene; POAG = primary open-angle glaucoma; TIGR = trabecular mesh-work inducible glucocorticoid response gene; PADI-4 = peptidylargininedeiminase-4; * references can be found in Shastry 2004.

SNPs and Medicinal Drug Response

Inter-individual variation in drug response among patients is well known and poses a serious problem in medicine. There are no biochemical markers at present that can predict which groups of patients respond positively, which patients are non-responders and who experience adverse reactions for the same medication and dose. SNPs in genes encoding drug metabolizing enzymes, drug transporters and receptors contribute, at least in part, to the inter-individual variability in drug response (Carlquist and Anderson 2011; Yoshida et al. 2011; Roden et al. 2011). For instance, the clinical outcomes regarding the use of clopidogrel and warfarin can be strongly influenced by genetic variability in cytochrome P450 (Wilke and Dolan 2011, Ingleman-Sundberg and Sim 2010; Simon et al. 2011; Rahmioglu et al. 2011). Similarly, clinical outcome of thiopurines (used for immune modulation and the management of the acute lymphoblastic leukemia), anticancer drugs (Deenen et al. 2011; Ramos et al. 2011), hepatitis C virus infection treatment in liver disease (Zimmer and Lammert 2011a; Yahoo et al. 2011), heart failure treatment with beta and alpha adrenergic receptors and endothelin receptors (Mestroni and Taylor 2011), colorectal tumors (Weidlich et al. 2011), response to psychosocial treatment in poststroke depression (Kohen et al. 2011) and eye disorders (Shastry 2010; 2011) depends upon genetic variability of an individual at loci of medical relevance. There are now several high-density SNP maps of genes encoding proteins of medical importance (Iida et al. 2003) and there is strong evidence that links SNPs to inter-individual difference in drug response (Nothen and Cichon 2002; Ansari and Krajinovic 2007). Over 10 million SNPs have been discovered to date and extensive investigations on SNPs have been performed to assess their clinical application (Laing et al. 2011). Therefore, based upon information gathered from individual genome sequence at loci of medical relevance, it is possible to develop individualized medicine with improved efficacy and safety (Offit 2011; Shastry 2006; 2005). For instance, particular SNPs in Vitamin K epoxide reductase complex subunit 1 (VKORC1), Cytochrome P450 2C9 (CYP2C9), uridine diphosphate glucoronosyltransferase 1A1 (UGT1A1) and thiopurine S-methyltransferase (TPMT) have now proven to be applicable (in different ethnic background) for optimizing the dosage for maximum efficiency and minimum adverse effects even though there is intra-ethnic difference in drug response and toxicity (Stojiljkovic et al. 2011; Suriapranata et al. 2011; Chan et al. 2011; Phan et al. 2011; Cheong et al. 2011). However, their immediate clinical application requires additional studies (see conclusion).

Copy Number Variation (CNV)

Recent studies have found that some individuals can have deletion or multiple copies of the same gene. These kinds of changes are called copy number variation (CNV) or copy number polymorphisms (CNPs) if CNV has a high frequency (Hind et al. 2006; McCarroll et al. 2006; Conard et al. 2006). The size of DNA variation in CNV could range from a kilobase to megabase. CNVs are not distributed uniformly in the genome but are enriched in segmental duplication. They are the main types of structural variations caused by genomic rearrangement. Large CNVs are often found to occur in regions containing large homologous repeats or high identity segmental duplication (Sharp et al. 2001) while smaller CNVs may occur because of non-homology driven mutational mechanisms. They represent an important class of human genetic diversity especially within segmental duplications.CNVs are abundant and their importance in human disease has become increasingly clear over the past five years. Although a true number of CNVs and their frequencies in human population are not known, they may result in altered level of gene expression. As a result they may account for a significant phenotypic variation within species (because they often encompass genes), susceptibility or resistance to disease, drug response, complex and common disorders and evolution of genome itself (Lupski 1998; 2007; Girirajan et al. 2011; Shastry 2009). Numerous CNVs have now been found to be associated with complex human diseases (Table 2). However, it is not necessary that all

Table 2. A partial list of diseases associated with CNVs *

Disease	Gene	Disease	Chromosomal region
ARMD	CFHR1, CFHR3	Autism	16p11.2
Color blindness	OPN1LW, OPN1MW	Autism	20p13
Coronary heart disease	LPA	Bipolar	3q13.3
Crohn disease	IRGM	Crohn disease	8p23.1
Diabetes type II	TSPAN 8	Charcot-Marie-tooth	17p12
Lupus	FCGR3b	DiGeorge syndrome	22q11.2
Neuroblastoma	NBPF23	Prader-Willi	15q11.3
Obesity	GPRC5B	Spinal muscular atrophy	5q13.2
Osteoporosis	UGT2B17		
Psoriasis	LCE3B, LCE3C		

* References can be found in Girirajan et al. 2011; Shastry 2009; ARMD = age-related macular degeneration.

CNVs are related to genome disorders and some of them are in fact present in healthy individuals with no obvious genetic disorders. Thus, it appears that the genome is more tolerant for duplications than deletions. Because of their mild effect on gene function, it is possible that some of the common CNVs alter phenotypes in complex and sporadic diseases including nervous system disorders (Lupski 2007; Lee and Lupski 2006) and inherited disorders. In this regard it is possible that CNVs can be modifiers of rare disease causing mutations. The relationship between the copy number of SMN1 (spinal muscular atrophy determining gene) and SMN2 also suggests that CNPs may interact with rare pathogenic variants and that may cause phenotypic variability (Girirajan et al. 2011).

Additionally, it is reported that at least 15% of human neurodevelopmental diseases is due to rare and large copy number changes that result in local dosage imbalance for dozens of genes. It was also observed that more than one CNV can contribute to severe developmental delay and often is responsible for phenotypic variability associated with genomic disorders. However, it is not clear to what extent genetic disease is caused by CNVs (Lee et al. 2007). Some studies did not find enrichment for large, rare CNVs in cases compared with controls

(Glessner et al. 2010; Ikeda et al. 2010; Need et al. 2009; Shi et al. 2008). According to the literature, their contribution to genetic disease appears to be smaller than that of SNPs. Moreover, their contribution to gene expression is also independent of SNPs. Therefore, it is possible that many of CNVs by themselves are not fully penetrant. However, in combination with other genetic factors they would provide a better explanation for the genetic complexity of disorders (Girirajan et al. 2011).

There is evidence that full complement of CNVs and SNPs may be important for understanding genetic diseases more broadly (O'Roak et al. 2011). However, such studies require genome sequencing and CNV characterization of a large number of normal genome to interpret potentially pathogenic CNVs. In order to understand the genetic diversity of these regions at the sequence level, it is essential that all classes of genomic variations (SNPs, CNVs and indeles) must be integrated into a more complete model of human disease. In short, rare CNVs may contribute to the human disease and genetic diversity. Although present in normal populations, they are enriched in individuals with the disease. In order to study complex diseases and to understand the genotype-phenotype relationship, it is important to integrate all genetic variations including SNPs and rare as well as common CNVs.

Epigenetic Variations

According to the whole-genome association studies variation in nucleotides alone does not account for all the heritable phenotypic variations (Rakyan et al. 2011). The individual variation in therapeutic outcome of a drug, disease susceptibility and even development may depend upon epigenetics (Candelaria et al. 2010; Dennis 2003; Reik et al. 2001). The term epigenetics refers to the study of mitotically and/or meiotically heritable variations (heritable chromatin modification) in gene expression arising in the absence of any alterations in the DNA sequence. These heritable changes include DNA methylation, modification of histone proteins, loss of imprinting and regulatory non-coding RNAs such as microRNAs (Peedicayl 2008; Leach et al. 2010; Esteller 2011). DNA cytosine methylation may alter accessibility for transcription factor complexes. Histone modification (phosphorylation, methylation, acetylation and ubiquitination) affects chromatin structure at regional level.

However, it is not clear whether histone modifications are inherited through the cell division or they only represent the transient indications of the transcriptional states. MicroRNAs on the other hand, act as important regulators of gene expression at the translational level and may become potential prognostic markers of certain diseases (Costa et al. 2011). Epigenetics may also form a new bridge between nutrition and age-related diseases (Park et al. 2011; Seddon et al. 2011). Therefore, epigenetic changes play an important role in the variation of predisposition to disease and drug response (Hamilton 2011). The importance of epigenetics in clinical medicine has been recently recognized but little is known regarding the role of epigenetics in the pathogenesis of diseases (Feinberg 2007). This is because some of the epigenetic studies have been conducted under the perspectives of gene regulation and not as a heritable change in gene activity (Zimmer and Lammert 2011b).

Epigenetic control mechanisms may differ among tissues and individuals and may also change overtime. Many genes encoding enzymes, drug transporters and drug targets are under epigenetic control. For instance, the development of drug resistance in cancer cells treated with physiological concentrations of doxorubicin was reported to be due to the hyperacetylation of histone 3 in close proximity to the promoter of the ABCG2 gene (the ATP dependent transporter) and reduced histone deacetylase binding in that region (Calcagno et al. 2008). This suggests the dysregulated transcription and translation of ABCG2 gene. Similarly, disease progression during doxorubicin treatment is correlated with the absence of methylation of ABCB1 promoter region (Dejeux et al. 2010).

Therefore, combining pharmacogenetics with epigenetics (pharmacoepigenetics) may provide a better opportunity to understand the individual variations in drug response and disease pathogenesis (Baer-Dubowska 2011; Ingleman-Sundberg and Gomez 2010; Gomez and Ingleman-Sundberg 2009a). Another example to support this view is the type II diabetes mellitus (T2DM) which is a prevalent disease. Although several drugs are available to treat T2DM, there is a significant variation in their clinical response among individuals. This variation could be due to polymorphisms in genes involved in drug metabolism, transport and absorption. However, polymorphisms (genetic diversity) in these genes alone cannot completely explain all of the drug effects (variability in their therapeutic and adverse drug reactions). Therefore, in addition to genetic diversity, studies on epigenetic modification (epigenomics) may provide a clearer picture and help in developing a personalized medicine more efficiently (Gomez and Ingelman-Sundberg 2009b; Monolopoulos et al. 2011). In this regard, it is necessary to know about many additional rare polymorphisms, structural variants and tissue specific epigenetic features of the human genome. Additionally, mechanisms of epigenetic silencing of shared promoters of adjacent genes are necessary to understand the disease pathogenesis and medicinal drug response (Damani and Topol 2011; Hitchins 2010; Bennett et al. 2010).

Conclusion

Recent progress in understanding of genetic contribution to individual variability in drug therapy and disease susceptibility may provide challenges to its translation into individualized medicine and disease pathogenesis in the future (Ma and Lu 2011; Ferraldesch and Newman 2011). This genomic medicine represents a new approach to health care and disease management (Boone et al. 2011).

However, studies of population genetics have demonstrated substantial variation in allele frequencies and haplotype structure at loci of medical importance. The genetic background of patient cohort may also often be complex (Tayo et al. 2011). In addition, cellular roles of many SNPs are not known at present and continue to be defined (Orr and Chanock 2008). The bulk of excess familial risk of many diseases is not accounted for by common variants (Maher 2008). Both the common and rare variants may contribute significantly to human disease susceptibility (Schork et al. 2009) and drug response.

Although CNVs in segmental duplication are likely important for human genetic diversity, accurate genotyping methods are essential to test these variants for their association to disease. Moreover, epistatic interactions between these variants as well as interaction with environmental factors are not yet known and they may also influence the drug response (Xu et al. 2011; Spencer et al. 2011) and disease susceptibility. It is also important to note that, before we design and conduct gene-environmental factor interactions in humans, it is necessary to employ various in vivo and in vitro models.

Additionally, most diseases are genetically and clinically heterogeneous (e.g. prostate Cancer) and genomic variants may play etiological roles for a spectrum of diseases. Therefore, we need an in depth molecular understanding of the pathophysiology and mechanism of drug action before developing individualized pharmacotherapy (Squire et al. 2011; Jorgensen 2011). It is also necessary to understand how polygenes and environmental factors interact to disturb cellular regulatory networks. We also need to know how this disturbance affects cellular phenotypes and determines the rationale for targeted presentation of disease (Conti et al. 2010); Schadt et al. 2009). Further understanding of chromatin and non-sequence based modifications of genome may also provide a great promise for the significant improvement in the development and management of therapeutics. Thus, a challenge in the transition from genetics to genomics is the complexity of information. In order to understand these aspects it may require a population size experiments. These as well as protection of consumers against potential harms may limit the immediate incorporation of genomics into the practice of medicine.

ACKNOWLEDGMENTS

My apologies to those whose work or original publications could not be cited in this article because of the short nature of the paper.

REFERENCE

Ansari M, Krajinovic M (2007) Pharmacogenomics in cancer treatment defining genetic basis for inter-individual differences in response to chemotherapy. *Curr. Opin. Pediatr.*, 19: 15-22.

Baer-Dubowska W, Majehrzak-Celiriska A, Cichocki M (2011) Pharmacoepigenetics: a new approach to predicting individual drug responses and targeting new drugs. *Pharmacol. Rep.*, 63: 293-304.

Bennett KL, Mester J, Eng C (2010) Germ-line epigenetic regulation of KILLIN in Cowden and Cowden-like syndrome. *JAMA,* 304: 2724-2731.

Boone PM, Wiszniewski W, Lupski JR (2011) Genome medicine and neurological disease. *Hum. Genet.*, 130: 103-121.

Calcagno AM, Fostel JM, To KK, Saleido CD, Martin SE, Chewning KJ, Wu CP, Varticorski L, Bates SE, Caplen NJ, Ambudkar SV (2008) Single step doxorubicin-selected cancer cells overexpress the ABCG2 drug transported through epigenetic changes. *Br. J. Cancer*, 98: 1515-1524.

Candelaria M, de la Cruz-Hernandez E, Perez-Cardenas E, Trejo-Bearril C, Gutierrez-Hernandez O, Duenas-Gonzalez A (2010). Pharmacogenetics and pharmacoepigenetics of gemcitabine. *Med. Oncol.*, 27: 1133-1143.

Carlquist JF, Anderson JL (2011) Pharmacogenetic mechanisms underlying unanticipated drug responses. *Discov. Med.*, 11: 469-478.

Chan SL, Suo C, Lee SC, Goh BC, Chia KS, Teo YY (2011) Translational aspects of genetic factors in the prediction of drug response variability: a case study of warfarin pharmacogenomics in a multi-ethnic cohort from Asia. *Pharmacogenomics J.*, (in Press).

Cheong HS, Kim HD, Na HS, Kim JO, Kim LH, Kim SH, Bar JS, Chung MW, Shin HD (2011) Screening of genetic variations of SLC 15A2, SLC 22A1, SLC 22A2 and SLC 22A6 genes. *J. Hum. Genet.*, 56: 666-670.

Conard DF, Andrews TD, Carter NP, Hurles ME, Pirtchard JK (2006) A high resolution survey of deletion polymorphism in the human genome. *Nat. Genet.*, 38: 75-81.

Conti R, Veenstra DL, Armstrong K, Lesko LJ, Grosse D (2010) Personalized medicine and genomics: challenges and opportunities in assessing effectiveness, cost-effectiveness and future research priorities. *Med. Decis. Mak.*, 30: 328-340.

Costa FF, Bischof JM, Vanin EF, Lulla RR, Wang M, Sredni ST, Rajaram V, de Fatima Bonaldo M, Wang D, Goldman S, Tomita T, Soares MB (2011) Identification of microRNAs as potential prognostic markers in ependymoma. *PLoS one*, 6: e25114.

Damani SB, Topol EJ (2011) Emerging genomic applications in coronary artery disease. *JACC Cardiovasc. Interv.*, 4: 473-482.

Deenen MJ, Cats A, Beijnen JH, Schellellen JH (2011) Pharmacogenetic variability in anticancer pharmacodynamic drug effects. *Oncologist*, 16: 1006-1020.

Dejeux E, Ronneberg JA, Solvang H, Bukholm I, Geisler S, Aas T, Gut IG, Borresen-Dale AL, Lonning PE, Kristensen VN, Tost J (2010) DNA methylation profiling in doxorubicin treated primary locally advanced breast tumors identifies novel genes associated with survival and treatment response. *Mol. Cancer*, 9: 68-81.

Dennis C (2003) Epigenetics and disease: altered states. *Nature*, 421: 686-688.

Emilien G, Ponchon M, Caldas C, Isacson Q, Maloteaux J-M (2000) Impact of genomics on drug discovery and clinical medicine. *Q. J. Med.*, 93: 391-423.

Esteller M (2011) Non-coding RNAs in human disease. *Nat. Rev. Genet.*, 12: 861-874.

Feinberg AP (2007) Phenotypic plasticity and the epigenetics of human disease. *Nature*, 447: 433-440.

Ferraldesch R, Newman WG (2011) Pharmacogenetics and pharmacogenomics: a clinical reality. *Ann. Clin. Biochem.*, 48: 410-417.

Gary IC, Campbell DA, Spurr NK (2000) Single nucleotide polymorphisms as tools in human genetics. *Hum. Mol. Genet.*, 9: 2403-2408.

Girirajan S, Campbell CD, Eichler EE (2011) Human copy number variation and complex genetic disease. *Ann. Rev. Genet.*, 45: 203-226.

Glessner JT, Reilly MP, Kim CE, Takahashi N, Albona A. et al. (2010) Strong synaptic transmission impact by copy number variation in Schizophrenia. *Proc. Natl. Acad. Sci. USA*, 107: 10584-10589.

Gomez A, Ingelman-Sundberg M (2009a) Epigenetic and microRNA – dependent control of cytochrome P450 expression: A gap between DNA and protein. *Pharmacogenomics.*, 10: 1067-1076.

Gomez A, Ingleman-Sundberg M (2009b) Pharmacoepigenetics: its role in interindividual differences in drug response. *Clin. Pharmacol. Ther.*, 85: 426-430.

Halushka MK, Fan JB, Bentley K, Hsie L, Shen NP, Weder A, Cooper R, Lipshutz R, Chakravarti A (1999) Patterns of single nucleotide polymorphisms in candidate genes for blood pressure homeostasis. *Nat. Genet.*, 22: 239-247.

Hamilton JP (2011) Epigenetics: principles and practice. *Dig. Dis.*, 29: 130-135.

Hind DA, Kloek AP, Jen M, Chen X, Frazer KA (2006) Common deletions and SNPs are in linkage disequilibrium in the human genome. *Nat. Genet.*, 38: 82-85.

Hitchins MP (2010) Inheritance of epigenetic aberrations (constitutional epimutations) in cancer susceptibility. *Adv. Genet.*, 70: 201-243.

Iida A, Saito S, Sekine A, Mishima C, Kitamura Y, Kondo K, Harigae S, Osawa S, Nakamura Y (2003) Catalog of 668 SNPs detected among 31 genes encoding potential drug targets on the cell surface. *J. Hum. Genet.*, 48: 23-36.

Ikeda M, Aleksic B, Kirov G, Kinoshita Y, Yamanouchi Y, Kitajima T, Kawashima K, Okochi T, Kishi T, Zaharieva I, Owen MJ, O'Donovan MC, Ozaki N, Iwata N (2010) Copy number variation in schizophrenia in the Japanese population. *Biol. Psychiatr.*, 67: 283-286.

Ingleman-Sundberg M, Gomez A (2010) The past, present and future of pharmacoepigenomics. *Pharmacogenomics*, 11: 625-627.

Ingleman-Sundberg M, Sim SC (2010) Intronic polymorphisms of cytochromes P450. *Hum. Genomics.*, 4: 402-405.

Jorgensen JT (2011) A challenging drug development processes in the era of personalized medicine. *Drug. Discov. Today*, 16: 891-897.

Kohen R, Cain KC, Buzaitis A, Johnson V, Becker KJ, Teri L, Tirschwell DL, Veith RC, Mitchell PH (2011) Response to psychosocial treatment in poststroke depression is associated with serotonin transporter polymorphism. *Stroke*, 42: 2068-2070.

Kimchi-Sarfaty C, Oh JM, Kim I-W, Sauna ZE, Calcagno AM, Ambudkar SV, Gottesman MM (2007) A silent polymorphism in the MDR1 gene changes substrate specificity. *Science*, 315: 525-528.

Krawezak M, Reiss J, Cooper DN (1992) The mutational spectrum of single base pair substitution in mRNA splice junctions of human genes: causes and consequences. *Hum. Genet.*, 90: 41-54.

Kruglyak L (1999) Prospects for the whole genome linkage disequilibrium mapping of common disease gene. *Nat. Genet.*, 22: 139-144.

Laing RE, Hess P, Wang J, Hu SX (2011) The role and impact of SNPs in pharmacogenomics and personalized medicine. *Curr. Drug. Metab.*, 12: 460-486.

Leach IM, van der Harst P, de Boer RA (2010) Pharmacoepigenetics in heart failure. *Curr. Heart Fail. Rep.*, 7: 83-90.

Lee JA, Lupski JR (2006) Genomic rearrangement and copy number alteration as a cause of nervous system disorders. *Neuron*, 52: 103-121.

Lee JA, Carvalho CM, Lupski JR (2007) A DNA replication mechanism for generating non-recurrent rearrangements associated with genomic disorders. *Cell*, 131: 1235-1247.

LaThangue NB, Kerr DJ (2011) Pedictive biomarkers: a paradigm shift towards personalized cancer medicine. *Nat. Rev. Clin. Oncol.*, 23: 587-596.

LeVan TD, Bloom JW, Bailey TJ, Karp CL, Halonen M, Martinez FD, Vercelli D (2001) A common single nucleotide polymorphism in the CD14 promoter decreases the affinity of Sp protein binding and enhances transcriptional activity. *J. Immunol.*, 167: 5838-5844.

Lohrer HD, Tangen U (2000) Investigations into the molecular effects of single nucleotide polymorphism. *Pathobiol.*, 68: 283-290.

Lupski JR (1998) Genomic disorders: structural features of the genome can lead to DNA rearrangement and human disease traits. *Trends Genet.*, 14: 417-422.

Lupski JR (2007) Genomic rearrangement and sporadic disease. *Nat. Genet.*, 39: S43-S47.

Ma Q, Lu AY (2011) Pharmacogenetics, pharmacogenomics and individualized medicine. *Pharmacol. Rev.*, 63: 437-459.

Maher B (2008) Personal genomes: the case of the missing heritability. *Nature*, 456: 18-21.

Martin N, Boomsma D, Machin G (1977) A twin pronged attacks on complex traits. *Nat. Genet.*, 17: 387-392.

McCarroll SA, Hadnott TN, Perry GH, Sabeti PC, Zody MC, Barrett JC, Dallaire S, Gabriel SB, Lee C, Daly MJ, Altshuler DM (2006) Common deletion polymorphism in the human genome. *Nat. Genet.*, 38: 86-92.

Mestroni L, Taylor MR (2011) Pharmacogenomics, personalized medicine and heart failure. *Discov. Med.*, 11: 551-561.

Monolopoulos VG, Ragia G, Tavridou A (2011) Pharmacogenomics of oral antidiabetic medication: current data and pharmacoepigenomics perspective. *Pharmacogenomics*, 12: 1161-1191.

Need AC, Ge D, Weale ME, Maia J, Feng S et al. (2009) A genome wide investigation of SNPs and CNVs in schizophrenia. *PLoS Genet.*, 5: e1000373.

Nothen MM, Cichon S (2002) Linking single nucleotide polymorphisms. *Pharmacogenetics*, 12: 89-90.

Offit K (2011) Personalized medicine: new genomics, old lessons. *Hum. Genet.*, 130: 3-4.

O'Roak BJ, Deriziotis P, Lee C, Vives I, Schwartz JJ, Girirajan S, Karakec E, Mackenzie AP, Ng SB, Baker C, Rieder MJ, Nickerson DA, Bemier R, Fisher SE, Shendure J, Eichler EE (2011) Exome sequencing in sporadic autism spectrum disorders identifies severe de novo mutations. *Nat. Genet.*, 43: 585-589.

Orr N, Chanock S (008) Common genetic variation and human disease. *Adv. Genet.*, 62: 1-32.

Park LK, Friso S, Choi SW (2011) Epigenetics is a new bridge between nutrition and age-related disease. *Proc. Nutr. Soc.*, 4: 1-9.

Peedicayl J (2008) Pharmacoepigenetics and pharmacoepigenomics. *Pharmacogenomics*, 9: 1785-1786.

Phan VH, Tan C, Rittau A, Xu H, McLachlan AJ, Clarke SJ (2011) An update on ethnic differences in drug metabolism and toxicity from anti-cancer drugs. *Expert. Opin. Drug. Metab. Toxicol.*, 7: 1395-1410.

Rahmioglu N, Heaton J, Clement G, Gill R, Surdnlescu G, Zlobecka K, Hodgkiss D, Ma Y, Hider RC, Smith NW, Ahmadi KR (2011) Genetic epidemiology of induced CYP 3A4 activity. *Pharmacogenet. Genomics.*, 21: 642-651.

Rakyan VK, Down TA, Balding DJ, Beck s (2011) Epigenome – wide association studies for common human diseases. *Nat. Rev. Genet.*,12: 529-541.

Ramos DL, Gaspar JF, Pingarilho M, Gil OM, Fernan des AS, Rueff J, Oliveira NG (2011) Genotoxic effects of doxorubicin in cultures human lymphocytes with different glutathione S-transferase genotypes. *Mutat. Res.*, 724: 28-34.

Reik W, Dean W, Walter J (2001) Epigenetic reprogramming in mammalian development. *Science*, 293: 1089-1093.

Riley JH, Allan CJ, Lai E, Roses A (2000) The use of single nucleotide polymorphism in the isolation of common disease gene. *Pharmacogenomics*, 1: 39-47.

Roden DM, Johnson JA, Kimmel SE, Krauss RM, Medina MW, Shuldiner A, Wilke RA (2011) Cardiovascular pharmacogenomics. *Circ. Res.*, 109: 807-820.

Schadt EE, Friend SH, Shaywitz DA (2009) A network view of disease and compound screening. *Nat. Rev. Drug. Discov.*, 8: 286-295.

Schork NJ, Fallin D, Lanchbury JS (2000) Single nucleotide polymorphisms and the future of genetic epidemiology. *Clin. Genet.*, 58: 250-264.

Schork NJ, Murray SS, Frazer KA, Topol EJ (2009) Common versus rare allele hypothesis for complex diseases. *Curr. Opin. Genet. Dev.*, 19: 212-219.

Seddon JM, Reynolds R, Shah HR, Rosner B (2011) Smoking, dietary betaine, methionine, and vitamin D in monozygotic twins with discordant macular degeneration: epigenetic implications. *Ophthalmol.*, 118: 1386-1394.

Sharp AJ, Locke DP, McGrath SD, Cheng Z, Bailey JA, Vallente RU, Pertz LM, Clark RA, Schwartz S, Segraves R, Oseroff VV, Albertson DG, Pinkel D, Eichler EE (2005) Segmental duplication and copy number variation in the human genome. *Am. J. Hum. Genet.*, 77: 78-88.

Shastry BS (2002) SNP alleles in human disease and evolution. *J. Hum. Genet.*, 47: 561-566.

Shastry BS (2003) SNPs and haplotypes: Genetic markers for diseases and drug response. *Int. J. Mol. Med.,* 11: 379-382.

Shastry BS (2004) Role of SNP/haplotype map in gene discovery and drug development: an overview. *Drug. Dev. Res.*, 62: 143-150.

Shastry BS (2005) Genetic diversity and new therapeutic concepts. *J. Hum. Genet.*, 50: 321-328.

Shastry BS (2006) Pharmacogenetics and the concept of individualized medicine. *Pharmacogenomics. J.*, 6: 16-21.

Shastry BS (2007) SNPs in disease gene mapping, medicinal drug development and evolution. *J. Hum. Genet.*, 52: 871-880.

Shastry BS (2009) Copy number variation and susceptibility to human disorders. *Mol. Med. Reports*, 2: 143-147.

Shastry BS (2010) Genetic diversity and medicinal drug response in eye care. *Graefes. Arch. Clin. Exp. Ophthalmol.*, 248: 1057-1061.

Shastry BS (2011) Pharmacogenomics in ophthalmology. *Discov. Med.*, 12: 159-167.

Shi YY, He G, Zhang Z, Tang W, Zhang J Jr., Zhao Q, Zhang J. Sr., Li XW, Xi ZR, Fang C, Zhao XZ, Feng GY, He L (2008) A study of rare structural variants in schizophrenia patients and normal controls from Chinese Han population. *Mol. Psychiatr.*, 13: 911-913.

Sim SC, Ingelman-Sundberg M (2011) Pharmacogenomic biomarkers: new tools in current and future drug therapy. *Trends Pharmacol. Sci.*, 32: 72-81.

Simon T, Bhatt DL, Bergougman L, Farenc C, Pearson K, Perrin L, Vicaut E, Lacreta F, Hurbin F, Dubar M (2011) Genetic polymorphisms and the impact of a higher clopidogrel dose regimen on active metabolite exposure and antiplatelet response in healthy subjects. *Curr. Pharmacol. Ther.*, 90: 287-295.

Spencer KL, Olson LM, Schnetz-Boutaud N, Gallins P, Agarwal A, Iannaccone A, Kritchevsky SB, Garcia M, Nalls MA, Newman AB, Scott WK, Pericak-Vance MA, Haines JL (2011) Using genetic and environmental risk factor data to identify individual at higher-risk for age-related macular degeneration. *PLoS One*, 6: e17784.

Squire JA, Park PC, Yoshimoto M, Alami J, Williams JL, Evans A, Joshua AM (2011) Prostate cancer as a model system for genetic diversity in tumors. *Adv. Cancer Res.*, 112: 183-216.

Stojiljkovic M, Patrinos GP, Pavlovic S (2011) Clinical applicability of sequence variations in genes related to drug metabolism. *Curr. Drug. Metab.*, 12: 445-454.

Suriapranata M, Tjong WY, Wang T, Utama A, Raharjo SB, Yuniadi Y, Tai SS (2011) Genetic factors associated with patient specific warfarin dose in ethnic Indonesians. *BMC Med. Genet.*, 12: 80.

Tayo BO, Teil M, Tong L, Qin H, Khitrov G, Zhang W, Song Q, Gottesman O, Zhu X, Pereira AC, Cooper RS, Bottinger EP (2011) Genetic background of patients from a university medical center in Manhattan: implications for personalized medicine. *PLoS One*, 6: e19166.

Weidlich S, Walsh K, Crowther D, Burczynski ME, Feuerstein G, Carey FA, Steele RJ, Wolf CR, Miele G, Smith G (2011) Pyrosequencing based methods revealed marked inter-individual differences in oncogene mutation burden in human colorectal tumors. *Br. J. Cancer*, 105: 246-254.

Weiss KM, Terwilliger JD (2000) How many diseases does it take to map a gene with SNPs. *Nat. Genet.*, 26: 151-157.

Wilke RA, Dolan ME (2011) Genetic and variable drug response. *JAMA* 306: 306-307.

Yoshida S, Sugawara T, Nishio T, Kanekos S (2011) Personalized medicine for epilepsy based on the pharmacogenomic testing. *Brain Nerve*, 63: 295-299.

Xu Z, Zhang Z, Shi Y, Pu M, Yuan Y, Zhang X, Li L, Reynolds GP (2011) Influence and interaction of genetic polymorphisms in the serotonin system and life stress on antidepressant drug response. *J. Psychopharmacol.*, (in Press).

Yahoo N, Sabahi F, Shahzamani K, Malboobi MA, Jabbari H, Sharifi H, Mousavi-Fard SH, Merat S (2011) Mutations in the E2 and NS 5A regions in patients infected with hepatitis C virus genotype 1A and their correlation with response to treatment. *J. Med. Virol.*, 83: 1332-1337.

Zillak D, O'Donnell PH, Im HK, Gamazon ER, Chen P, Delaney S, Shukla S, Das S, Cox NJ, Vokes EE, Kohen EE, Dolan ME, Huang RS (2011) Germ line polymorphisms discovered via a cell based genome wide approach predicts platinum response in head and neck cancers. *Transl. Res.*, 157: 265-272.

Zimmer V, Lammert F (2011a) Genetics in liver disease: New Concepts. *Curr. Opin. Gastroenterol.*, 27: 231-239.

Zimmer V, Lammert F (2011b) Genetics and epigenetics in the fibrogenic evolution of chronic liver diseases. *Best Pract. Res. Clin. Gastroenterol.*, 25: 269-280.

In: Genetic Diversity: New Research
Editors: J. A. C. Amaya and M. M. F. Jimenez
ISBN: 978-1-62081-598-4
© 2012 Nova Science Publishers, Inc.

Chapter 9

NEW RESEARCH ON THE GENETIC DIVERSITY OF NON-RHIZOBIAL ENDOPHYTES INHABITING LEGUME NODULES

Alvaro Peix[1,3], Lorena Carro[2], Eugenia Cerda-Castillo[2], Carmen Tejedor[2], Martha-Helena Ramírez-Bahena[1,3] and Encarna Velázquez[2,3]

[1]Instituto de Recursos Naturales y Agrobiología, IRNASA-CSIC, Salamanca, Spain
[2]Departamento de Microbiología y Genética, Universidad de Salamanca, Edificio Departamental de Biología, Campus M. Unamuno. Salamanca, Spain
[3]Unidad Asociada Grupo de Interacción Planta-Microorganismo Universidad de Salamanca-IRNASA-CSIC, Salamanca, Spain

ABSTRACT

Rhizobia form nitrogen fixing nodules in legumes but they are not the unique microbial nodular inhabitants. In the last years the endophytic bacteria of different legume nodules have been studied and the results revealed a high genetic diversity. Members of Proteobacteria, Firmicutes and Actinobacteria have been identified in the inner of nodules from different legumes. Moreover several of these endophytes are classified into novel genera and species of bacteria, revealing nodules as an important source of diversity that should be further studied, with special emphasis in their role in the formation and functionality of legume nodules.

1. INTRODUCTION

Legume nodules are structures formed in the roots or stems in response to the presence of bacteria that are able to induce a controlled development of plant tissues and to carry out the nitrogen fixation process. At the end of the XIX century Beijerinck obtained the first pure bacterial culture from a nodule suspension. Frank (1889) named this isolate *Rhizobium leguminosarum* and from this date ahead the name "rhizobia" was applied to all bacteria capable of forming nodules on leguminous plants. Since this moment research on legume nitrogen fixing symbiosis focused exclusively on "classic" rhizobia whose colonies with abundant polysaccharide are visually identifiable on YMA plates. When the strains forming these colonies can reproduce nodule formation in the host from which they were isolated (Koch's postulates), they are considered to be rhizobia and therefore good candidates for characterization and study of infection process. Bacteria isolated from nodules having morphology different than the typical rhizobia were commonly considered as contaminants and then traditionally discarded. In this way an enormous amount of information has been lost through the years because these bacteria were not considered relevant in the formation and functioning of nodules. Some of these bacteria were later discovered as able to induce legume nodules although they do not belong to the "classic" rhizobial genera *Rhizobium, Sinorhizobium, Mesorhizobium, Bradyrhizobium* or *Azorhizobium*. This way, Alpha-Proteobacteria such as *Methylobacterium nodulans* nodulating *Crotalaria*, *Devosia neptuniae* nodulating *Neptunia natans*, *Phyllobacterium trifolii* nodulating *Trifolium pratense* or *Ochrobactrum lupini* and *O. cytisi* nodulating *Lupinus* and *Cytisus scoparius*, respectively, were described (revised by Velázquez et al., 2011). Also strains from genus *Agrobacterium* were reported to be able to nodulate soybean (Chen et al., 2000), *Sesbania* (Cummings et al., 2009) and *Medicago* (Djedidi et al., 2011). Within beta-Proteobacteria, nodulation of *Mimosa* by *Burkholderia* and later by *Cupriavidus* has been reported (revised by Velázquez et al., 2011). Finally, nodulation of *Robinia pseudoacacia* by the gamma-Proteobacteria *Pseudomonas* has been recently reported (Shiraishi et al., 2010).

All these findings give rise to the search of "non-rhizobia" able to nodulate legumes and the investigations focused on this together with diversity analysis of rhizobia nodulating different legumes have also evidenced the existence of a high number of nodular endophytes unable to reproduce nodules. In the last decades research on the genetic diversity of these endophytes has been increasing and currently it is an attractive field of research complementary to that of rhizobial diversity. This is mainly focused on their identification by sequencing and analysis of the small ribosomal subunit (16S rRNA gene) that showed legume nodule endophytes known up to date belong to phyla Proteobacteria, Firmicutes and Actinobacteria. Some of these endophytes have *in vitro* mechanisms to promote the plant growth but there is still a lack of information about their role in the infection process by rhizobia and mainly about their possible contribution to nitrogen fixation. Further investigations on nodular endophytic diversity will highlight their importance in rhizobia-legume symbiosis.

2. Nodular Endophytes within Phylum Proteobacteria

Phylum Proteobacteria is one of the widest groups of bacteria present in nature. Constituted by Gram negative bacteria, it includes organisms aerobic and facultatively anaerobic widespread in soil and water and many of the most virulent plant, animal and human pathogens are classified within Proteobacteria. Nevertheless, this phylum also includes legume endosymbionts that belongs to the class alpha-Proteobacteria within order *Rhizobiales*, beta-Proteobacteria within order *Burkholderiales* (revised by Velázquez et al., 2011) and gamma-Proteobacteria within order *Pseudomonadales* (Shiraishi et al., 2010). Interestingly, the classes of Proteobacteria mentioned above comprise both the bacteria nodulating legumes (rhizobia) as well as most nodule endophytes found up to date.

2.1. Alpha-Proteobacteria Endophytes

Rhizobia, which belong to alpha-Proteobacteria, were traditionally considered the microorganisms responsible of legume infection processes and then only the strains able to reproduce symptoms in plant are commonly considered of interest by the rhizobiologists. However rhizobia can also behave as endophytes in nodules and it is frequent the isolation of rhizobial strains from nodules that despite their inability to reinfect the legume from which they were isolated, they often promote its growth.

For example *R. leguminosarum* bv phaseoli and *Mesorhizobium loti* were isolated from nodules of *Trifolium pratense*. Combinations of endophytic *R. leguminosarum* bv phaseoli strains and *R. leguminosarum* bv trifolii resulted in the promotion of clover growth (Sturtz et al., 1997). Other endophytic rhizobia were isolated from *Vicia* nodules that after sequencing of different genes were classified into genera *Ensifer* and *Shinella* as well as in the species *Rhizobium tropici* (Lei et al., 2008).

These findings showed that rhizobia are not only the major bacterial group originating nodules in legumes, but probably the most abundant nodular endophytes mainly considering that genus *Agrobacterium* has been reclassified into genus *Rhizobium* (Young et al., 2001). "*Agrobacterium*" strains are routinely isolated from nodules together with nodulating rhizobia and, since their colony morphologies are similar, this bacterium was commonly not discarded in the initial steps of the isolation of nodulating strains. For this reason *Agrobacterium* was one of the first reported and best studied endophytes of legume nodules up to date. One of the first reports was the isolation of *A. rhizogenes* from *Trifolium* nodules, being able to promote the nodulation of this legume by *R. leguminosarum* bv trifoli (Sturtz et al. 1997). Later, other strains were found in nodules of several legumes in different countries from Asia and Africa. For example, they have been widely isolated from *Phaseolus vulgaris* in China, Tunisia and Morocco (Han et al., 2005; Mhamdi et al., 2005; Faghire et al., 2012). The diversity of these endophytes has been recently analysed by Saïdi et al. (2011) in Tunisia concluding that most of them belong to the former species *Agrobacterium radiobacter* that is currently named *Rhizobium radiobacter* and includes the former species *Agrobacterium tumefaciens*. Some of these strains cause a reduction of nodulation of *P. vulgaris* by *R. gallicum* (Mrabet et al., 2006).

"*Agrobacterium*" endophytes have been found in nodules of both herbaceous and woody Papillionoideae. For example, they were isolated from nodules of herbaceous legumes such as *Melilotus dentatus* (Wang et al., 2006), *Crotalaria pallida* and *Trifolium fragiferum* (Kan et al., 2007), different species of *Vicia* (Kan et al., 2007; Lei et al. 2008) and in medicinal species from genus *Glycyrrhiza* (Li et al., 2011). Even two strains of *Agrobacterium tumefaciens* (currently *Rhizobium radiobacter*) isolated from *Vicia faba* nodules in India have been proposed as a new biovar of this species (Tiwary et al., 2007). From Papillionoideae woody legumes, *Agrobacterium* strains were found in *Sophora alopecuroides* (Zhao et al. 2010) and *Erythrina brucei* (Wolde-Meskel et al., 2005) and in shrubs such as *Amorpha fruticosa* (Liu et al., 2005), *Campylotropis* (Han et al., 2005), *Lespedeza cuneata* (Yu et al., 2007), *Ononis viciaefolia* (Kan et al., 2007) and *Argyrolobium uniflorum* (Mahdhi et al., 2008).

Genus *Agrobacterium* has also been found in nodules of woody legumes from Mimosoideae such as *Mimosa pudica* (Kan et al., 2007) and *Acacia seyal* (Diouf et al., 2007) and different species such as *A. tumefaciens* and *A. vitis*, currently named *Rhizobium radiobacter* and *Rhizobium vitis*, respectively, were isolated from *Acacia abyssinica* and *Acacia saligna* (Wolde-Meskel et al., 2005). Finally *Agrobacterium* was also found in nodules of the Cesalpinioideae genus *Cassia* (Han et al., 2005).

Several other genera within alpha-Proteobacteria contain species able to nodulate legumes and inhabiting legume nodules as endophytes. This is the case of the genera *Devosia* from family *Hyphomicrobiaceae* and *Phyllobacterium* from family *Phyllobacteraceae* that have been found as endophytes in nodules of *Acacia salicina* and *A. stenophylla* from Mimosoideae in Australia (Hoque et al., 2011). *Phyllobacterium* was also found in nodules of *Vicia* and *Oxytropis* in China (Kan et al., 2007; Lei et al. 2008) and recently, our group found a new species of this genus, named *P. endophyticum*, in nodules of *Phaseolus vulgaris* in Spain (Flores-Felix et al., unpublished). Also a new species from family *Brucellaceae*, *Ochrobactrum cicer*, was found in nodules of *Cicer arietinum* in Pakistan (Imran et al., 2010).

The abundance of endophytic alpha-Proteobacteria in legume nodules was pointed out by Zakhia et al. (2006) whose study is a good example of the genetic diversity of legume nodule endophytes. Strains belonging to very diverse families, genera and species were identified on the basis of their 16S rRNA gene sequences in nodules of different legumes growing spontaneously in Tunisia. From family *Nitrobacteraceae* (erroneously named *Bradyrizobiaceae*) two species of genus *Bosea*, *B. thiooxidans* and *B. mexicana*, were found in nodules of *Retama raetam, Ononis vaginalis* and *Astragalus gombiformis* and from genus *Rhodopseudomonas*, *R. rhenobacensis*, was found in nodules from *Ononis natrix*. From family *Sphingomonadacae*, the species *Sphingomonas asaccharolytica* was found in nodules of *Calycotome villosa, Astragalus gombiformis* and *Lotus argenteus. Paracraurococcus ruber* (family *Acetobacteraceae)* was found in nodules of *Retama raetam* together with *Ochrobactrum grignonense* (family *Brucellaceae). Inquilinus* (family *Rhodospirillaceae)* was isolated from nodules of *Astragalus algerianus* and *Calycotome villosa. Phyllobacterium myrsinacearum* (family *Phyllobacteriaceae)* was found in nodules of *Astragalus algerianus, Astragalus uniflorus, Calycotome villosa* and *Lathyrus numidicus*. Although these strains did not form nodules, some of them have *nifH* genes close to those of *Sinorhizobium meliloti* (currently *Ensifer meliloti*).

The study of Deng et al. (2011a) also showed the high diversity of endophytes present in nodules of *Sphaerophysa salsula* in China identifying several alpha-Proteobacteria such as *Sphingomonas pruni* (family *Sphingomonadacae), Inquilinus limosus* (family *Rhodospirillaceae)* and two species of genus *Paracoccus* (family *Rhodobacteraceae) Paracoccus halophilus* and the novel species *Paracoccus sphaerophisae* (Deng et al., 2011b).

2.2. Beta-Proteobacteria Endophytes

As was previously mentioned, beta-Proteobacteria are the new rhizobia discovered at the beginning of the present century. Before 2001, beta Proteobacteria isolated from legume nodules were considered as endophytic bacteria because non-nodulating genera such as *Burkholderia* are commonly isolated from nodules.

Endophytic *Burkholderia* strains have been reported in nodules of *Mimosa pudica* in India (Pandey et al., 2005), *Glycine max* in China (Li et al., 2008), *Acacia seyal* in Senegal (Diouf et al., 2007) and recently in two species of *Acacia* in Australia (Hoque et al., 2011). Although some of these strains are able to induce ineffective nodules (Diouf et al., 2007), most of them are endophytes presenting *in vitro* plant growth mechanisms in some cases such as phytohormone secretion, ACC deaminase, phosphate solubilization and biocontrol activity (Pandey et al., 2005).

Despite the frecuent occurrence of *Burkholderia* in nodules, there are still very few reports about endophytic *Burkholderia* because the current research on nodular *Burkholderia* strains is focused on the ability for nodulation of these strains, being the non-nodulating strains little studied yet.

In addition to *Burkholderia*, within beta-Proteobacteria the species *Herbaspirillum lusitanum* from family *Oxalobacteraceae* has been isolated in Portugal from nodules of *Phaseolus vulgaris* by our research group (Valverde et al. 2003). Curiously, also some alpha-Proteobacteria uncommonly found in plant tissues such as *Bordetella avium* have been identified in nodules of clover (*Trifolium pratense*), (Sturz et al., 1997). Interestingly this species was able to promote clover growth either individually and in combination with *R. leguminosarum* bv. trifolii (Sturz et al., 1997).

2.3. Gamma-Proteobacteria Endophytes

Members of gamma-Proteobacteria have been widely isolated from nodules of different legumes, most of them classified in the families *Enterobacteriaceae, Pseudomonadaceae* and *Moraxellaceae.*

Within *Enterobacteriaceae, Pantoea agglomerans, Enterobacter kobei, Enterobacter cloacae, Leclercia adecarboxylata* and *Escherichia vulneris* were identified after 16S rRNA gene sequencing in nodules of *Hedysarum carnosum, Hedysarum spinosissimum subsp. capitatum* and *Hedysarum pallidum* in Algeria (Benhizia et al., 2004; Muresu et al., 2008). *Enterobacter* and *Klebsiella* were identified in nodules of *Arachis hypogaea* in Argentina (Ibáñez et al., 2009). Besides, there are several studies in China reporting the occurrence of *Enterobacteriaceae* strains in nodules of *Vicia faba, Vicia angustifolia, Oxytropis glabra* and *Oxytropis ochrocephala* (Kan et al., 2007). *Serratia plymuthica* was found in nodules of *Sphaerophysa salsula* (Deng et al., 2011a) and *Pantoea* and *Serratia* were also found in *Glycine max* (Li et al., 2008). Although some of these strains present *in vitro* plant growth promotion mechanisms such as the production of indol acetic acid, phosphate solubilization and free-living nitrogen fixation ability, no effect on growth and nodulation of soybean was reported.

The other wide group of gamma-proteobacteria found as nodular endophytes are included in family *Pseudomonadaceae* and specifically into genus *Pseudomonas* which was firstly described in nodules of different species from *Hedysarum* by Benhizia et al. (2004) in Algeria. Later, *P. syringae* and *P. brassicacearum* identified by 16S rRNA gene sequencing were isolated from nodules of *Medicago truncatula* and *Hedysarum carnosum*, respectively, in Tunisia (Zakhia et al., 2006). *Pseudomonas* strains were also found in nodules of *Arachis hypogaea* in Argentina and their ability to promote plant growth and to colonize the pre-nodules was reported (Ibáñez et al., 2009). A strain of *Pseudomonas fluorescens* was found in nodules of *Sphaerophysa salsula* in China (Deng et al., 2011a).

From family *Moraxellaceae,* the genus *Acinetobacter* has been identified in legume nodules of soybean in China (Li et al., 2008).

Finally, within family *Xanthomonadaceae* three genera, *Stenotrophomonas, Starkeya* and *Labrys* were found in legume nodules. Concretely, *Stenotrophomonas* was isolated from nodules of *Vicia angustifolia* in China (Kan et al., 2007). *Starkeya novella* was found in nodules of *Retama raetam* in Tunisia (Zakhia et al., 2006) and *Labrys neptuniae* was isolated from *Neptunia oleracea* in Taiwan (Chou et al., 2007).

3. Firmicutes Endophytes

The phylum Firmicutes comprises Gram positive bacteria with low G+C content including the class bacilli that comprises sporulated bacteria. These bacteria are commonly isolated from plant tissues including legume nodules, and the most frequent genera are *Bacillus* and *Paenibacillus* belonging to families *Bacillaceae* and *Paenibacillaceae*, respectively. For example, *Bacillus megaterium* isolated from *Trifolium pratense* nodules was able to promote its growth in absence of *Rhizobium* and *B. insolitus* and *B. brevis* isolated from the same nodules were able to promote nodulation by *R. leguminosarum* bv trifolii (Sturz et al., 1997). *Bacillus subtilis* and *Bacillus thuringiensis* were isolated from *Glycine max* (soybean) in China promoting plant growth when they were coinoculated with *Bradyrhizobium* nodulating this host (Bai et al., 2007). The same results were found with *B. pumilus* when it was coinoculated with *Bradyrhizobium* in soybean (Li et al., 2008). *B. thuringiensis* strains able to promote plant growth of wheat were also isolated in China from nodules of *Pueraria thunbergiana* (Selvakumar et al., 2008).

Bacillus and *Paenibacillus* strains were also found in nodules of wild legumes in Tunisia (Zakhia et al., 2006), *Robinia pseudoacacia* in Japan (Shiraishi et al., 2010) and medicinal species of *Glycyrrhiza* in China (Li et al., 2011).

Strains from several Firmicutes genera were identified after sequencing of 16S rRNA gene in nodules of *Sphaerophysa salsula* in China (Deng et al., 2011a). These strains belong to *Staphylococcus saprophyticus* from family *Staphylococcaceae*, *Bacillus simplex, Bacillus licheniformis, B. cereus, B. pumilus* and *Lysinibacillus fusiformis* from family *Bacillaceae* and *Paenibacillus amylolyticus* and *Brevibacillus borstelensis* from family *Paenibacillaceae* (Deng et al., 2011a).

Finally, some recently described new species were isolated from legume nodules such as *Paenibacillus prosopidis* isolated from *Prosopis farcta* in Tunisia (Valverde et al., 2010) and *Cohnella phaseoli* from *Phaseolus vulgaris* in Spain (García-Fraile *et al.*, 2009).

4. Actinobacteria Endophytes

Phylum Actinobacteria includes gram positive non-sporulated bacteria with high G+C content. This phylum is one of the most widely represented in legume nodules from which several new species have been described, mainly within genus *Micromonospora*. This genus, commonly considered as an external contaminant, is routinely isolated from nodules when the plates are incubated for more than one week and it is easily identifiable by its characteristic colour and morphology. In the last years several reports on the occurrence of *Micromonospora* in nodules have been published concluding that this genus is widespread in legume nodules (Trujillo et al., 2010).

Also, some new species of *Micromonospora* isolated from legume nodules have been described such as *Micromonospora lupini* and *Micromonospora saelicesensis* from *Lupinus angustifolius* (Trujillo et al., 2007) and *Micromonospora pisi* from *Pisum sativum* nodules (García et al., 2010).

The species *Curtobacterium luteum* (family *Microbacteriaceae)* was found in clover nodules (Sturz et al., 1997). Members of the same family were also found in nodules of different legumes in Tunisia (Zakhia et al., 2006). *Microbacterium flavescens* was found in *Retama raetam* and *Ononis natrix*, *Microbacterium barkeri* in *Retama retam* and *Agromyces cerinum* was found in nodules of *Argyrolobium uniflorum*. In the same study *Mycobacterium frederiksbergensis* (family *Mycobacteraceae)* was found in nodules of *Astragalus armatus* and *Ornithicoccus hortensis* (family *Intrasporangiaceae)* was found in nodules of *Medicago truncatula*.

Deng et al. (2011a) in China also found several species of Actinobacteria in nodules of *Sphaerophysa salsula* such as *Mycobacterium sacrum* (family *Mycobacteraceae)*, *Nocardia uniformis* (family *Nocardiaceae)* and *Streptomyces bottropensis* (family *Streptomycetaceae)*. *Streptomyces lydicus*, found in *Pisum sativum* nodules was able to promote growth of pea increasing nodule size and improving the vigor of bacteroids within the nodules by enhancing nodular assimilation of iron and possibly other soil nutrients (Tokala et al., 2002).

Conclusion

All the studies revised in this chapter pointed out the high genetic diversity of the nodular legume endophytes, they only showed a very small fraction of the endophytic microbiota existent within legume nodules. This clearly shows the need to increase knowledge of the diversity of non-nodulating bacteria in nodules but also in the interactions among endophytes-rhizobia-plants. This will allow us to find new tools for biofertilization of legumes exploiting the natural resources in benefit of humans without damaging the environment or health.

Acknowledgments

The authors would like to thank our numerous collaborators and students involved in this research over the years. Funding was provided by Ministerio de Ciencia e Innovación (MICINN) and Junta de Castilla y León from Spain.

REFERENCES

Bai, Y., D'Aoust, F., Smith, D. L. and Driscoll, B. T. (2002). Isolation of plant-growth-promoting *Bacillus strains* from soybean root. *Canadian Journal of Microbiology* 48, 230-238.

Benhizia, Y., Benhizia, H., Benguedouar, A., Muresu, R., Giacomini, A. and Squartini, A. (2004). Gamma proteobacteria can nodulate legumes of the genus *Hedysarum*. *Systematic and Applied Microbiology* 27, 462-468.

Chen, L. S., Figueredo, A., Pedrosa, F. O. and Hungria, M. (2000). Genetic characterization of soybean rhizobia in Paraguay. *Applied Environmental Microbiology* 66, 5099-5103.

Chou, Y. J., Elliott, G. N., James, E. K., Lin, K. Y., Chou, J. H., Sheu, S. Y., Sheu, D. S., Sprent, J. I. and Chen, W. M. (2007). *Labrys neptuniae* sp. nov., isolated from root nodules of the aquatic legume *Neptunia oleracea*. *International Journal of Systematic and Evolutionary Microbiology* 57, 577-581.

Cummings, S. P., Gyaneshwar, P., Vinuesa, P., Farruggia, F. T., Andrews, M., Humphry, D., Elliott, G. N., Nelson, A., Orr, C., Pettitt, D., Shah, G. R., Santos, S. R., Krishnan, H. B., Odee, D., Moreira, F. M., Sprent, J. I., Young, J. P. and James, E. K. (2009). Nodulation of *Sesbania* species by *Rhizobium* (*Agrobacterium*) strain IRBG74 and other rhizobia. *Environmental Microbiology* 11, 2510-2525.

Deng, Z. S., Zhao, L. F., Kong, Z. Y., Yang, W. Q., Lindström, K., Wang, E. T. and Wei, G. H. (2011a). Diversity of endophytic bacteria within nodules of the *Sphaerophysa salsula* in different regions of Loess Plateau in China. *FEMS Microbiology Ecology* 76, 463-475.

Deng, Z. S., Zhao, L. F., Xu, L., Kong, Z. Y., Zhao, P., Qin, W., Chang, J. L. and Wei, G. H. (2011b). *Paracoccus sphaerophysae* sp. nov., a siderophore-producing, endophytic bacterium isolated from root nodules of *Sphaerophysa salsula*. *International Journal of Systematic and Evolutionary Microbiology* 61, 665-669.

Diouf, D., Samba-Mbaye, R., Lesueur, D., Ba, A. T., Dreyfus, B., de Lajudie, P. and Neyra, M. (2007). Genetic diversity of *Acacia seyal* Del. rhizobial populations indigenous to Senegalese soils in relation to salinity and pH of the sampling sites. *Microbial Ecology* 54, 553-566.

Djedidi, S., Yokoyama, T., Ohkama-Ohtsu, N., Risal, C. P., Abdelly, C. and Sekimoto, H. (2011). Stress tolerance and symbiotic and phylogenic features of root nodule bacteria associated with *Medicago* species in different bioclimatic regions of Tunisia. *Microbes Environment* 26, 36-45.

Faghire, M., Mandri, B., Oufdou, K., Bargaz, A., Ghoulam, C., Ramírez-Bahena, M. H., Velázquez, E. and Peix, A. (2012). Identification at the species and symbiovar levels of strains nodulating *Phaseolus vulgaris* in saline soils of the Marrakech region (Morocco) and analysis of the *otsA* gene putatively involved in osmotolerance. *Systematic and Applied Microbiology doi*:10.1016/j.syapm.2012.02.003.

García, L. C., Martínez-Molina, E. and Trujillo, M. E. (2010). *Micromonospora pisi* sp. nov., isolated from root nodules of *Pisum sativum*. International *Journal of Systematic and Evolutionary Microbiology* 60, 331-337.

García-Fraile, P., Velázquez, E., Mateos, P. F., Martínez-Molina, E. and Rivas, R. (2008). *Cohnella phaseoli* sp. nov., isolated from root nodules of *Phaseolus coccineus* in Spain, and emended description of the genus *Cohnella*. *International Journal of Systematic and Evolutionary Microbiology* 58, 1855-1859.

Han, S. Z., Wang, E. T. and Chen, W. X. (2005). Diverse bacteria isolated from root nodules of *Phaseolus vulgaris* and species within the genera *Campylotropis* and *Cassia* grown in China. *Systematic and Applied Microbiology* 28, 265-276.

Hoque, M. S., Broadhurst, L. M. and Thrall, P. H. (2011). Genetic characterization of root-nodule bacteria associated with *Acacia salicina* and *A. stenophylla* (Mimosaceae) across south-eastern Australia. *International Journal of Systematic and Evolutionary Microbiology* 61, 299-309.

Ibáñez, F., Angelini, J., Taurian, T., Tonelli, M. L. and Fabra, A. (2009). Endophytic occupation of peanut root nodules by opportunistic Gammaproteobacteria. *Systematic and Applied Microbiology* 32, 49-55.

Imran, A., Hafeez, F. Y., Frühling, A., Schumann, P., Malik, K. A. and Stackebrandt, E. (2010). *Ochrobactrum ciceri* sp. nov., isolated from nodules of *Cicer arietinum*. *International Journal of Systematic and Evolutionary Microbiology* 60, 1548-1553.

Kan, F. L., Chen, Z. Y., Wang, E. T., Tian, C. F., Sui, X. H. and Chen, W. X. (2007). Characterization of symbiotic and endophytic bacteria isolated from root nodules of herbaceous legumes grown in Qinghai-Tibet plateau and in other zones of China. *Archives of Microbiology* 188, 103-115.

Lei, X., Wang, E. T., Chen, W. F., Sui, X. H. and Chen, W. X. (2008). Diverse bacteria isolated from root nodules of wild *Vicia* species grown in temperate region of China. *Archives of Microbiology* 190, 657-671.

Li, L., Sinkko, H., Montonen, L., Wei, G., Lindström, K. and Räsänen, L. A. (2011). Biogeography of symbiotic and other endophytic bacteria isolated from medicinal *Glycyrrhiza* species in China. *FEMS Microbiology Ecology* 79, 46-68.

Li, J. H., Wang, E. T., Chea, W. F. and Chen, W. X. (2008). Genetic diversity and potential for promotion of plant growth detected in nodule endophytic bacteria of soybean grown in Heilongjiang province of China. *Soil Biology and Biochemistry*. 40, 238-246.

Liu, J., Wang, E. T. & Chen, W. X. (2005). Diverse rhizobia associated with woody legumes *Wisteria sinensis*, *Cercis racemosa* and *Amorpha fruticosa* grown in the temperate zone of China. *Systematic and Applied Microbiology* 28, 465-477.

Mahdhi, M., de Lajudie, P. and Mars, M. (2008). Phylogenetic and symbiotic characterization of rhizobial bacteria nodulating *Argyrolobium uniflorum* in Tunisian arid soils. *Canadian Journal of Microbiology* 54, 209-217.

Mhamdi, R., Mrabet, M., Laguerre, G., Tiwari, R. and Aouani, M. E. (2005). Colonization of *Phaseolus vulgaris* nodules by *Agrobacterium*-like strains. *Canadian Journal of Microbiology* 51, 105-111.

Mrabet, M., Mnasri, B., Romdhane, S. B., Laguerre, G., Aouani, M. E. and Mhamdi, R. (2006). *Agrobacterium* strains isolated from root nodules of common bean specifically reduce nodulation by *Rhizobium gallicum*. *FEMS Microbiology Ecology* 56, 304-309.

Muresu, R., Polone, E., Sulas, L., Baldan, B., Tondello, A., Delogu, G., Cappuccinelli, P., Alberghini, S., Benhizia, Y., Benhizia, H., Benguedouar, A., Mori, B., Calamassi, R., Dazzo, F. B. and Squartini, A. (2008). Coexistence of predominantly nonculturable rhizobia with diverse, endophytic bacterial taxa within nodules of wild legumes. *FEMS Microbiology Ecology* 63, 383-400.

Pandey, A., Kang, S. C. and Maheshwari, D. K. (2005). Isolation of endophytic plant growth promoting *Burkholderia* sp. MSSP from root nodules of *Mimosa pudica. Current Science* 89, 177-180.

Saïdi, S., Mnasri, B. and Mhamdi, R. (2011). Diversity of nodule-endophytic agrobacteria-like strains associated with different grain legumes in Tunisia. *Systematic and Applied Microbiology* 34, 524-530.

Shiraishi A, Matsushita N, Hougetsu T. (2010). Nodulation in black locust by the Gammaproteobacteria *Pseudomonas* sp. and the Betaproteobacteria *Burkholderia* sp. *Systematic and Applied Microbiology* 33, 269-274.

Sturz, A. V., Christie, B. R., Matheson, B. G. and Nowak, J. (1997). Biodiversity of endophytic bacteria which colonize red clover nodules, roots, stems and foliage and their influence on host growth. *Biology and Fertility of Soils* 25, 13-19.

Tiwary, B. N., Prasad, B., Ghosh, A., Kumar, S. and Jain, R. K. (2007). Characterization of two novel biovar of *Agrobacterium tumefaciens* isolated from root nodules of *Vicia faba. Current Microbiology* 55, 328-333.

Tokala, R. K., Strap, J. L., Jung, C. M., Crawford, D.L., Salove, M. H., Deobald, L. A., Bailey, J. F. and Morra, M. J. (2002). Novel plant-microbe rhizosphere interaction involving *Streptomyces lydicus* WYEC108 and the pea plant (*Pisum sativum*). *Applied and Environmental Microbiology* 68, 2161-2171.

Trujillo, M. E., Alonso-Vega, P., Rodríguez, R., Carro, L., Cerda, E., Alonso, P. and Martínez-Molina, E. (2010). The genus *Micromonospora* is widespread in legume root nodules: the example of *Lupinus angustifolius. ISME Journal* 4, 1265-1281.

Trujillo, M. E., Kroppenstedt, R. M., Fernández-Molinero, C., Schumann, P. and Martínez-Molina E. (2007). *Micromonospora lupini* sp. nov. and *Micromonospora saelicesensis* sp. nov., isolated from root nodules of *Lupinus angustifolius. International Journal of Systematic and Evolutionary Microbiology* 57, 2799-2804.

Valverde, A., Fterich, A., Mahdhi, M., Ramírez-Bahena, M. H., Caviedes, M. A., Mars, M., Velázquez, E. and Rodríguez-Llorente, I. D. (2010). *Paenibacillus prosopidis* sp. nov., isolated from the nodules of *Prosopis farcta. International Journal of Systematic and Evolutionary Microbiology* 60, 2182-2186.

Valverde, A., Velázquez, E., Gutiérrez, C., Cervantes, E., Ventosa, A. and Igual, J. M. (2003). *Herbaspirillum lusitanum* sp. nov., a novel nitrogen-fixing bacterium associated with root nodules of *Phaseolus vulgaris. International Journal of Systematic and Evolutionary Microbiology* 53, 1979-1983.

Velázquez, E., García-Fraile, P., Ramírez-Bahena, M.H., Peix, A. and Rivas, R. (2011). Proteobacteria forming nitrogen fixing symbiosis with higher plants. In: *M. L. Sezenna (Ed.). Proteobacteria: Phylogeny, Metabolic Diversity and Ecological Effects* (first edition, pp. 37-56). New York, USA: Nova Science Publishers Inc.

Wang, L. L., Wang, E. T., Liu, J., Li, Y. & Chen, W. X. (2006). Endophytic occupation of root nodules and roots of *Melilotus dentatus* by *Agrobacterium tumefaciens. Microbial Ecology* 52, 436-443.

Wolde-Meskel, E., Terefework, Z., Frostegård, A. and Lindström, K. (2005). Genetic diversity and phylogeny of rhizobia isolated from agroforestry legume species in southern Ethiopia. *International Journal of Systematic and Evolutionary Microbiology* 55, 1439-1452.

Young, J. M., Kuykendall, L. D., Martínez-Romero, E., Kerr, A. and Sawada, H. (2001). A revision of *Rhizobium* Frank 1889, with an emended description of the genus, and the inclusion of all species of *Agrobacterium* Conn 1942 and *Allorhizobium undicola* de Lajudie et al. 1998 as new combinations: *Rhizobium radiobacter*, *R. rhizogenes*, *R. rubi, R. undicola* and *R. vitis*. *International Journal of Systematic and Evolutionary Microbiology* 51, 89-103.

Yu, J., Fan, L., Yang, S., Tang, M., Yang, W., Li, H. and Wei G. (2007). Characterization of copper-resistant agrobacterium isolated from legume nodule in mining tailings. *Bulletin of Environmental Contamination and Toxicology* 82, 354-357.

Zakhia, F., Jeder, H., Willems, A., Gillis, M., Dreyfus, B. and de Lajudie, P. (2006). Diverse bacteria associated with root nodules of spontaneous legumes in Tunisia and first report for *nifH*-like gene within the genera *Microbacterium* and *Starkeya*. *Microbial Ecology* 51, 375-393.

Zhao L., Deng Z, Yang W., Cao Y, Wang E., Wei G. (2010). Diverse rhizobia associated with *Sophora alopecuroides* grown in different regions of Loess Plateau in China. *Systematic and Applied Microbiology* 33, 468-477.

In: Genetic Diversity: New Research
Editors: J. A. C. Amaya and M. M. F. Jimenez
ISBN: 978-1-62081-598-4
© 2012 Nova Science Publishers, Inc.

Chapter 10

GENETIC DIVERSITY AND GENETIC RESCUE WITHIN WILDLIFE POPULATIONS

Shawn Larson
Curator of Conservation Research, Seattle Aquarium,
Seattle, WA, US

ABSTRACT

The importance of genetic diversity within populations has been debated since the study of genetics began. This debate was brought to a head when O'Brien et al., 1985 suggested that population viability problems within the cheetah (*Acinonyx jubatus*) could be linked to low genetic diversity thus supporting the balanced school. Since then there has been a significant increase in conservation genetics studies within threatened and endangered species focusing on genetic variability and it's importance in wildlife conservation. Populations that experience bottlenecks through extirpation, isolation or fragmentation are thought to lose genetic diversity through genetic drift and inbreeding. Small population sizes may lead to inbreeding where related individuals produce offspring. Inbreeding may lead to the buildup of deleterious recessive genes, termed inbreeding depression, that may cause decreased fecundity, increased mortality, slowed growth, developmental defects, increased susceptibility to disease, decreased ability to withstand stress, and decreased ability to compete. Genetic load is the longer term loss of population fitness over time due to an increase in detrimental mutations becoming fixed in small populations through inbreeding. Genetic load may be documented as low estimates of fitness related variables when compared to other populations that have not experience bottlenecks or have no evidence of loss of fitness associated with loss of genetic diversity. Genetic rescue is a management technique that has been used to manage critically endangered populations with high genetic load. This strategy uses unrelated individuals from one population that are introduced into the population with apparent low fitness. This introduction of new genetic material acts to reduce genetic load by lowering or eliminating the frequency of detrimental gene variants in the population. This management technique has been used successfully within both captive and wild critically endangered populations. Due to the rapid increase in the study of genetic variation within wildlife populations there is a growing body of evidence linking fitness variables with genetic methods, suggesting meaningful trends and management practices like genetic rescue that may affect long-term population survival and viability. In this chapter we highlight case studies that describe the consequences of low genetic diversity, inbreeding, genetic load and genetic rescue within several endangered wildlife populations.

INTRODUCTION

The importance of genetic diversity for optimal population health and long term viability has been debated since the study of genetics began. The debate was brought to public attention when O'Brien et al., 1985 suggested that population problems such as low reproduction and high incidence of disease within the cheetah (*Acinonyx jubatus*) could be linked to low genetic diversity. Since then there has been an increase in genetics studies within threatened and endangered species focusing on genetic variability and it's importance in wildlife conservation and population viability.

Most threatened and endangered wildlife populations have experience significant loss of individuals through extirpations and population fragmentation creating local population bottlenecks characterized by relatively small populations isolated from other groups preventing genetic exchange. These small, isolated populations are thought to lose genetic diversity through genetic drift and inbreeding (Charlesworth and Charlesworth, 1999; Crnokrak and Roff, 1999; Hedrick and Kalinowski, 2000; Lacy, 1997; Lynch, 1996; Ralls et al., 1988). Small population sizes out of necessity may lead to inbreeding where related individuals produce offspring. Inbreeding may then lead to the buildup of deleterious recessive genes, termed inbreeding depression that may cause decreased fecundity, increased mortality, slowed growth, developmental defects, increased susceptibility to disease, decreased ability to withstand stress, and decreased ability to compete. Inbreeding and the potential effects on population viability are important because populations that experience a bottleneck are thought to suffer from decreased genetic variation for hundreds of thousands of years (in the absence of immigration) due to the low locus mutation rate that prevails at most genetic locations or loci (10-8 to 10–5 mutations per year; Driscoll et al., 2002; Lande andBarrowclough, 1987; Lynch, 1996). Recovery of lost genetic variation in the absence of immigration is the inverse of the mutation rate (Driscoll et al., 2002; Nei et al., 1975). Small populations that remain small and isolated from immigration over several generations run the risk of fixation of alleles. Most individuals are assumed to contain unique deleterious genes at several loci, all subject to chance fixation in a small and isolated founder population (Lynch, 1996). The probability of fixation of deleterious alleles in founder populations of 30 or fewer individuals is close to the alleles' initial frequencies in the population because genetic drift overwhelms natural selection, and may lead to relatively rapid extinction of the population even in the absence of any catastrophic events (Lynch, 1996; Shaffer, 1981). Thus if the deleterious allele frequency is 10% that is the probability thah the allele will become fixed in the population and that is the probability that the negative effects on population viability will be realized.

Fixing of deleterious genes over time through inbreeding is termed genetic load which is the longer term loss of population fitness due to an increase in detrimental genes becoming fixed. Genetic load may be documented as low estimates of fitness related variables when compared to populations that have not experience bottlenecks or have no evidence of loss of fitness associated with low genetic diversity. One way to manage inbreeding and genetic load is through a management technique called genetic rescue. This strategy uses unrelated individuals from a similar population that are introduced into the population with apparent low fitness and assumed genetic load. This introduction of new genetic diversity acts to reduce genetic load by lowering or eliminating the frequency of detrimental genes in the population. This management technique has been used successfully within both captive and wild endangered populations (Hedrick and Fredrickson, 2010). There are clearly recognized guidelines for the successful use of genetic rescue as a conservation management strategy. Ideal candidate populations for genetic rescue must have several of the following: Evidence of low fitness; the existence of a closely related donor population from the same species and from similar habitat; experimental data in a captive situation such as successful mating, good survival of progeny and molecular data; a translocation protocol; a detailed monitoring plan and a commitment for long term management (Hedrick and Fredrickson, 2010).

Due to the rapid increase in the study of genetic variation within wildlife populations, there is a growing body of evidence linking fitness variables with genetic diversity suggesting management practices like genetic rescue may positively affect long-term population survival and viability. In this chapter we highlight case studies that describe the consequences of low genetic diversity, inbreeding, genetic load and genetic rescue within several endangered wildlife populations.

CASE STUDIES

Sea Otter

Sea otters, *Enhydra lutris*, were once abundant across their range in the north Pacific Rim from northern Japan to the central Pacific Coast of Baja California, Mexico (Lensink, 1962; Kenyon, 1969; Riedman and Estes 1990). During the 18th and 19th centuries, the fur trade eliminated or greatly reduced sea otter populations throughout this area, resulting in 11 scattered populations after protection in 1911, with a combined population totaling approximately 1% of the original abundance of sea otters throughout the North Pacific (Lensink, 1962; Kenyon, 1969).

By the 1970's, few sea otter populations had recovered to pre-exploitation levels, with the majority of historic sea otter habitat vacant along the west coast of North America from Prince William Sound, Alaska, southward to California (Estes, 1990; Kenyon, 1969; Riedman and Estes, 1990). In an effort to re-establish sea otter populations throughout their former range, several translocations were made from the 1950's through the 1970's (Jameson et al., 1982). In total, 715 otters were captured at Amchitka Island in the Aleutian chain and Prince William Sound, Alaska, and then released at various unoccupied habitats in Alaska, British Columbia, Washington and Oregon (Jameson et al., 1982). The translocations to Washington, Oregon and the Pribilof Islands included only animals captured at Amchitka, and only the Washington effort was successful (Bodkin et al., 1999; Jameson et al., 1982). The translocations to Southeast Alaska and British Columbia (off the west coast of Vancouver Island) included a mix of Amchitka and Prince William Sound animals, and both were successful (Bodkin et al., 1999; Jameson et al., 1982). This mixture of sea otters from two different populations resulted in groups of sea otters with the highest genetic diversity and the highest population growth rates, a case of genetic rescue (Larson et al, 2009). However, in spite of these successful translocation efforts, sea otter populations today remain fragmented with extant populations geographically separated and, in most cases, reproductively isolated (Bodkin et al., 1999).

Investigation into the genetic variation remaining in sea otters post exploitation found only half the original genetic variation left within sea otter populations (Larson et al., 2002b).

Genetic diversity estimates for all extant populations were similar, with expected heterozygosities ranging from 40% within California to 47% within Southeast Alaska and the allelic richness (mean number of alleles corrected for sample size) ranging from 3.20 in Prince William Sound to 3.70 in Amichitka and Washington (Larson et al., 2002a).

The question is do sea otter suffer inbreeding depression because of their loss of genetic diversity and population fragmentation and reduction? California sea otters historically have had the lowest population growth rate of any sea otter population as well as the highest reported incidence of disease within adult animals, those that make up the breeding population (Gerber et al., 2004; Hanni et al., 2003; Kreuder et al., 2003; Miller et al., 2004; Thomas and Cole, 1996). In addition, a high incidence of infectious diseases among this population suggests a weakened immune system or immunosuppression (Gerber et al., 2004; Hanni et al., 2003; Kreuder et al., 2003; Miller et al., 2004).

It seems at least conceivable that the low genetic diversity levels in this population have played a role in susceptibility to infectious disease via immunosuppression.

The higher incidence of disease eventually leads to elevated mortality of prime-aged animals and low overall population growth rates. Many extant sea otter populations appear to be in a precarious balance. Because they inhabit coastal areas, they are often in contact with humans and can suffer negative interactions associated with fishing activities and exposure to shoreline and near-shore pollution sources.

Add to these difficulties the potential for negative effects associated with the loss of genetic diversity and buildup of deleterious alleles called genetic loading, and we may improve our understanding of why some sea otter populations fail to thrive, like the threatened California population. Genetic rescue success stories like the thriving South East Alaskan sea otter population made up of individuals from two different populations may be the key to managing inbreeding depression in sea otters.

Florida Panther

The American mountain lion, cougar or puma (*Puma concolor*) once was widely distributed throughout the Americas. There are currently seven recognized subspecies based on genetic data: *P. c. cougar*: North America, *P. c. corryi*: Florida, *P. c. costaricensis*: Central America, *P. c. capricornensis*: eastern South America, *P. c. concolor*: northern South America, *P. c. cabrerae*: central South America, and *P. c. puma*: southern South America (Culver et al., 2000).

The cougar subspecies found in southern Florida swamplands, *P. c. corryi,* is commonly referred to as "panther" and is listed as critically endangered by the International Union for the Conservation of Nature (IUCN) (2008 IUCN red list) and endangered by the Endangered Species Act (1967 ESA list).

The Florida panther is restricted to a small area that includes the Big Cypress National Preserve, Everglades National Park, and the Florida Panther National Wildlife Refuge, the entire area thought to represent only 5% of its historic range. The number of living Florida panthers was estimated to be approximately 100 in 2010 (Johnson et al. 2010). Extirpation from hunting is responsible for the population bottleneck which created a small, isolated population that eventually became inbred.

This genetic isolation and inbreeding in the Florida panther population is thought to have resulted in several problems associated with reproduction (such as cryporchidism in males and high cub mortality), as well as widespread genetic abnormalities such as heart defects and kinked tails (Roelke et al., 1983).

This prompted a genetic rescue program to protect and enhance the genetic diversity of the remaining panthers. In 1995, conservation managers translocated eight unrelated female cougars from Texas into the Florida panther population to increase genetic diversity, improve population growth, and reverse indications of inbreeding depression (Hedrick, 1995). The effect of this genetic rescue program was dramatic.

Since the introduction of those unrelated females from Texas the Florida panther numbers increased threefold with the panther population growth rate increasing from 0 to 12% annually, genetic heterozygosity doubled, individual survival and fitness measures improved, and inbreeding correlates declined significantly (Hedrick and Fredrickson, 2010; Johnson et al. 2010).

The remarkable success of the genetic rescue experiment within Florida panthers makes it a model for other genetic rescue programs.

Mexican Wolf

The Mexican wolf (*Canis lupis baileyi*) is an endangered subspecies of the grey wolf (*Canis lupus*) (Leonard et al, 2005). The population was driven to critically low numbers because of hunting resulting in their extinction in the United States and only a few small isolated populations in Mexico by 1980. These small, isolated populations were thought to have been negatively affected by inbreeding because of zero population growth over time due primarily to low reproductive rate (litter sizes) and high levels of mortality. Based on these population parameters this wolf was listed as endangered by the ESA in 1976 (Hedrick and Fredrickson, 2010). Because there were so few left in the wild the surviving individuals were eventually caught and placed in captivity. All Mexican wolves surviving today thus are derived from three captive lineages descended from a total of seven wild founders caught between 1960 and 1980 (Hedrick et al., 1997). The captive wolves were kept and bred within and between founder lineages. These captive groups were monitored over several years for fitness related traits to determine potential effects of inbreeding and out breeding. Initial results of fitness variables found that crossed lineages between the three captive lineages had significantly higher numbers of pups survive in each litter (Hedrick and Fredrickson, 2010). In 1998 a pure lineage group was reintroduced into the wild. Then in 2009 a group from another lineage was introduced resulting in the first genetically mixed population in the wild (Hedrick and Fredrickson, 2010). Because of the mixing of these two lineages, almost immediately the population rate started growing in the wild with the average litter size doubling and a significant decline in the measured inbreeding coefficient (Hedrick and Fredrickson, 2010). Based on these reproductive indices (litter size and inbreeding coefficients in pups) there appeared to be successful genetic rescue in wild Mexican wolves after only two different lineages were allowed to interbreed (Hedrick and Fredrickson, 2010).

Big Horn Sheep

Rocky Mountain bighorn sheep (*Ovis canadensis*) in the National Bison Range in Montana were established by translocation in 1922. Since the initial introduction it remained isolated until 1985, when bighorn sheep derived from two other herds were introduced. Population demographics ,genetic and life-history data were collected from 1979–2003 to describe the population and the effects of the bottleneck, to quantify changes in genetic variation during isolation and following introductions, and to compare the relative fitness of individuals (Hogg et al. 2006).

The change in population size from founding became slightly negative over time. Genetic variability was extremely low for the founders alive in 1985, observed heterozygosity and allelic diversity averaged only 0.44 and 2.2 alleles per locus. In addition the distribution of alleles across allele frequency classes was the same as that expected for a populations subjected to strong genetic drift with rare alleles lost and others fixedwith potentially deleterious results.

After the introduction of new founders in 1985 the beneficial effects of genetic rescue on fitness indicies was striking. For example, m

ale annual reproductive success in outbred rams was significantly higher than that of inbred rams with outbred rams having 2.6 times more reproductive success relative to inbred rams of the same age. Outbreeding was also associated with greater annual female reproductive success with outbred females 2.2 times more reproductive success than inbred ewes of the same age. In addition more outbred females gave birth to larger lambs and had much earlier initial breeding dates (Hogg et al. 2006; Miller et al. 2012).

Genetic variation was also rapidly restored by this case of genetic rescue. Observed heterozygosity increased from 0.50 in 1984 to 0.67 in 2003. Allelic diversity also increased from 2.4 alleles per locus in 1984 to 5.4 alleles per locus by 2003. These values post genetic rescue are comparable to those reported for large, outbred bighorn populations. Finally annual population growth rates followed a strong positive trend after decades of negative population growth rates (Tallmon et al. 2004; Hogg et al. 2006; Miller et al. 2012).

Northern Elephant Seals

There are some populations that experience severe bottlenecks, have very little genetic variation, yet seem not to suffer from inbreeding depression.

One case of this is the northern elephant seal (*Mirounga angustirostris*) which was exploited for its blubber in the 19th century resulting in only 10-30 founders (Hoelzel, 1997). Furthermore a genetic study of pre-exploitation northern elephant seal genetic diversity revealed a significant loss of genetic diversity (55% loss) due to the bottleneck (Weber et al.,

2000). Even though numbers were so low and genetic analysis revealed little genetic variation the population recovered quickly and was estimated to be close to 127,000 in 1991 (Bonnell and Selander, 1974; Hoelzel, 1997).

However even though the northern elephant seal recovery has been remarkable, new evidence suggests the population may be exhibiting decreased reproductive fitness. Specifically paternity success of male northern elephant seals was lower than expected when compared to paternity success of the southern elephant seals, *M. leonine*, which has not experienced similar severe exploitation reduction (Hoelzel, 1997).

CONCLUSION

Low diversity combined with inbreeding depression increase a small population's vulnerability to extinction from stochastic events due to fixation of deleterious alleles and potential the loss of adaptive variation (Lacy, 1997; Lynch, 1996).

Population sizes of at least 1000 are suggested to protect against the fixation of deleterious alleles while a breeding population of 10,000 is suggested to protect adaptive genetic variation (Lynch, 1996). Unfortunately, these population sizes are rarely found in wildlife populations and certainly not in endangered species, specifically those that have been fragmented by human activities such as sea otters, Florida panthers and Mexican wolves. The fragmentation of wildlife life populations resulting in small, isolated populations, along with fixation of deleterious alleles, genetic drift, loss of adaptive genetic variation and inbreeding depression all dramatically increase a small population's vulnerability to extinction (Charlesworth and Charlesworth 1999; Crnokrak and Roff, 1999; Fowler and Whitlock, 1999; Hedrick and Kalinowski, 2000; Lacy, 1997; Lynch, 1996; Ralls et al., 1988). In a world where wildlife individuals have little opportunity to regularly migrate freely between isolated populations to maintain geneflow and thus overall effective population size, it becomes imperative for management policies to encourage genetic diversity through establishment of migration routes and encouraging translocations. If the ultimate goal is to maintain maximal diversity and eliminate the problems associated with small, inbred wildlife populations within wildlife population then management decisions such as genetic rescue may need to be made based on maintaining genetic diversity rather than maintaining unique populations such as subspecies.

Genetic rescue may require the mixing of unique stocks or subspecies to retain lost diversity and reduce the presence of deleterious gene variations. Managers must decide what is most important to preserve, the species with maximal genetic diversity for future stochastic changes or a genetically unique race that may be negatively affected by isolation and inbreeding and incapable of responding to future challenges.

REFERENCES

Bodkin J.L., Ballachey, B.E., Cronin, M.A., and Scribner, K.T. (1999). Population demographics and genetic diversity in remnant and translocated populations of sea otters. *Conservation Biology* 13:1378-1385.

Bonnell M.L., and Selander, R.K. (1974). Elephant seals: Genetic variation and near extinction. *Science* 184:908-909.

Charlesworth, B., and Charlesworth, D. (1999). The genetic basis of inbreeding depression. *Genetics Research* 74: 329-340.

Crnokrak, P., and Roff, D.A. (1999). Inbreeding depression in the wild. *Heredity* 83: 260-270.

Culver, M., Johnson, W.E., Pecon-Slatterym, J., and O'Brien, S.J. (2000). Genetic ancestry of the American Puma (*Puma concolor*). *Journal of Heredity* 91:186-197.

Driscoll, C.A., Raymond, M.M., Nelson, G., Goldstein, D., and O'Brien, S.J. (2002). Genomic Microsatellites as Evolutionary Chronometers: A Test in Wild Cats. *Genome Research* 12: 414-423.

Gerber, L. R., Tinker, T. M., Doak, D. F., Estes, J. A., and D.A. Jessup. 2004. Mortality sensitivity in life-stage simulation analysis: A case study of southern sea otters. *Ecological Applications* 14:1554–1565.

Hanni, K. D., Mazet, J. A. K., Gulland, F. M. D., Estes, J., Staedler, M., Murray, M.J., Miller, M., and Jessup, D.A. (2003). Clinical pathology and assessment of pathogen exposure in southern and Alaskan sea otters. *Journal of Wildlife Diseases* 39:837–850.

Hedrick P.W., and S.T. Kalinowski. (2000). Inbreeding depression in conservation biology. *Annual Review of Ecology, Evolution, and Systematics* 31:139-162.

Hedrick, P.W., and Fredrickson, R. (2010). Genetic rescue guidelines with examples from Mexican wolves and Florida panthers. *Conservation Genetics* 11:615-626.

Hedrick, P.W., Miller, P.S., Geffen, E., and Wayne, R. (1997). Genetic evaluation of the three captive Mexican wolf lineages. *Zoo Biology* 16:47-69.

Hedrick, P.W. (1995). Gene flow and genetic restoration: the Florida panther as a case study. *Conservation Biology* 9:996-1007.

Hoelzel, A.R. (1997). Molecular ecology of pinnipeds. In: *Molecular genetics of marine mammals*. Dizon, A.E., Chivers, S.J., and Perrin, W.F., eds. Pp. 147-157 Special Publication, 3. Society for Marine Mammalogy.

Hogg, J.T., Forbes, S.H., Steele, B.M., and Luikart, G (2006) Genetic rescue of an insular population of large mammals. Proceedings of the Royal Society of Biology 273(1593) 1491-1499. doi: 10.1098/rspb.2006.3477.

Jameson, R.J., Kenyon, K.W., Johnson, A.M., and Wright, H.M. (1982). History and status of translocated sea otter populations in North America. *Wildlife Society Bulletin* 10:100-107.

Johnson, W.E., Onorato, D. P., Roelke, M.E., Land, E. D., Cunningham, M., Belden, R.C, McBride, R., Jansen, D., Lotz, M., Shindle, D., Howard, J., Wildt, D.E., Penfold, L.M. , Hostetler, J.A. , Oli, M.K., and O'Brien, S.J. (2010). Genetic Restoration of the Florida Panther. *Science* 49:329.

Kenyon, K.W. (1969). *The sea otter in the eastern Pacific Ocean*. North American Fauna 68. U.S. Department of the Interior, Washington, D.C.

Kreuder, C., Miller, M.A., Jessup, D.A., Lowenstine, L.J., Harris, M.D. , Ames, J.A. , Carpenter, T.E., Conrad, P.A., and Mazet, J.A.K. (2003). Patterns of mortality in southern sea otters (*Enhydra lutris nereis*) from 1998–2001. *Journal of Wildlife Diseases*. 39:495–509.

Lacy, R.C. (1997). Importance of genetic variation to the viability of mammalian populations. *Journal of Mammalogy* 78(2):320-335.

Lande, R., and Barrowclough, G.F. (1987). Effective population size, genetic variation, and their use in population management. In: *Viable populations for conservation*. Soule, M.E., ed. Pp. 87-123 Cambridge University Press, New York, NY.

Larson, S.E., Jameson, R.J., Bodkin, J.L., Staedler, M., and Bentzen, P. (2002a). Microsatellite DNA and mtDNA variation within and among remnant and translocated sea otter, *Enhydra lutris*, populations. *Journal of Mammalogy* 83 (3): 893-906.

Larson, S.E., Jameson, R.J., Etnier, M., Fleming, M., and Bentzen, P. (2002b). Loss of genetic diversity in sea otters (*Enhydra lutris*) associated with the fur trade of the 18th and 19th centuries. *Molecular Ecology* 11: 1899-1903.

Larson, S.E., Monson, D. Ballachey, B., Jameson, R., Wasser, S.K. (2009). Stress-related hormones and genetic diversity
in sea otters (*Enhydra lutris*). *Marine Mammal Science* 25(2): 351–372.

Lensink, C.J. (1962). *The history and status of sea otters in Alaska*. PhD dissertation. Purdue University, New York.

Leonard, J.A., Vila, C., and Wayne, R.K. (2005). Legacy lost: genetic variability and population size of extirpated US grey wolves (*Canis lupus*). *Molecular Ecology* 14:9-17.

Lynch, M. (1996). A quantitative-genetic perspective on conservation issues. In: *Conservation genetics. Case histories from nature* Avise, J. C., and Hamrick, J. L., eds. Pp. 471-501. Chapman and Hall. New York.

Miller, M. A., Grigg, M. E., Kreuder, C., James, E. R., Melli, A. C., Crosbie, P. R., Jessup, D.A., Boothroyd, J. C., Brownstein, D., and Conrad, P. A. (2004). An unusualgenotype of Toxoplasma gondii is common in California sea otters (*Enhydra lutris nereis*) and is a cause of mortality. *International Journal for Parasitology* 34:275–284.

Miller, J.M., Poissant, J., Hogg, J.T., and Coltman, D.W. (2012) Genomic consequences of genetic rescue in an insular population of bighorn sheep (Ovis canadensis). Molecular Ecology DOI: 10.1111/j.1365-294X.2011.05427.x.

Nei, M., Maruyama, T., and Chakraborty, R. (1975). The bottleneck effect and genetic variability in populations. *Evolution* 29:1-10.

O'Brien, S.J., Roelke, M.E., Marker, L., Newman, A., Winkler, C.A., Meltzer, D., Colly, L., Evermann, J.F., Bush, M., and Wildt, D.E. (1985). Genetic basis for species vulnerability in the cheetah. *Science* 227: 1428-1434.

Ralls, K., Ballou, J.D., and Templeton, A. (1988). Estimates of lethal equivalents and the cost of inbreeding in mammals. *Conservation Biology* 2 (2): 185-193.

Riedman, M.L., and Estes, J.A. (1990). The sea otter (*Enhydra lutris*): Behavior, ecology and natural history. U.S. Department of the Interior, Fish and Wildlife Service. *Biological Report 90* (14).

Roelke, M.E., Martenson, J.S., and O'Brien, S.J. (1993). The consequence of demographic reduction in the endangered Florida panther. *Current Biology* 3:304-350.

Tallmon D.A, Luikart G, andWaples R.S (2004) The alluring simplicity and complex reality of genetic rescue. *Trends Ecological Evolution* 1: 489–496. doi:10.1016/j.tree.2004.07.003.

Thomas, N. J., and Cole, R. A. (1996). The risk of disease and threats to the wild population. *Endangered Species Update* 13:23–27.

Weber D.S., Stewart, B.S., Garza, J.C., and Lehman, N. (2000). An empirical genetic assessment of the severity of the northern elephant seal population bottleneck. *Current Biology* 10: 1287-1290.

In: Genetic Diversity: New Research
Editors: J. A. C. Amaya and M. M. F. Jimenez
ISBN: 978-1-62081-598-4
© 2012 Nova Science Publishers, Inc.

Chapter 11

The Genus *Cunila* D. Royen Ex. L. (Lamiaceae): A Review on Genetic Variability and Advances in Taxonomy

G. Agostini,[1,*] T. S. Ribeiro,[2] S. Echeverrigaray[2] and T. T. Souza-Chies[1]

[1]Programa de Pós-Graduação em Botânica, Universidade Federal do Rio Grande do Sul, Porto Alegre, RS, Brazil. Departamento de Botânica, UFRGS. Instituto de Biociências Av. Bento Gonçalves, Porto Alegre, RS, Brazil
[2]Instituto de Biotecnologia, Universidade de Caxias do Sul, Caxias do Sul, RS, Brazil. Universidade de Caxias do Sul. R. Francisco Getúlio Vargas, Caxias do Sul, RS, Brazil

Abstract

Several native species are used in South American folk medicine justifying the scientific surveys and earning increasing attention from pharmacological and cosmetic industries among others. *Cunila* is an important aromatic and medicinal genus with increasing researches regarding chemical and genetic variability due its interesting essential oil composition. Along with the increasing scope of chemical studies, concerns about the maintenance of natural populations and its genetic variability have raised toward to the sustainable and rational use of these plants. The current chapter encompasses an actual review about genetic variability among South American *Cunila* species and an introduction to the genetic relationship among South and North American species.

* Corresponding author: Agostini G.: sabonim.radan@gmail.com. Tel.: 55 51 3308-7578; Fax: 55 51 3308-7755.

1. INTRODUCTION

The genus *Cunila* D. Royen ex L. belongs to Lamiaceae family, subfamily Nepetoideae, tribe Mentheae, subtribe Menthinae (Cantino et al., 1992). This genus is distributed in North America, which includes nine species, and 11 species in southern South America (Bordignon, 1997; Garcia-Peña, 1989).

The North American group is composed by one species native to the west of the United States (*C. origanoides* (L.) Britton). Mexico group is composed by eight species; seven of which are found in the western and central regions (*C. longiflora* A. Gray, *C. secunda* S. Watson, *C. crenata* M.R. Garcia-Peña amd P. Tenorio Lezama, *C. lythrifolia* Benth., *C. polyantha* Benth., *C. pycnantha* B.L. Rob. amd Greenm., and *C. ramamoorthiana* M.R. Garcia-Peña) and one species (*C. leucantha* Kunth ex Schltdl. et Cham.) found from southern Mexico to Panama.

South American species (southern Brazil, Paraguay, Uruguay, and Argentina), are classified into three botanical sections based on growth habits and inflorescence morphology: *Incanae*, *Incisae*, and *Spicatae* (Epling, 1936). Section *Incanae* is represented by *C. incana* Benth. and is characterized by shrubs with hairy pale yellow flowers at the upper axils. Section *Incisae* is composed by *C. incisa* Benth. and *C. angustifolia* Benth, and is characterized by shrubs with inflorescences formed by small glabrous white flowers. Section *Spicatae* is composed by eight species: *C. menthoides* Benth., *C. platyphylla* Epling, *C. menthiformis* Epling, *C. spicata* Benth., *C. galioides* Benth., *C. fasciculata* Benth. (Figure 1), *C. microcephala* Benth., and *C. tenuifolia* Epling. This section is characterized by subshrubs or evergreen-perennial herbs with terminal spikes or globular inflorescences formed by small white, almost glabrous flowers.

Both northern and southern species of *Cunila* are used in popular medicine as stimulants, aromatics, antispasmodics, emenagogues, antithermics, and in the treatment of chronic coughs and respiratory infections (Simões et al. 1994).

Figure 1. *Cunila fasciculata* Benth.

Additionaly, *Cunila* essential oils have compounds responsible for antibacterial, antifungal and insecticidal activity (Duke 1994; Sandri et al., 2007). A wide array of different essential oil has been reported for *Cunila* (Manjarrez e Mendoza, 1966; Moreira e Krambeck, 1976; Bordignon *et al.*, 1997, 1998, 1998b; 1999; Cicció e Poveda, 1999; Echeverrigaray *et al.*, 2003; 2009; Agostini et al., 2006; 2010a; 2010b, 2012).

The aim of the present review is to provide information about the inter- and intraspecific genetic variability of the genus *Cunila*, as well, to introduce advances in the taxonomic study of this genus.

2. Genetic Relationships among South American *Cunila* Species

Agostini et al. (2008) analyzed the genetic relationship among South American *Cunila* species based on ISSR (Inter Simple Sequence Repeat), using seven primers selected based on the number of amplification products, quality and level of polymorphism and reproducibility. These primers displayed high discriminating ability, substantiated by Simpson's indexes, which demonstrates that a small number of primers are sufficient for this methodology.

Samples from each species of *Cunila* were collected at different locations in the Rio Grande do Sul state (Brazil). *Cunila fasciculata* (Figure 1) is an endangered species and was not found during the fild trips, in this case, a dried herbarium sample was used. The authors included one herbarium sample of *C. origanoides* (L.) Britton (North American species) to estimate the genetic divergence between North and South American species. *Mentha piperita* L. and *Rosmarinus officinalis* L., were also included in the analysis as outgroups.

Considering *Cunila* species, the polymorphism found was 96.3%, but excluding the North American species (*C. origanoides*), the polymorphism was slightly lower (95.3%).

Cunila microcephala and *C. menthiformis* showed the highest similarity values between species, followed by *C. incisa* and *C. angustifolia*. As expected, the lowest similarity values were found between *Cunila* species and the outgroup *Mentha* and *Rosmarinus*.

The results based on ISSR analyses separated the *Cunila* species from the outgroup. The North American species *C. origanoides* clustered at the base of the South American species, and formed an independent cluster.

Among South American species two groups were formed, one composed by the shrubs species of the sections *Incisae* and *Incanae*: *C. incisa*, *C. angustifolia* and *C. Incana*; and the second composed by the subshrubs species of the section *Spicatae*: *C. galioides*, *C. spicata*, *C. fasciculata*, *C. menthiformis*, *C. menthoides*, *C. platyphylla*, and *C. microcephala* (Figure 2).

Following Agostini et al (2008), the genetic similarity observed among the shrub species *C. incisa*, *C. angustifolia* (*Incisae*) and *C. incana* (*Incanae*) does not support their separation into two different sections.

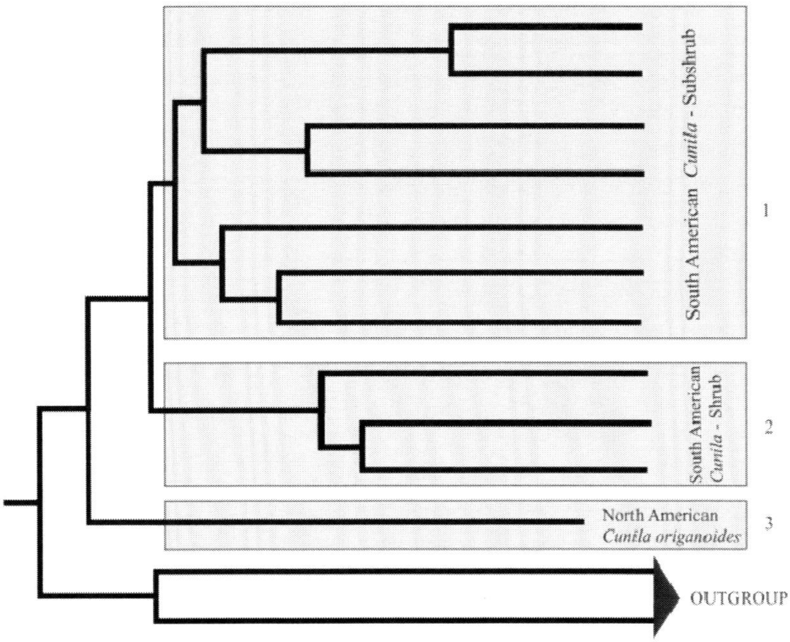

Modified from Agostini et al. (2008).

Figure 2. Dendrogram representing the genetic similarity among *Cunila* species based on ISSR markers. Group 1 – South American *Cunila* subshrubs (section *Spicatae*), Group 2 – South American *Cunila* shrub (*Incanae* and *Incisae*), Group 3 - North American species (*C. origanoides*), and OUTGROUP – *Mentha piperita* and *Rosmarinus officinalis.*

3. Phylogenetic Relationships among *Cunila* Species

The genus *Cunila* is distributed in North America - nine species, and in southern South America -11 species (Bordignon, 1997; Garcia-Peña, 1989). The South American species are classified into three botanical sections: *Incanae*, *Incisae*, and *Spicatae* (Epling, 1936).

Considering the distance between the centers of distribution, Agostini (2008; data not published) used nuclear and plastid sequence data to produce molecular phylogenies for *Cunila* and to test the genus' monophyly. Seventeen species of the genus *Cunila* were included in the study: ten species from South America, six species from Mexico, and one from the United States. Other Lamiaceae species belonging to the subtribe Menthinae were used as outgroup (Table 1).

All results obtained by the authors strongly supported subtribe Menthinae as monophyletic with respect to the representatives of subtribe Salviinae and Nepetinae included in the analyses. Regardless of the phylogenetic method, the results enabled separating the South American from the North American *Cunila* species. Moreover, the South American species were divided in two groups: shrub species of the sections *Incisae* and *Incanae*; and subshrub species of the section *Spicatae* (Figure 3).

Mexican *Cunila* species clustered together with high support in all analyses performed by Agostini (2008; data not published).

Table 1. Plant material, geographical coordinates and GenBank accession numbers

Taxon	Latitude	Longitude	GenBank *trn*L-F	GenBank ITS
Cunila fasciculata Benth.	S 29°44′56,9″	W 53°02′48,3″	-	-
Cunila galioides Benth.	S 28°38′39″*	W 50°38′14″*	-	-
Cunila menthiformis Epl.	S 28°03′13″*	W 51°11′25″*	-	-
Cunila menthoides Benth.	S 30°32′28.2″	W 53°33′18.7″	-	-
Cunila microcephala Benth.	S 30°53′13.9″	W 52°14′58.3″	-	-
Cunila platyphylla Epl.	S 28°30′44″*	W 50°56′02″*	-	-
Cunila spicata Benth.	S 28°03′13″*	W 51°11′25″*	-	-
Cunila angustifolia Benth.	S 29°02′52″*	W 50°08′41″*	-	-
Cunila incisa Benth.	S 28°56′10″*	W 51°32′8″*	-	-
Cunila incana Benth.	S 30°30′44″*	W 53°29′29″*	-	-
Cunila origanoides (L.) Britton	-	-	-	-
Cunila leucantha Kunth ex Schltdl. et Cham.	-	-	-	-
Cunila lythrifolia Benth.	-	-	-	-
Cunila polyantha Benth.	-	-	-	-
Cunila pycnantha B.L. Rob. amd Greenm.	-	-	-	-
Cunila ramamoorthiana M.R. Garcia-Peña	-	-	-	-
Cunila secunda S. Watson	-	-	-	-
Blephilia ciliata Raf.	-	-	-	AF369167
Clinopodium coccineum Kuntze	-	-	DQ667433	DQ667233
Conradina glabra Shinners	-	-	DQ860318	DQ860349
Dicerandra thinicola H.A. Mill.	-	-	DQ860346	DQ860377
Glechon marifolia Benth.	-	-	DQ667489	DQ667303
Glechon thymoides Spreng.	-	-	DQ667496	DQ667310
Hesperozygis ringens (Benth.) Epl.	S 29°46′59″*	W 55°47′31″*	-	-
Hesperozygis nitida (Benth.) Epl.	-	-	-	-
Hedeoma costata A. Gray	-	-	DQ667436	DQ667236
Hoehnea epilobioides (Epl.) Epl.	-	-	DQ667497	-
Mentha spicata L.	-	-	AY570461	DQ667244
Monarda fistulosa L.	-	-	AY506603	DQ667318
Origanum vulgare L.	-	-	AY570463	DQ667243
Pycnanthemum incanum (L.) Michx.	-	-	AY506604	AY506640
Rhabdocaulon gracile (Benth.) Epl.	-	-	-	-
Rhabdocaulon stenodontum (Briq.) Epl.	-	-	-	-
Satureja hortensis L.	-	-	AY506611	AY227143
Thymus vulgaris L.	-	-	AY506613	EU785939
Hyptis alata (Raf.) Shinners	-	-	DQ667435	DQ667235
Agastache foeniculum Kuntze	-	-	AY506626	AY506660
Lepechinia lancifolia (Rusby) Epl.	-	-	DQ667492	DQ667306
Rosmarinus officinalis L.	S 29°10′05″*	W 51°10′46″*	-	-
Salvia austriaca Jacq.	-	-	DQ667512	DQ667323

*Coordinates of nearby cities.
Modified from Agostini (2008).

The US species *C. origanoides* nested weakly supported, with Mexican *Cunila* species and with other North American species, depending on the analyses performed. To best clarify the placement of *C. origanoides*, more samples are being added to the analyses.

Classifications including non-monophyletic genera are not supported by evolutionary rules and obscure phylogenetic relationships. Taxon sampling is currently being expanded to address more South American species with the aim of increasing the resolution within New World groups.

Figure 3. Maximum parsimony analysis based on nuclear dataset.

4. Genetic Variability among and within *Cunila* Populations

4.1. *Cunila galioides* Benth

Cunila galioides belongs to the section *Spicatae* and it is a popular aromatic and medicinal plant in South Brazil. This species is characterized by three chemotypes: 1) citral (neral and geranial); 2) trans-b-ocimene; and 3) menthene (Fracaro et al., 2001; Echeverrigaray et al., 2003).

Twenty populations collected at Rio Grande do Sul and Santa Catarina States, were analyzed based on RAPD markers by Fracaro et al. (2005). *Cunila microcephala* Benth and *C. incisa* Benth were also included in the analysis as outgroups. From a set of 60 primers, 20 were selected for the analyses, based on band resolution and the reproducibility of band patterns. Including *C. microcephala* and *C. incisa*, 92.5% of the bands were polymorphic, but considering only *C. galioides*, the percentage of polymorphic bands was 59.8%. Cluster analysis anabled separating the populations in three groups: Group 1 was formed by the 20 populations of *C. galioides*; Group 2 was formed by *C. microcephala*; and Group 3 by *C. incisa*.

Following Fracaro et al. (2005), RAPD markers analyses revealed wide variation within *C. galioides* that reflected a high level of diversity within this species. Cluster and canonical discriminating analysis indicated that the populations of the citral and menthene chemotypes are more closely related to each other, and the populations of the ocimene chemotype represent a different genetic pool (Figure 4).

4.2. *Cunila incisa* Benth

Cunila incisa belongs to the section *Incisae* and it is a highly aromatic species commonly used in south Brazilian folk medicine. This species is characterized by high contents of 1.8-cineole (50-60%) in its essential oil (Bordignon *et al.*, 1996; Agostini *et al.*, 2006). Sixty five plant samples, collected from 13 *Cunila incisa* populations in different regions of Rio Grande do Sul state, Brazil, were analyzed based on RAPD markers by Agostini (2003). One sample belonging to *C. incana* and one to *C. angustifolia* was used as outgroup. Sixty primers were previously tested for RAPD amplification, but only 14 primers were selected for further analyses, based on the quality of profiles. Including the outgroup, 89.1% of the bands were polymorphic, but considering only *C. incisa*, the polymorphism reached 42.2%.

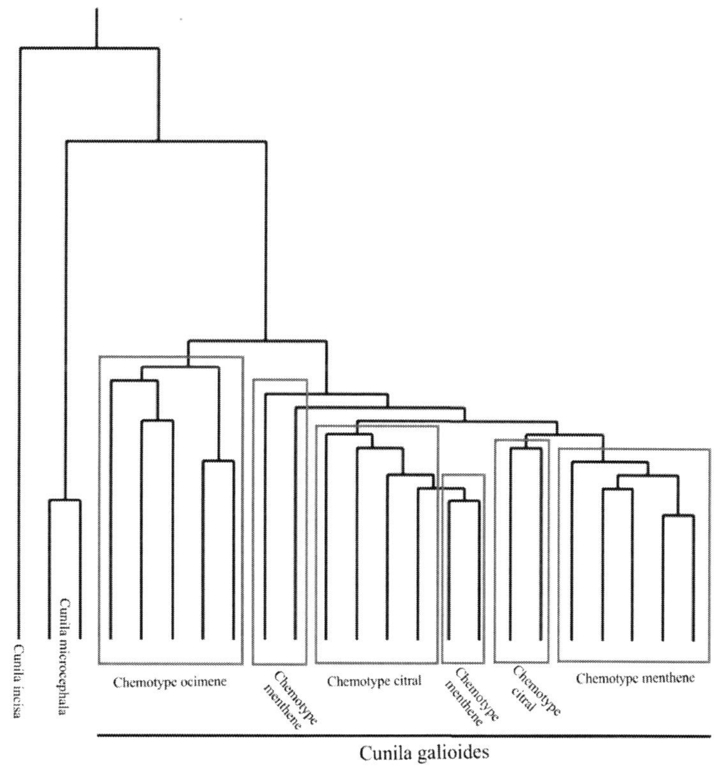

Modified from Fracaro et al. (2005).

Figure 4. Dendrogram representing the genetic similarity among *Cunila galioides* chemotypes based on RAPD markers.

The analyses enabled discriminating all *C. incisa* populations, and the amplification pattern revealed the existence of species-specific markers for the three species analyzed. Cluster analysis afforded to discriminate populations under three groups based mainly in its geographical origin.

A correlation was found between the distances obtained based on the essential oil analysis (Agostini *et al.*, 2006) and the similarity index calculated based on RAPD markers. This correlation suggests that the essential oil variation observed in *C. incisa* is under genetic control rather than environmental control.

Following Agostini (2003), *C. incisa* populations presented low intrapopulational variability, indicating that the populations form distinct genetic pools possible due to founder effects and low interpopulation gene flow.

4.3. *Cunila menthoides* Benth

Cunila menthoides (Figure 5) belongs to the section *Spicatae* and it is an endangered species found in fragmented environments threatened by local agriculture and pasture activities. This species presents strong and pleasant smell, and is commonly used in south Brazilian folk medicine. Three chemotypes have been reported in *C. menthoides*: pulegone, linalool (Agostini et al., 2010b) and isomenthone (Bordignon et al., 1998).

Following Bordignon (1997) and Coelho de Souza (1997), *C. menthoides* is restricted to a few locations in Rio Grande do Sul state (Brazil). Agostini et al. (2011) found only six populations in their fields trips, three of which in sites never reported before. Based on these six populations Agostini et al. (2010c) conducted genetic variability analyzes using ISSR markers. Twenty primers were previously tested for ISSR amplification, but only 10 primers were selected, based on the profiles. The selected primers generated 55.4% of polymorphic bands and population-specific markers were revealed by the amplification pattern.

Following the authors, high similarity values were obtained within populations suggesting that the populations are well-defined and structured as distinct genetic pools. Cluster analysis afforded to discriminate populations under four groups, with low genetic variability, formed based on the geographical origin of the populations. The authors associate the low genetic diversity observed for *C. menthoides* mainly to habitat degradation, since the number of individuals per population was relatively small, being found in fragmented environments threatened by local agriculture and pasture activities.

Regarding *C. menthoides*, *in situ* and *ex situ* strategies are considered as an alternative of conservation, by the authors.

4.4. *Cunila spicata* Benth

Cunila spicata belongs to the section *Spicatae* and it is an endangered aromatic and medicinal species of South Brazil, Argentina, Paraguay and Uruguay. In the last decades *C. spicata* populations are becoming rarer due to intense anthropic interference as agriculture, associated with drier summers. This species presents pleasant smell, and is commonly used in south Brazilian folk medicine for the treatment of respiratory disorders (Pio Correa, 1974; Toursarkissian, 1980).

Figure 5. *Cunila menthoides* Benth.

Four chemotypes have been reported in *C. spicata*: linalool/1,8-cineole; 1,8-cineole; carvone/carveol; and 1,8-cineol/limonene (Echeverrigaray et al., 2009). Van Baren et al. (2001) found three other main constituents from an Argentinean accession: limonene; geranyl acetate; and linalool.

Ten *C. spicata* populations (10 samples per population) collected in different regions of Rio Grande do Sul state, Brazil, were analyzed based on ISSR markers by Albuquerque (2004). Sample belonging to *C. galioides*, *C. menthoides* and *C. origanoides* (US) were used as outgroup. Twenty primers were previously tested for ISSR amplification, but only nine primers were selected for further analyses, based on the robustness and reproducibility of the profiles. Including the outgroup, 99.3% of the bands were polymorphic, but considering only *C. spicata*, this percentage fell to 84.4%.

Cluster analysis confirmed the genetic structure of *C. spicata,* separating all individuals by their original population, but no significant grouping was evidenced among populations. Low genetic diversity at populational and species level, and high differentiation among populations were detected by Albuquerque (2004). The genetic diversity, low estimated genetic flow and absence of correlation between genetic and geographical distances and chemical composition, indicates that genetic drift and inbreeding may be the main factors involved in the genetic structure of *C. spicata* (Albuquerque, 2004).

References

Agostini, G., 2003. Micropropagação, variabilidade química e genética em *Cunila incisa* Benth. *MSc. Thesis*, Universidade de Caxias do Sul. Caxias do Sul, RS. Brazil.

Agostini, G., 2008. Filogenia e diversidade genética do gênero Cunila D. Royen ex L.,(Lamiaceae). *PhD Thesis*. Universidade Federal do Rio Grande do Sul. Porto Alegre, RS, Brazil.

Agostini, G., Agostini, F., Atti-Serafini, L., Echeverrigaray, S., 2006. Essential oil variability within and among populations of *Cunila incisa* Benth. *Biochem. Syst. Ecol.* 34: 802–808.

Agostini, G., Agostini, F., Atti-Serafini, L., Echeverrigaray, S., Souza-Chies, T. T., 2010a. Essential Oil Composition of *Cunila incana* Benth. (Lamiaceae). *J. Essent. Oil Res.* 22: 432-434.

Agostini, G., Agostini, F., Bertolazzi, M., Echeverrigaray, S., Souza-Chies, T. T., 2010b. Variation of the chemical composition of essential oils in Brazilian populations of *Cunila menthoides* Benth. (Lamiaceae). *Biochem. Syst. Ecol.* 38: 906-910.

Agostini, G., Echeverrigaray, S., Souza-Chies, T. T., 2010c. Genetic diversity of the endangered Brazilian endemic herb *Cunila menthoides* Benth. (Lamiaceae) and its implications for conservation. *Biochem. Syst. Ecol.* 38, 1111-1115.

Agostini, G., Echeverrigaray, S., Souza-Chies, T. T., 2008. Genetic relationships among South American species of *Cunila* D. Royen ex L. based on ISSR. Pl. *Syst. Evol.* 274, 135-141.

Agostini, G., Ribeiro, T., Moura, S., Echeverrigaray, S., Souza-Chies, T. T., 2012. *Cunila* D. Royen ex. L., *Glechon* Epl. and *Hesperozygis* Epl. (Lamiaceae) in South America: an ethnobotanical and phytochemical review. In: Shen Ten Ma; Zhi Yu Su. (Org.). Herbs: Cultivation, Medicinal Uses and Environmental Concerns. v. 1.

Albuquerque, M. R. (2004) Avaliacão de variabilidade química e genética em *Cunila spicata* Benth.. *MSc. Thesis*. Universidade de Caxias do Sul. Caxias do Sul, RS. Brazil.

Bordignon, S. A. L., 1997. Estudo botânico e químico de espécies de *Cunila* Royen ex L. (Lamiaceae) nativas do sul do Brasil. *PhD. Thesis*. Universidade Federal do Rio Grande do Sul. Porto Alegre, RS. Brazil.

Bordignon, S. A. L., Schenkel, E. P., Spitzer, V., 1996. The essential oil of *Cunila incisa* (Lamiaceae). A rich source of 1,8-cineole. *Quím. Nova* 19: 105–107.

Bordignon, S. A. L., Schenkel, E. P., Spitzer, V., 1997. The essential oil composition of *Cunila microcephala* and *Cunila fasciculata*. *Phytochemistry* 44: 1283– 1286.

Bordignon, S. A. L., Schenkel, E. P., Spitzer, V., 1998. The essential oil of *Cunila menthoides* Bentham (Lamiaceae). *J. Essent. Oil Res.* 10: 317–320.

Bordignon, S. A. L., Schenkel, E. P., Spitzer, V., 1998b. The essential oil composition of *Cunila platyphylla* Epling (Lamiaceae). *Acta Farm. Bonaerense* 17: 143– 146.

Bordignon, S. A. L., Schenkel, E. P., Spitzer, V., 1999. The essential oil composition of *Cunila angustifolia* Bentham (Lamiaceae). *J. Essent. Oil Res.* 11: 145–148.

Cantino, P. D., Harley, R. M., Wagstaff, S. J., 1992. Genera of Labiate: status and classification, In: Harley, R. M., Reynolds, T., (Eds), Advances in Labiate Science. *Royal Botanic Gardens, Kew, UK*, pp. 511-522.

Cicció, J. F., Poveda, L. J., 1999.Volatile constituents of *Cunila Polyantha* (Lamiaceae) from Costa Rica. A rich source of menthone. *Rev. Biol. Trop.* 47, 377.

Coelho de Souza, G. P., 1997. Estudo etnobotânico da família Lamiaceae no Rio Grande do Sul, com ênfase na busca de espécies com propriedades anticonvulsantes. *MSc Thesis*. Universidade federal do Rio Grande do Sul. Porto Alegre, RS.

Duke, J. A., 1994. Biologically-active compounds in important spicies, in: Charalembours, G., (Ed), Spices, herbs, and edible fungi. *Amsterdam: Elsevier Publishers*. P. 595

Echeverrigaray, S., Albuquerque, M., Zacaria, J., Atti dos Santos, A. C., Atti-Serafini, L., 2009. Chemical variations on the essential oils of *Cunila spicata* Benth. (Lamiaceae), an aromatic and medicinal plant from South Brazil. *J. Essent. Oil Res.* 21: 241-245.

Echeverrigaray, S., Fracaro, F., Atti dos Santos, A. C., Paroul, N., Wasum, R., Atti-Serafini, L., 2003. Essential oil composition of south Brazilian populations of *Cunila galioides* and its relation with the geographic distribution. *Biochem. Syst. Ecol.* 31, 467-475.

Epling, C., 1936. Synopsis of the South American Labiatae. Repert. Spec. Nov. *Regni Veg. Beih*. II. 85, 97-192.

Fracaro, F., Atti-Serafini, L., Dos Santos, A. C. A., Paroul, N., Echeverrigaray, S., Wasum, R., 2001. Analysis of the essential oil composition of *Cunila galioides* Benth. *J. Essent. Oil Res*. 11, 27e30.

Fracaro, F., Zacaria, J., Echeverrigaray, S., 2005. RAPD based genetic relationship between populations of three chemotypes of *Cunila galioides* Benth. *Biochem. Syst. Ecol.* 33, 409-417.

Garcia-Peña, M. R., 1989. A new species of *Cunila* (Lamiaceae) from southwestern Mexico. *Kew Bulletin*. 44, 727-730.

Manjarrez, A., Mendoza, V., 1966. The volatile oils of *Agastache mexicana* (Benth.) Epling and *Cunila lythrifolia* Benth. Perfum. *Essent. Oil Rec*., 57, 561-562.

Moreira, E. A., Krambeck, R., 1976. Análise cromatográfica do óleo essencial das folhas de *Cunila angustifolia* Bentham, vol. 44. *Tribuna Farmacêutica*, pp. 50-59.

Pio Correa, M., 1974. Dicionário das plantas úteis do Brasil e das exóticas cultivadas. Vol. 5. Instituto Brasileiro de Desenvolvimento Florestal, Rio de Janeiro, Brasil.

Sandri, I. G., Zacaria, J., Fracaro, F., Delamare, A. P. L., Echeverrigaray, S., 2007 Antimicrobial activity of the essential oils of Brazilian species of the genus *Cunila* against foodborne pathogens and spoiling bacteria. *Food Chem* 103:823–828

Simões, C. M. O., Mentz, L. A., Schenkel, E. P., Irgang, B. E., Stehmann, J. R., 1994. *Plantas da medicina popular no Rio Grande do Sul*, fourth ed. Editora da Universidade/ UFRGS, Porto Alegre, Brazil.

Toursarkissian, M., 1980. Plantas Medicinales de la Argentina. Hemisferio Sur. Buenos Aires, Argentina.

Van Baren, C. M., Muschietti, L. V., Lira, P. L., Coussio, J. D., Bandoni, A. L., 2001. Volatile Constituents from Aerial Parts of *Cunila spicata*. *J. Essent. Oil Res*. 13: 351-353.

In: Genetic Diversity: New Research
Editors: J. A. C. Amaya and M. M. F. Jimenez
ISBN: 978-1-62081-598-4
© 2012 Nova Science Publishers, Inc.

Chapter 12

MODELLING MINIMIZATION OF GENETIC DIVERSITY LOSS IN JAPANESE MARINE FINFISH STOCK ENHANCEMENT

Maria del Mar Ortega-Villaizan[1,*] and Nobuhiko Taniguchi[2]
[1]Institute of Molecular and Cell Biology,
Miguel Hernandez University, Elche, Spain
[2]Department of Marine Biological Science, Faculty of Biotechnology,
Fukuyama University, Fukuyama City, Hiroshima Pref., Japan

ABSTRACT

Japanese stock enhancement has been carried out since the early sixties by releasing artificially bred seed of several economically important fish and shellfish species. Even though the aim of stocking is to increase the amount of resources in the wild, breeding part of the native population in a captive environment, to achieve this goal also implies the reduction of the genetic variability even in first generation offspring, due to the small genetic contribution of just a few breeders. Consequently, stocking may comprise a potential risk to the wild population if no action is taken with respect to the loss of the genetic variability of the hatchery stock. Hence, in the last two decades, when referring to conservation programs of rare and endangered species, strong attention has been paid on conservation genetics of the captive population, in addition to the primary protection of the species and the maintenance of an adequate natural habitat.

Specifically, in the case of Japanese marine finfish stocked species, such as barfin flounder, spotted halibut or red sea bream -which are migratory species, and thus not easily traceable- it is especially complicated to obtain feasible genetic population data to establish the genetic variability of the native population. Consequently, Japanese marine finfish stock enhancement becomes a novel model for marine finfish genetic diversity conservation programs.

For this purpose, computational modelling simulation of selective breeding scenarios is being applied in Japan in order to minimize genetic diversity loss of especially rare and

* Corresponding author: Dr. Maria del Mar Ortega-Villaizan; Romo, Institute of Molecular and Cell Biology, Miguel Hernandez University, 03202 Elche, Spain. E-mail: mortega-villaizan@umh.es. Tel: 34-658131423.

endangered finfish species and thus to conserve the genetic variability of the native population.

INTRODUCTION

In the last two decades, when referring to conservation programs of rare and endangered species, strong attention has been paid on the conservation genetics of the captive population, together with the primary protection of the species and maintenance of an adequate natural habitat (Hedrick & Miller 1992).

In order to ensure high survival and growth rate of the released stock, three factors have been established for the effectiveness of stocking: the quality of the fish released, environmental conditions, and releasing techniques (Svåsand 2004). In relation to the quality of the fish, which comprises morphological, physiological, and behavioural factors (Svåsand 2004), the genetic variability of the hatchery stock plays an important role, since genetic variability is extremely important in allowing to maintain the fitness of the stock as well as the capability of adapting to changing environments (Hindar et al. 1991, Perez-Enriquez et al. 1999, Ryman 1991, Taniguchi et al. 2003, Utter 1998). These facts become especially important in the case of stock enhancement programs intended for rare and endangered species, where the wild stock is close to or already suffering the consequences of an effect of decline caused by the small population size (Primack 1993).

Even though the aim of supportive breeding is to increase the amount of resources in the wild by means of breeding part of the native population in a captive environment, this process also implies reducing the genetic variability even in the first generation offspring as a result of the small genetic contribution of just a few breeders (Jackson et al. 2003, Sekino et al. 2002, Taniguchi et al. 1983, Verspoor 1988). Therefore, if the level of genetic variation of the released stock is reduced, the success of the stock enhancement program could fail, and, even more, the integrity and future of the wild stock could be threatened by introgressive hybridization of hatchery releases with indigenous populations (Utter 1998). Hence, in the last two decades, when referring to conservation programs of rare and endangered species, careful attention is devoted to the conservation genetics of the captive population, as well as on the primary protection of the species and maintenance of an adequate natural habitat (Hedrick and Miller 1992).

For that purpose, the number of unrelated captive broodstock used to produce the juveniles for stock enhancement should be kept above a minimal threshold in order to preserve the genetic variability of the population. However, even in the case where the number of parental individuals is above a minimal threshold, different factors as the way of fertilization, suitable fertilization time and hatchery management procedures may reduce the effective number of contributing parentals far below the current census number. Even when parental participation is under control, the fertilization rate depends on the time of fertilization, which varies among individuals. Besides, the hatch-out rate as well as the larvae and juvenile survival rate may vary from one family to another. Therefore, to know exactly which parents contribute to the juvenile stock is of exceptional importance in order to maintain a high level of genetic variability within the hatchery population used for stock enhancement. FAO (1980) recommended genetic monitoring of cultured species to assess the genetic changes that result from a particular culture scheme, and advised that a minimum threshold of the effective population size (N_e) should not be below 50 for short-time fitness preservation; however, a minimum effective size of 500 is desirable to guarantee a useful genetic variation for long-time fitness preservation.

This rule can be applied without difficulty in the case of abundant fish species, such as the red sea bream, the Japanese flounder, or the Japanese smelt (ayu). However, it is not easily applicable to endangered and rare fish species, such as barfin flounder, since the current number of wild individuals has diminished to a great extent and does no longer exceed the minimum recommended value. The policy recommended in these situations is to retain the genetic diversity in the ancestral population by compensatory mating, in which founder lineages that are under-represented and/or less related are selected for mating preferentially so as to increase their contribution (Doyle et al., 2001).

Molecular Markers, Population Genetic Analysis and Pedigree Tracing

The utility of molecular markers has been widely proved to be an exceptional indicator of genetic variation within and between populations, especially microsatellite DNA (msDNA) markers, which compared with other molecular markers exhibit very high levels of polymorphism (Ferguson et al. 1995, O'Reilly and Wright 1995, Taniguchi et al. 2003). MsDNA markers have been efficiently used in the determination of parentage and kinship in aquaculture (O'Reilly et al. 1998, Perez-Enriquez et al. 1999, Perez-Enriquez & Taniguchi 1999, Sekino et al. 2003, Sugaya et al.2002, Ortega-Villaizan et al.2006a), as well as in the estimation of relatedness and the reconstruction of the pedigree of natural populations (Blouin et al. 1996; Blouin 2003, Van de Casteele et al. 2001, Ortega-Villaizan et al. 2005, Ortega-Villaizan et al. 2006b).

Estimation of the Effective Population Size in Recaptured Stock by Pedigree Evaluation

The stock enhancement programs have the aim of supplementing the decreasing wild population resources with hatchery-produced offspring. For this purpose, a reduced number of adult individuals captured from the wild are bred in captivity. This procedure is commonly referred to supportive breeding (Ryman 1991). However, the introgression of the hatchery gene pool into the wild one can result in: i) the loss of adapted genes because of interbreeding between the hatchery stock and the wild stock; ii) the homogenization of genetically distinct populations; and iii) the modification of the genetically effective population size of the wild stock (Taniguchi 2003). Therefore, Taniguchi (2004) stated that artificially propagated seed fish should meet the following criteria: i) their broodstock should come from the same local population into which the offspring will be released; ii) they should have a high genetic variability equal to that of the wild population; iii) the effective population size of the broodstock should be large; and iv) the level of inbreeding of the seed fish should be low at both the population and the individual level. The genetic variability of the recaptured stock has been evaluated for two Japanese flatfish species, barfin flounder and spotted halibut

(Ortega-Villaizan et al. 2005, Ortega-Villaizan et al. 2006b, respectively). In both studies, genetic tagging successfully identified the recaptured stock in the wild. Barfin flounder's and spotted halibut's recaptured stocks, which could be called "future pseudo-wild" due to the low population size of these species, showed that their Ne was far below 50, which is the minimum viable population size recommended for short term fitness preservation of the species (FAO 1980). The reason for this was recognized to be the lack of policy for broodstock management, which included a lack of effective breeding designs to enhance Ne and lack of offspring equalization.

Minimization of Genetic Diversity Loss of Captive Broodstocks by Means of Minimal Kinship Selective Crossbreeding

The minimal kinship (MK) selective crossbreeding approach, which means selecting for mating those individuals which show a lower level of kinship, has proven to be the best way to decrease the loss of genetic variation in captive populations which is caused by founder lineage bias and drift (Ballou and Lacy, 1995; Doyle et al., 2001; Sekino et al., 2004; Sriphairoj et al., 2007).

However, in practice, MK selection requires genetic background information of the founder pedigree, which is not available in the early hatchery generations. Different relatedness estimators and similarity indexes have been used to estimate inter-individual relatedness by means of microsatellite DNA markers (Blouin, 2003; Toro et al., 2002, Toro et al., 2003). The Bowcock's similarity index (P_s), which estimates the proportion of shared alleles between individuals based on microsatellite DNA genotypes (Bowcock et al., 1994), has depicted good performance in identifying related individuals (Ortega-Villaizan et al., 2006a) as well as estimating interindividual relatedness to maximize the genetic variability of future hatchery generations of the rare fish species barfin flounder (Ortega-Villaizan et al. 2011). The latter show that the best policy for maintaining or increasing genetic and allelic diversity, while minimizing the effective number of contributing breeders (N_c) of the base population (G_{-1}), was to apply the MK selective crossbreeding. They also state that after implementing the MK breeding scenario, an N_c of more than 150, and therefore under the recommended N_e of 500, can preserve about 80% of the allelic diversity (number of alleles, N_a) and 100% of the gene diversity (expected heterozygosity, H_e) for thirty generations, thus demonstrating that the MK selective crossbreeding scenario as analyzed with the algorithm of Noguchi and Taniguchi (2006) proved to be a highly advantageous approach to retain both gene and allelic diversities in the stock enhancement program of rare and endangered species.

Conclusion

When talking about genetic monitoring on the stock enhancement programs, two main subjects are considered to be essential (Ryman and Laikre 1991; Cross 2000; Gold 2004, Taniguchi 2004):

1. The need to identify the genetic structure of the species so as to assure that the appropriate population is targeted for propagation and release in a particular area, as well as for matching the broodstock selection and supplementation with the population structure of the species.
2. The importance of maintaining genetic variability in the captive broodstock population, in the subsequent hatchery generation, and, after reintroduction into the wild, of monitoring any changes in genetic composition or variability. For this purpose, "genetic tags" are required for identifying the fish at family level.

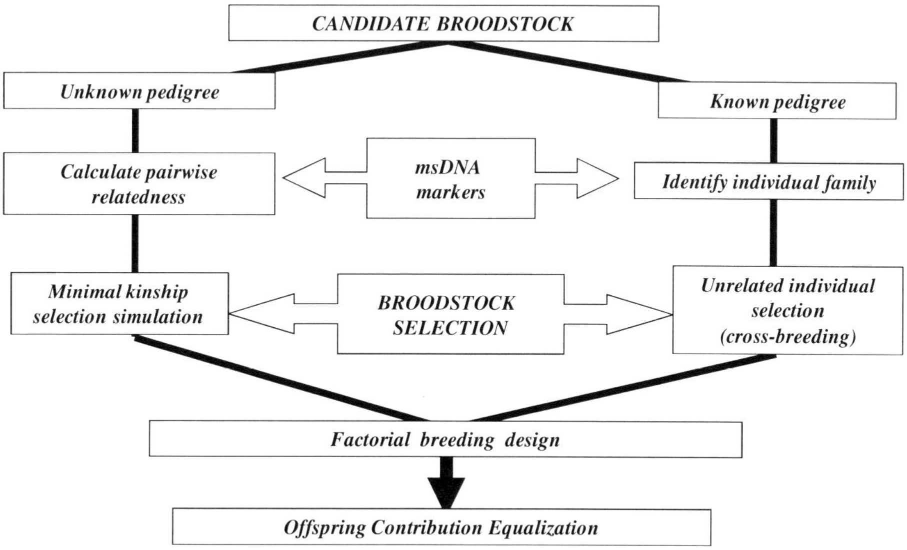

Figure 1. Diagram of the proposal for broodstock management to minimize loss of genetic variation in the stock enhancement program.

Nonetheless, the effects of the majority of the enhancement stocking exercises have not been adequately monitored after their release (Cross 2000). In Japan, genetic monitoring has been fulfilled before and after their release for several finfish species such as spotted halibut (Ortega-Villaizan et al. 2006b, Ortega-Villaizan 2006c) and barfin flounder (Ortega-Villaizan et al, 2005, Ortega-Villaizan 2006c), and gilthead seabream (Perez-Enriquez et al. 1999, Perez-Enriquez et al. 2001).

A proposal for future broodstock management which should minimize the loss of genetic variation in the stock enhancement program of Japanese marine finfish is described in Figure 1. Depending on whether the pedigree of the candidate broodstock is known or unknown, either of the following procedures should be carried out:

- If the pedigree of the candidate broodstock is unknown, a coefficient to determine inter-individual category of relationship, like Bowcock's similarity index (P_s), (Bowcock et al., 1994), Queller and Goodnight (1989) relatedness (R_{XY}), and the inter-individual genetic identity (GI) (Sugaya et al. 2002), or any other estimator (Van de Casteele et al. 2001), should be calculated using molecular markers in order to estimate the degree of relationship among the available broodstock. Afterwards, minimal kinship selection procedures following the model of Doyle et al. (2001), which has been demonstrated to increase genetic diversity in captive populations (Sekino et al. 2004, Noguchi and Taniguchi 2006, Ortega-Villaizan et al. 2011), should be carried out in order to select those individuals which are less related with the rest. Afterwards, factorial breeding designs could be applied to mate the less related individuals. Factorial breeding designs, although time- and cost-consuming, have proved to maintain genetic diversity in supplemental breeding programs (Fiumera et al. 2004), an aspect that is extremely necessary to be taken into account in the case of stock enhancement programs intended for rare or endangered species.

- On the other hand, if the pedigree of the candidate broodstock is known, direct crossbreeding between unrelated individuals is known to give better results than applying minimal kinship selection procedures.

ACKNOWLEDGMENTS

Special thanks are due to Prof. Henny Gevers for helping with English editing.

REFERENCES

Ballou, J. D. and R. C. Lacy. 1995. Identifying genetically important individuals for management of genetic diversity in pedigreed populations. Pages 76-111 in J.D. Ballou, M. Gilpin, and T.J. Foose (eds.), *Population Management for Survival and Recovery*, Analytical Methods and Strategies in Small Population Conservation, Columbia University Press, New York.

Blouin M. S., Parsons M., Lacaille V, Lotz S. 1996; Use of microsatellite loci to classify individuals by relatedness; *Mol. Ecol.* 5: 393-401.

Blouin M. S. 2003. DNA-based methods for pedigree reconstruction and kinship analysis in natural populations. *Trends Ecol. Evol.* 18: 503-511.

Bowcock A. M., Ruiz-Linares A., Tomfohrde J., Minch E., Kidd J. R., Cavalli-Sforza L. L. 1994. High resolution of human evolutionary trees with polymorphic microsatellites. *Nature.* 368: 455-457.

Cross T. F. 2000. Genetic implications of translocation and stocking of fish species, with particular reference to Western Australia. *Aquaculture Res.* 31: 83-94.

Doyle R. W., Pérez-Enríquez R., Takagi M., Taniguchi N. 2001. Selective recovery of founder genetic diversity in aquacultural broodstocks and captive, endangered fish populations. *Genetica.* 111: 291-304.

FAO. 1980. Conservation of the genetic resources of fish: problems and recommendations. *Report of the expert consultation on the genetic resources of fish*. Rome, 9-13 June 1980. FAO Fisheries Technical Paper No. 217.

Ferguson A., Taggart J. B., Prodöhl P. A., McMeel O., Thompson C., Stone C., McGinnity P., Hynes R. A. 1995. The application of molecular markers to the study and conservation of fish populations, with special reference to *Salmo*. *J. Fish Biol.* 47: 103-126.

Fiumera A. C., Porter B. A., Looney G., Asmussen M. A., Avise J. C. 2004. Maximizing offspring production while maintaining genetic diversity in supplemental breeding programs of highly fecund managed species. *Cons. Biol.* 18: 94-101.

Gold J. R. 2004. Stock structure and effective size of the red drum (*Sciaenops ocellatus*) in the northern Gulf of Mexico and implications relative to stock enhancement and recruitment. In: *Stock enhancement and sea ranching*. Developments, pitfalls and opportunities (eds. Leber K.M, Kitada S, Blankenship H.L and Svåsand T). Blackwell Publishing. pp. 353-370.

Hedrick P. W. and Miller P. S. 1992. Conservations genetics: techniques and fundamentals. *Ecol. Appl.* 2: 30-46.

Hindar K., Ryman N., Utter F. 1991. Genetic effects of cultured fish on natural fish populations. *Can. J. Fish. Aquat. Sci.* 48: 945-957.

Jackson T. R., Martin-Robichaud D. J., Reith M. E. 2003. Application of DNA markers to the management of Atlantic halibut (*Hippoglossus hippoglossus*) broodstock. *Aquaculture.* 220: 245-359.

Noguchi D., Taniguchi N. 2006. Evaluation of Minimal Kinship Selective Breeding to Maintain Genetic Diversity in Masu salmon, *Oncorhynchus masou masou*, Based on Computer Simulation. *Fish Gen. Breed. Sci.* 35: 165-170 (in Japanese).

O'Reilly P., Wright J. M. 1995. The evolving technology of DNA fingerprinting and its application to fisheries and aquaculture. *J. Fish Biol.* 47: 29-55.

O'Reilly P. T., Herbinger C., Wright J. M. 1998. Analysis of parentage determination in Atlantic salmon (*Salmo salar*) using microsatellites. *Animal Genet.* 29: 363-370.

Ortega-Villaizan M., Suzuki S., Ikeda M., Nakajima M., Taniguchi N. 2005. Monitoring of the genetic variability of the hatchery and recaptured fish in the stock enhancement program of the rare species barfin flounder Verasper moseri. *Fish. Sci.* 71: 1120-1130.

Ortega-Villaizan M., Suzuki S., Nakajima M., Taniguchi N. 2006a. Genetic evaluation of interindividual relatedness for broodstock management of the rare species barfin flounder *Verasper moseri* using microsatellite DNA markers. *Fish. Sci.* 72: 33–39.

Ortega-Villaizan M., Aritaki M., Taniguchi N. 2006b. Pedigree analysis of recaptured fish in the stock enhancement program of spotted halibut Verasper variegatus. *Fish. Sci.* 72: 48-52.

Ortega-Villaizan M., Aritaki M., Suzuki S., Ikeda M., Asahida T., Taniguchi N. 2006c. Genetic population evaluation of two closely related flatfish species, the rare barfin flounder and spotted halibut, along the Japanese coast. *Fish. Sci.* 72: 556–567.

Ortega-Villaizan M., Noguchi D., Taniguchi N. 2011. Minimization of genetic diversity loss of endangered fish species captive broodstocks by means of minimal kinship selective crossbreeding. *Aquaculture* 318: 239-243.

Perez-Enriquez R., Takagi M., Taniguchi N. 1999. Genetic variability and pedigree tracing of a hatchery-reared stock of red sea bream (*Pagrus major*) used for stock enhancement, based on microsatellite DNA markers. *Aquaculture.* 173: 413-423.

Perez-Enriquez R., Taniguchi N. 1999. Use of microsatellite DNA as genetic tags for the assessment of a stock enhancement program of red sea bream. *Fish. Sci.* 65: 374-379.

Perez-Enriquez R., Takemura M., Tabata K., Taniguchi N. 2001. Genetic diversity of red sea bream Pagrus major in western Japan in relation to stock enhancement. *Fish. Sci.* 67: 71-78.

Primack R. B. 1993. *Essentials of conservation biology*. Sinauer Associates, Inc.

Queller D. C., Goodnight K. F. 1989. Estimating relatedness using genetic markers. *Evolution.* 43: 258-275.

Ryman N. 1991. Conservation genetics considerations in fishery management. *J. Fish Biol.* 39: 211-224.

Ryman N., Laikre L. 1991. Effects of supportive breeding on the genetically effective population size. *Cons. Biol.* 5.: 325-329.

Sekino M., Hara M., Taniguchi N. 2002. Loss of microsatellite and mitochondrial DNA variation in hatchery strains of Japanese flounder Paralichthys olivaceus. *Aquaculture.* 213: 101-122.

Sekino M., Saitoh K., Yamada T., Kumagai A., Hara M., Yamashita Y. 2003. Microsatellite-based pedigree tracing in a Japanese flounder *Paralichthys olivaceus* hatchery strain: implications for hatchery management related to stock enhancement program. *Aquaculture.* 221: 255-263.

Sekino M., Sugaya T., Hara M., Taniguchi N. 2004. Microsatellite relatedness estimator used for minimal kinship selection *Paralichthys olivaceus. Aquaculture.* 233:163-172.

Sriphairoj K., Kamonrat W., Na-Nakorn U. 2007. Genetic aspect in broodstock management of the critically endangered Mekong giant catfish, Pangasianodon gigas in Thailand. *Aquaculture.* 264: 36–46.

Sugaya T., Ikeda M., Mori H., Taniguchi N. 2002. Inheritance mode of microsatellite DNA markers and their use for kinship estimation in kuruma prawn *Penaeus japonicus. Fish. Sci.* 68: 299-305.

Svåsand T. 2004. Why Juvenile Quality and Release Strategies are Important Factors for Success in Stock Enhancement and Sea Ranch. In: *Stock enhancement and sea ranching.* Developments, pitfalls and opportunities (eds. Leber K.M, Kitada S, Blankenship H.L and Svåsand T). Blackwell Publishing.

Taniguchi N., Sumantadinata K., Iyama S. 1983. Genetic change in the first and second generations of hatchery stock of black sea-bream. *Aquaculture.* 35: 309-320.

Taniguchi N. 2003. Genetic factors in broodstock management for seed production. *Rev. Fish. Biol. Fish.* 13: 175-185.

Taniguchi N. 2004. Broodstock management for stock enhancement programs of marine fish with assistance of DNA markers (a review). In: *Stock enhancement and sea ranching.* Developments, pitfalls and opportunities (eds. Leber K.M, Kitada S, Blankenship H.L and Svåsand T). Blackwell Publishing. pp. 329-338.

Toro M., Barragán C., Ovilo C., Rodrigáñez J., Rodríguez C., Silió L. 2002. Estimation of coancestry in Iberian pigs using molecular markers. *Cons. Genet.* 3: 309-320.

Toro M., Barragán C., Ovilo C. 2003 Estimation of genetic variability of the founder population in a conservation scheme using microsatellites. *Anim. Genet.* 34: 226–228.

Utter F. 1998. Genetic problems of hatchery-reared progeny released into the wild, and how to deal with them. *Bull. Mar. Sci.* 62: 623-640.

Van de Casteele T., Galbusera P., Matthysen E. 2001. A comparison of microsatellite-based pairwise relatedness estimators. *Mol. Ecol.* 10: 1539-1549.

Verspoor E. 1988. Reduced genetic variability in first-generation hatchery populations of Atlantic salmon (*Salmo salar*). *Can. J. Fish. Aquat. Sci.* 45: 1686-1690.

INDEX

B

C

D

E

I

J

K

L

M

N

O

P

Q

R

S